Photosynthesis Bibliography

volume 4 1973

References no. 15368-18420 / ABD - ZUT

Editors Z. Šesták & J. Čatský

Dr. W. Junk b.v. - Publishers - The Hague 1978

ISBN-13:978-90-6193-043-3   e-ISBN-13:978-94-009-9949-7
DOI: 10.1007/978-94-009-9949-7

PREFACE

The bibliography includes papers in all fields of photosynthesis research - from studies of model biochemical and biophysical systems of the photosynthesis mechanism to primary production studied by the so-called growth analysis. In addition to papers devoted entirely to photosynthesis, papers on other topics are included if they contain data on photosynthetic activity, photorespiration, chloroplast structure, chlorophyll and carotenoid synthesis and destruction, *etc.*, or if they contain valuable methodological information (measurement of selected environmental factors, leaf area, *etc.*). In many branches it has been very difficult to define the limits of interest for photosynthesis researchers. This problem has arisen *e.g.* in topics dealing with the transport of gases, where - in addition to the papers on CO_2 transfer - some papers on water vapour transfer are included, these being of general application. On the other hand, many papers dealing with the anatomy and physiology of stomata have been omitted, if the aspect of carbon dioxide or water vapour exchange has not been discussed.

This volume contains references to papers published in the year 1973, and, similarly to Vol. 3, also addenda including references published in the preceding period (*i.e.* 1966 - 1972). The numbers of these additional references are labelled with an asterisk in the list of references as well as in indexes.

To maximize the value of the bibliography the references are arranged alphabetically by authors' names, and each volume is provided with three indexes. The authors' index contains all names of authors, co-authors and editors. The subject index covers only primary items chosen according to their interest for photosynthesis researchers. In this volume its preparation was based mainly on the paper titles, key words and abstracts. In the plant index, only important crop plants and selected plant types and groups are indexed.

Cumulative indexes accompany Volumes 1, 5, and then every fifth volume, *i.e.* Vols. 10, 15, *etc.*

We have tried to cover fully the relevant papers which have appeared in the most important scientific periodicals and books. Articles published in local journals, mimeographed booklets, abstracts of theses and of symposia contributions, *etc.*, were chosen mostly from reprints and lists of publications received directly from the authors.

Since some 3000 relevant papers are currently published every year and included in this bibliography, and since the majority of citations have been checked with the originals, collecting and preparing for publication of such a large amount of material would have been impossible without the collaboration of the authors of the relevant publications. The courtesy of those authors who have already supplied us with reprints is highly appreciated.

We acknowledge with thanks the cooperation of our colleagues from the Department of Physiology of Photosynthesis and Water Relations of the Institute of Experimental Botany of the Czechoslovak Academy of Sciences in Prague, especially Mrs. DRAHOMÍRA TĚŽKÁ who helped in preparing the card material and typing the manuscript, Dr. INGRID TICHÁ who helped with completing and checking the references, and Mr. PETR ZÁZVORKA who supplied us with rare periodicals. In addition, the former librarian of the Institute, Mrs. ALENA ŠTĚTINOVÁ, helped us with checking the references.

Dr. Z. ŠESTÁK and Dr. J. ČATSKÝ

Institute of Experimental Botany
Czechoslovak Academy of Sciences

Flemingovo n. 2

160 00   PRAHA 6

Czechoslovakia

INSTRUCTIONS FOR USE

All references are arranged alphabetically according to the authors' names and the year of publication. They are numbered and these numbers are used in the indexes. In case of a book title, the number is preceded by B. An asterisk preceding the number denotes the reference published in the preceding period (1966 - 1972).

The references contain the original unshortened title of the paper (book). English, French and German titles are cited in the original language. Titles in other languages are supplemented with a translation in English (sometimes using the title of the respective English abstract or a shortened title with omitted deadweight words). Titles of Japanese, Chinese *etc.* papers are given in English translation only. The journals' names are abbreviated mainly according to the "Style Manual for Biological Journals" (Second Edition, Amer. Institute of Biological Sciences, Washington, D.C. 1964), *e.g.* :

Abhandlungen	chinese	Industry	Publishers
Abstract	Chromatography	inorganic	quantitative
Abteilung.	Commission	Institute	Quarterly
Academy	Communication	international	Radiation
Acta	comparative	Investigation	Radiobiology
Africa	Comptes rendus	italian	Rastenii̯
agricultural	Conference	Izvestiya	Recherche
Agriculture	Congress	Jahrbuch	Report
Agronomy	Contribution	japanese	Research
Akademie (-emiya)	Cytochemistry	Japan	Review
Algology	Cytology	Journal	royal
allgemeine	czechoslovak	Klasse	russian
american	Dendrology	Laboratory	russkii̯
America	Department	Landwirtschaft	scandinavicus
analytical	Deutschland	Letters	Science
Anatomy	Disease	Limnology	Section
angewandte	Dissertation	Magazin	Series (-iya)
Annals	Doklady	marine	Society
annual	Dopovidi	Mathematics	sovetskii̯
anorganisch (-nic)	Ecology	Microbiology	soviet
applied	Education	miscellaneous	special
Arbeit	Embryology	molecular	SSSR
Archiv	Encyclopedia	Monograph	Station
Atmosphere	Engineer	moskovskii̯	Supplement
atomic	Enzymology	Mycology	Survey
Australia	european	national	Symposium
Beiheft	experimental	natural	technical
Belgique	Experiment	Naturforschung	Technology
Bericht	Faculty	neerlandicus	Tijdschrift
biochemical	Federation	Netherland	Transaction
Biochemistry	Fizika	New Zealand	Travail (-aux)
biokhimicheskii̯	Fiziologiya	nuclear	tropical
Biokhimiya	Forestry	Oceanography	Trudy
biological(-ogicheskii̯)	Forschung	Optics	ukrainian
Biology (-ogiya)	Foundation	organic	UK
biophysical	France	original	US, USA
Biophysics	Gazette	Otdelenie	USSR
Bodenkunde	general	Pathology	University
bolgarskii̯	genetical	Pflanzen-	végétal
botanical (-anicheskii̯)	Genetics	Philosophy	Virology
Botany	Gesellschaft	physical	Virusforschung
british	Giornale	Physics	Volume
Bulletin	helveticus	physiological	Weekblad
Canada	Histochemistry	Physiology	Wetenschappen
cellular (-ulaire)	Histology	Phytopathology	Wissenschaft
central	Horticulture	Plant (-arum)	Zeitschrift
chemical	hungaricus	polish	Zeitung
Chemistry	Husbandry	Proceedings	Zentralblatt
chimicus	imperial	Publication	Zhurnal

The numbers at the end of each reference of a journal article denote : volume (issue) : first page - last page, year of publication. The number of issue is given only in the journal where each issue is paginated separately.

Book titles are cited according to the title page, not to the book jacket or cover (if the names of the editors are not given on the title page, they are not cited in the reference). The publishing house, place and year of publication are included.

Brackets at the end of the reference give bibliographic details and explanations to the contents, not given in the original. The following abbreviations are used most often :

ab	abstract	Jap.	Japanese
Arm.	Armenian	Latv.	Latvian
Belorus.	Belorussian	Lithu.	Lithuanian
Bulg.	Bulgarian	Norweg.	Norwegian
Car	carotenoids	PC	paper chromatography
CC	column chromatography	PhAR	photosynthetically active radiation
Chin	Chinese		
Chl	chlorophyll	Pol.	Polish
Croa .	Croatian	Ps	photosynthesis
E	English	R	Russian
F	French	Roum.	Roumanian
G	German	Span.	Spanish
GC	gas chromatography	Swed.	Swedish
Georg.	Georgian	TLC	thin-layer chromatography
Hung.	Hungarian	Tr	transpiration
IRGA	infra-red gas analyser	Ukr.	Ukrainian
Ital.	Italian	Uz.	Uzbeg

The transliteration of Cyrillic characters is in accordance with the BSI-ASA/ /SC-Z39 draft table, *i.e.* :

Translit.	Cyrill.	Translit.	Cyrill.
a	а	p	п
b	б	r	р
ch	ч	s	с
d	д	sh	ш
e	е	shch	щ
ė	э	t	т
f	ф	ts	ц
g	г	u	у
i	и	v	в
ĭ	й	y	ы
k	к	ya	я
kh	х	yu	ю
l	л	z	з
m	м	zh	ж
n	н	"	ъ
o	о	'	ь

Several exceptions apply for Ukrainian, Belorussian and Serbian :

Translit.		Cyrill.		Translit.		Cyrill.
Ukr.	y	и	Serbian :	ć		ħ
	i	i		dj		ђ
	ï	ï		dž		џ
Beloruss.				h		х
	ŭ	ў		j		j
				lj		љ
				nj		њ

Authors' names are presented in spelling used in the original paper. If this spelling does not correspond to the original spelling used by the author (*e.g.* Russian papers of English authors), one spelling is referred to the other in the Authors' Index.

Printers' errors in the original papers are marked by underlining the respective words (letters).

E R R A T A

Ref. no.	For	Read
Volume 1, Part 1		
2226	Life Sci.	Life Sci. Part II.
	915	913
2227	reduttase	reduttasi
Volume 1, Part 2		
6480	PUSCAS	PUŞCAŞ
	OTARASANU	OTĂRĂŞANU
	PUSCASU	PUŞCAŞU
	toxica	toxică
	Timisoara	Timişoara
	agron.	Agron.
		[In Roum., ab : E, R.]
SUBJECT INDEX		
Age of tissue and plant, effect on chlorophyll	7911-2	7211-2
Volume 2		
SUBJECT INDEX		
Growth of regulators, effect on chlorophyll	12944-5	12044-5
Volume 3		
15192	permaplasts	permeaplasts
AUTHORS' INDEX		
DINGER, B.E.	126	12672

15368 - **ABDULLAEV, H.A., USMANOV, P.D., TAGEEVA, S.V., NASYROV, Yu.S.** : On the homological variability of the chloroplast membrane system. - *Arabidopsis* Inform. Serv. *10* : 18 - 19, 1973.

*15369 - **ABDULLAEV, Kh.A., TAGEEVA, S.V.** : Gomologicheskaya izmenchivost' ul'trastrukturnoĭ organizatsii khloroplastov. [Homological variability of the chloroplast ultrastructural organization.] - In : Geneticheskie Aspekty Fotosinteza. Tezisy Dokladov. Pp. 7 - 8. Donish, Dushanbe 1972. [In R.]

15370 - **ABDULLAEVA, S.K., GILLER, Yu.E.** : O zakonomernostyakh obrazovaniya iskusstvennogo kompleksa khlorofilla *a* s belkom. [Regularities of formation of artificial complex of chlorophyll *a* with protein.] - Dokl. Akad. Nauk tadzh. SSR *16* (10) : 59 - 63, 1973. [In R, ab : Tajik.]

*15371 - **ABDURAKHMANOVA, Z.N., BELAN, N.F.** : Issledovanie metabolizma ugleroda u pigmentnykh mutantov *Arabidopsis thaliana*. [Carbon metabolism in chlorophyll mutants of *Arabidopsis thaliana*.] - In : Geneticheskie Aspekty Fotosinteza. Tezisy Dokladov. Pp. 40 - 41. Donish, Dushanbe 1972. [In R.]

15372 - **ABILOV, Z.K., GAZANCHYAN, R.M., GASANOV, R.A., KURBANOVA, I.M.** : O znachenii fosfolipidov v stabilizatsii fotokhimicheskikh tsentrov pigmentnykh sistem khloroplastov. [Importance of phospholipids in the stabilization of photochemical centres of chloroplast pigment systems.] - Tr. mosk. Obshch. Ispytateleĭ Prirody *49* (Problemy Biofotokhimii) : 225 - 233, 1973. [In R, ab : E.]

15373 - **ABILOV, Z.K., GAZANCHYAN, R.M., GASANOV, R.A., KURBANOVA, I.M.** : Spektral'nye izmeneniya khloroplastov, vyzvannye ekstraktsieĭ fosfolipidnoĭ fraktsii. [Chloroplast spectral changes induced by the extraction of phospholipid fraction.] - Izv. Akad. Nauk azerb. SSR, Ser. biol. Nauk *1973* (3) : 36 - 42, 1973. [In R, ab : Azerb.]

15374 - **ABOU-ZIED, E.N.** : Effects of morphactin on growth and alkaloid formation in *Datura metel* L. - Egypt. J. Bot. *16* : 137 - 144, 1973. [Chl, Car.]

*15375 - **ACZEL, A.** : Einfluß chemischer Vorgänge auf die Farbe von Tomatenmark während der Lagerung. II. Die Veränderung von Carotinoiden während der Lagerung. - Ind. Obst- Gemüseverwert. *57* : 436 - 440, 1972. [TLC, CC.]

15376 - **ADAMS, M.S., STONE, W.** : Field studies on photosynthesis of *Cladophora glomerata (Chlorophyta)* in Green Bay, Lake Michigan. - Ecology *54* : 853 - 862, 1973.

15377 - **ADEDIPE, N.O.** : Effects of 2-chloroethylphosphonic acid on growth, plastid pigments, and sucrose translocation in radish. - J. exp. Bot. *24* : 124 - 129, 1973.

15378 - **ADEDIPE, N.O., FLETCHER, R.A., ORMROD, D.P.** : Ozone lesions in relation to senescence of attached and detached leaves of tobacco. - Atmos. Environ. *7* : 357 - 361, 1973. [Chl.]

15379 - **ADEDIPE, N.O., KHATAMIAN, H., ORMROD, D.P.** : Stomatal regulation of ozone phytotoxicity in tomato. - Z. Pflanzenphysiol. *68* : 323 - 328, 1973. [Chl.]

15380 - **ADEY, W.H.** : Temperature control of reproduction and productivity in a subarctic coralline alga. - Phycologia *12* : 111 - 118, 1973.

15381 - **AGALIDIS, I., JAUNEAU, E., REISS-HUSSON, F.** : Cytochrome composition in *Rhodopseudomonas spheroides* grown under iron limited conditions. - In : Ninth International Congress of Biochemistry, Stockholm, July 1-7, 1973. Abstract Book. P. 229. IUB, Stockholm 1973.

15382 - **AHMADI, N., TING, I.P.** : Activation of NADP-malate dehydrogenase by lipoate. - Plant Sci. Lett. *1* : 11 - 14, 1973.

15383 - **AHMED, A.M., OSMAN, M.E.H.** : The influence of light on $^{14}CO_2$-fixation by synchronous cultures of *Chlorella pyrenoidosa*. - Egypt. J. Bot. *16* : 319 - 327, 1973.

15384 - **AIKAZYAN, V. Ts., MUTUSKIN, A.A., PSHENOVA, K.V., NALBANDYAN, R.M.** : The effect of denaturating agents on the EPR and optical absorption spectra of plastocyanins. - FEBS Lett. *34* : 103 - 105, 1973.

*15385 - **AKHMEDOV, B., ISHMUKHAMEDOVA, S.G.** : Osobennosti fotosinteza u nekotorykh sortov i mutantov khlopchatnika. [Specificity of photosynthesis in some cotton cultivars and mutants.] - In : Geneticheskie Aspekty Fotosinteza. Tezisy Dokladov. Pp. 52 - 53. Donish, Dushanbe 1972. [In R.]

15386 - AKITA, S., MOSS, D.N. : Photosynthetic responses to CO_2 and light by maize and wheat leaves adjusted for constant stomatal apertures. - Crop Sci. *13* : 234 - 237, 1973.

15387 - AKITA, S., MOSS, D.N. : The effect of an oxygen-free atmosphere on net photosynthesis and transpiration of barley (*Hordeum vulgare* L.) and wheat (*Triticum aestivum* L.) leaves. - Plant Physiol. *52* : 601 - 603, 1973.

15388 - AKITA, S., TANAKA, I. : [Studies on the mechanism of differences in photosynthesis among species. III. Influence of low oxygen concentration on dry matter production and grain fertility of rice plant.] - Proc. Crop Sci. Soc. Jap. *42* : 18 - 23, 1973. [In Jap., ab : E.]

15389 - AKITA, S., TANAKA, I. : Studies on the mechanism of differences in photosynthesis among species. IV. The differential response in dry matter production between C_3 and C_4 species to atmospheric carbon dioxide enrichment. - Proc. Crop Sci. Soc. Jap. *42* : 288 - 295, 1973.

15390 - AKOYUNOGLOU, G., MANETAS, J., MICHELINAKI, M., MICHALOPOULOS, G. : Studies on the fate of "exogenous" protochlorophyllide. - Plant Physiol. *51* (Suppl.) : 28, 1973.

15391 - AKULOVA, A.E., MURZAEVA, S.V. : Deĭstvie katalazy, peroksidazy i perekisi vodoroda na svetovye reaktsii izolirovannykh khloroplastov gorokha. [Influence of catalase, peroxidase and hydrogen peroxide on the light reactions of the isolated chloroplasts of pea.] - Biokhimiya *38* : 901 - 905, 1973. [In R, ab : E.]

15392 - AKULOVICH, N.K., ARLOŬSKAYA, K.I., PARSHYKAVA, T.A. : Dasledavanne rêaktsyi adnaŭlennya protakhlarafilidu ètyyaliravanykh listsyaŭ z chastkova razburanaĭ nagravannem formaĭ 650. [Reduction of protochlorophyllide in etiolated leaves with partially heat-degraded form 650.] - Vestsi Akad. Navuk belarus. SSR, Ser. biyal. Navuk *1973* (2) : 52 - 60, 139, 1973. [In Belorus., ab : R.]

15393 - AKULOVICH, N.K., ORLOVSKAYA, K.I., PARSHIKOVA, T.A. : Issledovanie obrazovaniya spektral'nykh form protokhlorofillida i ego nakopleniya v protsesse razvitiya rasteniĭ v temnote. [Formation of spectral forms of protochlorophyllide and its accumulation in the process of plant development in darkness.] - In : Upravlenie Skorost'yu i Napravlennost'yu Biosinteza u Rasteniĭ. P. 5. Krasnoyarsk 1973. [In R.]

15394 - AKULOVICH, N.K., ORLOVSKAYA, K.I., PARSHIKOVA, T.A. : Vzaimosvyaz' sostoyaniya i funktsii form protokhlorofillovogo pigmenta ètiolirovannykh rasteniĭ. [Interrelation of shape and function of forms of the protochlorophyll pigment in etiolated plants.] - In : Formirovanie Pigmentnogo Apparata Fotosinteza. Pp. 3 - 29. Nauka i Tekhnika, Minsk 1973. [In R.]

15395 - ALBERTE, R.S., THORNBER, J.P., NAYLOR, A.W. : Biosynthesis of the photosystem I chlorophyll-protein complex in greening leaves of higher plants. - Proc. nat. Acad. Sci. USA *70* : 134 - 137, 1973.

15396 - ALDERFER, R.G. : Interaction of solar radiation with plant systems. - Solar Energy *15* : 77 - 82, 1973. [Energy budget.]

15397 - ALIKOV, Kh. K. : Fotokolorimetricheskiĭ metod opredeleniya soderzhaniya ugleroda v list'yakh mokrym szhiganiem v khromovoĭ smesi. [Photocolorimetric method of carbon contents determination in leaves by wet combustion in bichromate-sulfuric acid mixture.] - In : Metody Kompleksnogo Izucheniya Fotosinteza. Vol. 2. Pp. 6 - 14. VASKHNIL, Leningrad 1973. [In R.]

15398 - ALLAWAY, W.G. : Accumulation of malate in guard cells of *Vicia faba* during stomatal opening. - Planta *110* : 63 - 70, 1973.

15399 - ALLAWAY, W.G., HSIAO, T.C. : Preparation of rolled epidermis of *Vicia faba* L. so that stomata are the only viable cells : Analysis of guard cell potassium by flame photometry. - Aust. J. biol. Sci. *26* : 309 - 318, 1973.

15400 - ALLEN, H.L. : Production and utilization of dissolved organic carbon during *in situ* phytoplankton photosynthesis measurements. - Int. Rev. ges. Hydrobiol. *58* : 843 - 849, 1973.

15401 - ALLEN,J.F., HALL, D.O. : Superoxide reduction as a mechanism of ascorbate-stimulated oxygen uptake by isolated chloroplasts. - Biochem. biophys. Res. Commun. *52* : 856 - 862, 1973.

15402 - ALLEN, T.D., HAIGH, M.V., HOWARD, A. : Ultrastructure of giant plastids in a radiation induced mutant of *Osmunda regalis*. - J. Ultrastruct. Res. *42* : 491 - - 501, 1973.

15403 - ALLEN, T.F.H., KOONCE, J.F. : Multivariate approaches to algal stratagems and tactics in systems analysis of phytoplankton. - Ecology *54* : 1234 - 1246, 1973. [Ps.]

15404 - ALLESSIO, M.L., TIESZEN, L.L. : Patterns of translocation and allocation of ^{14}C-photoassimilate *in situ* studies with *Dupontia fischeri* R.BR. Barrow, Alaska. - In : BLISS, L.C., WIELGOLASKI, F.E. (ed.) : Primary Production and Production Processes, Tundra Biome. Pp. 219 - 229. Tundra Biome Steering Comm., Edmonton 1973.

15405 - AMBROSAŬ, A.L., SHCHUTSKAYA, V.V. : Uplyŭ virusnaĭ infektsyi na intensiŭnasts' fotasintezu i dykhannya ŭ raslin bul'by. [Effect of virus infection on the rates of photosynthesis and respiration in potato plants.] - Vestsi Akad. Navuk belarus. SSR, Ser. biyal. Navuk *1973* (6) : 55 - 58, 1973. [In Belorus., ab : R.]

15406 - AMESZ, J. : III. Photosynthesis. Biophysical aspects. - Fortschr. Bot. *35* : 89 - 102, 1973.

15407 - AMESZ, J. : Spectrophotometric methods in photobiology. - In : CHECCUCCI, A., WEALE, R.A. (ed.) : Primary Molecular Events in Photobiology. Pp. 21 - 43. Elsevier, Amsterdam - London - New York 1973. [Chl.]

15408 - AMESZ, J. : Primary photochemistry of systems 1 and 2 of photosynthesis. - In: CHECCUCCI, A., WEALE, R.A. (ed.) : Primary Molecular Events in Photobiology. Pp. 125 - 145. Elsevier, Amsterdam - London - New York 1973.

15409 - AMESZ, J. : The function of plastoquinone in photosynthetic electron transport. - Biochim. biophys. Acta *301* : 35 - 51, 1973.

15410 - AMESZ, J., 't MANNETJE, A.H., de GROOTH, B.G. : Light-induced shifts in pigment absorption and changes of fluorescence yield of bacteriochlorophyll in *Rhodopseudomonas spheroides*. - In : Abstr. Symp. Prokaryotic Photosynthetic Organisms, September 19 - 23, 1973. Pp. 34 - 35. Freiburg i. Br. 1973.

15411 - AMESZ, J., PULLES, M.P.J., VELTHUYS, B.R. : Light-induced changes of fluorescence and absorbance in spinach chloroplasts at - 40 °C. - Biochim. biophys. Acta *325* : 472 - 482, 1973.

15412 - AMIRDZHANOV, A.G., POTAPOV, N.S., VELIEV, S. Ya., KIRPICHEV, I.V. : Opredelenie pogloshcheniya solnechnoĭ radiatsii vinogradnikom. [Determination of absorption of solar radiation by vineyard.] - Fiziol. Rast. *20* : 1198 - 1203, 1973. [In R, ab : E.]

15413 - ANDERSON, J.E., McNAUGHTON, S.J. : Effects of low soil temperature on transpiration, photosynthesis, leaf relative water content, and growth among elevationally diverse plant populations. - Ecology *54* : 1220 - 1233, 1973.

15414 - ANDERSON, J.L., THOMSON, W.W. : The effects of herbicides on the ultrastructure of plant cells. - Residue Rev. *47* : 167 - 189, 1973. [Chloroplast.]

15415 - ANDERSON, J.L., THOMSON, W.W., SWADER, J.A. : Fine structure of *Wolffia arrhiza*. - Can. J. Bot. *51* : 1619 - 1622, 1973. [Chloroplast.]

15416 - ANDERSON, J.M., BOARDMAN, N.K. : Localization of low potential cytochrome b-559 in photosystem I. - FEBS Lett. *32* : 157 - 160, 1973.

15417 - ANDERSON, J.M., GOODCHILD, D.J., BOARDMAN, N.K. : Composition of the photosystems and chloroplast structure in extreme shade plants. - Biochim. biophys. Acta *325* : 573 - 585, 1973.

15418 - ANDERSON, J.M., THAN-NYUNT, BOARDMAN, N.K. : Light-induced redox changes of cytochrome b-559. - Arch. Biochem. Biophys. *155* : 436 - 444, 1973.

15419 - **ANDERSON, L.E.** : Dithiothreitol activation of some chloroplast enzymes in extracts of etiolated pea seedlings. - Plant Sci. Lett. *1* : 331 - 334, 1973.

15420 - **ANDERSON, M.M.** : A study on the role of galactolipids in spinach chloroplast lamellar membranes. - Diss. Abstr. Int. B *33* : 5641-B, 1973.

15421 - **ANDERSON, S.M., KRINSKY, N.I.** : Protective action of carotenoid pigments against photodynamic damage to liposomes. - Photochem. Photobiol. *18* : 403 - - 408, 1973.

15422 - **ANDERSSON, G., CRONBERG, C., GELIN, C.** : Planktonic changes following the restoration of Lake Trummen, Sweden. - Ambio *2* : 44 - 47, 1973. [Production.]

15423 - **ANDREENKO, T.I., PYT'EVA, N.F., RIZNICHENKO, G. Yu., RUBIN, L.B., RUBIN, A.B.** : O nekotorykh osobennostyakh stroeniya élektron-transportnoĭ tsepi bakterial'-nogo fotosinteza. [Some characteristics of the structure of the electron-transport chain of bacterial photosynthesis.] - Nauch. Dokl. vyssh. Shkoly, biol. Nauki *16* (10) : 57 - 62, 1973. [In R.]

15424 - **ANDREEVA, T.F., PERSANOV, V.M.** : Vliyanie fosfornogo golodaniya na fotosinteticheskuyu aktivnost' list'ev, ottok i ispol'zovanie assimilyatov v svyazi s rostom i produktivnost'yu rasteniĭ. [Effect of phosphorus deficiency on the photosynthetic activity of leaves, outflow and using of photosynthates in connection with growth and productivity of plants.] - Tr. biol.-pochv. Inst., dal'nevost. nauch. Tsentr Akad. Nauk SSSR *20* (123) : 179 - 185, 1973. [In R, ab : E.]

*15425 - **ANDREI, M.** : La production primaire de quelques associations de macrophytes aquatiques du complexe des lacs Crapina-Jijila (Dobroudja). - Rev. roum. Biol., Sér. Bot. *16* : 335 - 339, 1971.

15426 - **ANDREO, C.S., VALLEJOS, R.H.** : Inhibition of energy transfer reactions in spinach chloroplasts by discarine *B*, a new peptide alkaloid. - FEBS Lett. *33* : 201 - 204, 1973.

15427 - **ANDREWS, T.J., LORIMER, G.H., TOLBERT, N.E.** : Ribulose diphosphate oxygenase. I. Synthesis of phosphoglycolate by fraction-1 protein of leaves. - Biochemistry *12* : 11 - 18, 1973.

15428 - **ANISIMOV, A., OLUNINA, L., BULATOVA, T.** : Influence of mineral nutrition on the translocation of assimilates and auxins in plant organisms. - Proc. Res. Inst. Pomol., Skierniewice, Ser. E *1973* (3) : 227 - 231, 1973.

15429 - **ANISIMOV, A.A.** : Faktory, opredelyayushchie intensivnost' i napravlenie transporta assimilyatov v raznykh usloviyakh mineral'nogo pitaniya. [Factors determining the rate and direction of photosynthate transport under different mineral nutrition.] - Tr. biol.-pochv. Inst., dal'nevost. nauch. Tsentr Akad. Nauk SSSR *20* (123) : 168 - 173, 1973. [In R, ab : E.]

15430 - **ANPILOGOVA, N.N.** : Obmen veshchestv u rasteniĭ ozimykh pshenits v zavisimosti ot yarovizatsii i dliny dnya. [Metabolism of winter wheat plants in dependence on vernalisation and day length.] - Byull. vses. nauch. issled. Inst. Rastenievod. N.I. Vavilova *32* : 19 - 23, 1973. [Ps; in R.]

15431 - **ANSTIS, P.J.P., NORTHCOTE, D.H.** : Development of chloroplasts from amyloplasts in potato tuber discs. - New Phytol. *72* : 449 - 463, 1973.

15432 - **ANTOSZEWSKI, R., DZIĘCIOŁ, U.** : Influence of growth inhibitors on the import of assimilates into strawberry receptacle. - Proc. Res. Inst. Pomol., Skierniewice, Ser. E *1973* (3) : 191 - 195, 1973.

15433 - **ANTOSZEWSKI, R., KAMIŃSKA, M.K., DZIĘCIOŁ, U.** : Translocation profile of assimilates in the strawberry petiole. - Proc. Res. Inst. Pomol., Skierniewice, Ser. E *1973* (3) : 31 - 40, 1973.

15434 - **D'AOUST, A.L., CANVIN, D.T.** : Effect of oxygen concentration on the rates of photosynthesis and photorespiration of some higher plants. - Can. J. Bot. *51* : 457 - 464, 1973.

*15435 - **D'AOUST, B.G.** : Natural resistance to oxygen poisoning. - In : **LAMBERTSEN,** **C.J.** (ed.) : Underwater Physiology. Pp. 23 - 34. Academic Press, New York - London 1971. [Ps.]

15436 - **APEL, P., LEHMANN, C.O., FRIEDRICH, A.** : Beziehungen zwischen Fahnenblattflä-
che, Photosyntheserate und Einzelährenertrag bei Sommerweizen. - Kulturpflanze
21 : 89 - 95, 1973.

15437 - **APEL, P., TSCHÄPE, M., SCHALLDACH, I., AURICH, O.** : Die Bedeutung der Karyop-
sen für die Photosynthese und Trockensubstanzproduktion bei Weizen. - Photosynthetica *7* : 132 - 139, 1973.

15438 - **ARENS, I.P., SAMPAT, T.V.** : Vliyanie raznykh doz i sposobov primeneniya molibdena na urozhaĭ i kachestvo lyupina. [Effect of different application methods
and doses of molybdenum on the yield and quality of lupine.] - Vest. mosk. U-
niv., Ser. VI - Biol., Pochvoved. *28* (2) : 72 - 80, 1973. [Chl; in R, ab : E.]

15439 - **ARGYROUDI-AKOYUNOGLOU, J.H., AKOYUNOGLOU, G.** : On the formation of photosynthetic membranes in bean plants. - Photochem. Photobiol. *18* : 219 - 228, 1973.

15440 - **ARGYROUDI-AKOYUNOGLOU, J.H., KONDYLAKI, S., AKOYUNOGLOU, G.** : Studies on the
formation of grana. - In : Ninth International Congress of Biochemistry,
Stockholm, July 1-7, 1973. Abstract Book. P. 292. IUB, Stockholm 1973.

15441   **ARKCOLL, D.B.** : The preservation and storage of leaf protein preparations.
- J. Sci. Food Agr. *24* : 437 - 445, 1973. [Chl, Car.]

15442 - **ARKCOLL, D.B., HOLDEN, M.** : Changes in chloroplast pigments during the preparation of leaf protein. - J. Sci. Food Agr. *24* : 1217 - 1227, 1973.

15443 - **ARMITAGE, B.J., SIMMONS, G.M. Jr.** : Phytoplankton and primary productivity
investigations on a new reservoir. - J. Phycol. *9* (Suppl.) : 9, 1973.

15444 - **ARNON, D.I., KNAFF, D.B.** : Primary electron acceptor in Photosystem II of
plant photosynthesis. - In : 4th International Biophysics Congress. Symposial
Papers. Vol. 2 (II). Pp. 491 - 503. Moskva 1973.

15445 - **ARNOTT, H.J., HARRIS, J.B.** : Development of chloroplast substructures with
apparent secretory roles in the young tobacco leaf. - Tissue Cell *5* : 337 -
- 347, 1973.

15446 - **ARNTZEN, C.J., HAUGH, M.F., BOBICK, S.** : Induction of stomatal closure by
Helminthosporium maydis pathotoxin. - Plant Physiol. *52* : 569 - 574, 1973.
[Ps.]

15447 - **ARTAMONOV, V.I., KURAMAGOMEDOV, M.K.** : Vliyanie gibberellina na razrushenie
khlorofilla v intaktnykh rasteniyakh i vysechkakh iz list'ev fasoli. [Effect
of gibberellin on the breakdown of chlorophyll in intact bean plants and leaf
discs.] - Nauch. Dokl. vyssh. Shkoly, biol. Nauki *16* (6) : 79 - 84, 1973.
[In R.]

15448 - **ARUGA, Y.** : Primary production in the Indian Ocean II. - In : ZEITSCHEL, B.
(ed.) : The Biology of the Indian Ocean. Pp. 127 - 130. Springer-Verlag, Berlin - Heidelberg - New York 1973.

15449 - **ARYA, S.K., WAYGOOD, E.R.** : Carbon dioxide fixation by wheat leaf chloroplasts
in the presence of P-enolpyruvate or ribulose-1,5-diphosphate. - Plant Physiol.
51 (Suppl.) : 7, 1973.

15450 - **ASADA, K., KISO, K.** : Initiation of aerobic oxidation of sulfite by illuminated spinach chloroplasts. - Europe. J. Biochem. *33* : 253 - 257, 1973.

15451 - **ASADA, K., KISO, K.** : The photo-oxidation of epinephrine by spinach chloroplasts and its inhibition by superoxide dismutase : evidence for the formation
of superoxide radicals in chloroplasts. - Agr. biol. Chem. *37* : 453 - 454,
1973.

15452 - **ASADA, K., KISO, K.** : The univalent reduction of oxygen in chloroplasts on
illumination. - In : Ninth International Congress of Biochemistry, Stockholm,
July 1-7, 1973. Abstract Book. P. 230. IUB, Stockholm 1973.

15453 - **ASADA, K., URANO, M., TAKAHASHI, M.** : Subcellular location of superoxide dismutase in spinach leaves and preparation and properties of crystalline spinach
superoxide dismutase. - Europe. J. Biochem. *36* : 257 - 266, 1973.

15454 - **ASHOUR, N.I., NEUMANN, D., zur NIEDEN, U.** : Gibberellic acid induced changes
in the ultrastructure of chloroplasts and the content of chlorophyll in leaves

of dwarf maize (*Zea mays* L.). - Biochem. Physiol. Pflanzen *164* : 402 - 413, 1973.

15455 - ASLAM, M., HUFFAKER, R.C. : Effect of DCMU, simazine and atrazine on nitrate reductase activity in *Hordeum vulgare in vitro* and *in vivo*. - Physiol. Plant. *28* : 400 - 404, 1973.

15456 - ASLAM, M., HUFFAKER, R.C., TRAVIS, R.L. : The interaction of respiration and photosynthesis in induction of nitrate reductase activity. - Plant Physiol. *52* : 137 - 141, 1973.

*15457 - ASOEVA, L.M., LIPKIND, B.I., GILLER, Yu.E. : Ob osobennostyakh énergeticheskogo vzaimodeĭstviya fotosinteticheskikh pigmentov v uproshchennykh sistemakh *in vivo* i *in vitro*. [The peculiarities of energetic interaction of photosynthetic pigments in the simplified systems *in vivo* and *in vitro*.] - In : Geneticheskie Aspekty Fotosinteza. Tezisy Dokladov. Pp. 81 - 82. Donish, Dushanbe 1972. [In R.]

*15458 - ASROROV, K.A. : Vidovye osobennosti produktivnosti rasteniĭ. [Species peculiarities of plant productivity.] - In : Geneticheskie Aspekty Fotosinteza. Tezisy Dokladov. Pp. 95 - 96. Donish, Dushanbe 1972. [In R.]

15459 - ASTON, A.R., MILLINGTON, R.J., PETERS, D.B. : The energy balance of leaves. - In : SLATYER, R.O. (ed.) : Plant Response to Climatic Factors. Pp. 37 - 44. Unesco, Paris 1973.

15460 - ASTON, M.J. : Changes in internal water status and the gas exchange of leaves in response to ambient evaporative demand. - In : SLATYER, R.O. (ed.) : Plant Response to Climatic Factors. Pp. 243 - 247. Unesco, Paris 1973.

15461 - AUBERT, B., GAILLARD, J.P., PY, C., LOSSOIS, P., MARCHAL, J. : Influence de l'altitude sur le comportement de l'ananas "Cayenne lisse": Essais réalisés au pied du Mont Cameroun. - Fruits *28* : 203 - 214, 1973. [Growth analysis.]

15462 - van AUKEN, O.W., LEE, R.T.S., MENDEZ, V.M., ROWLANDS, J.R. : Effects of SO_2 on photosynthesis and respiration in bean leaves and cellular organelles. - Plant Physiol. *51* (Suppl.) : 21, 1973.

15463 - AULENBACH, B.B., WORTHINGTON, J.T. : New portable colorimeter to evaluate external fruit color of tomato and peach. - HortScience *8* : 92 - 94, 1973.

15464 - AUSTIN, M.E., AUNG, L.H. : Patterns of dry matter distribution during development of sweet potato (*Ipomoea batatas*). - J. hort. Sci. *48* : 11 - 17, 1973. [Growth analysis.]

15465 - AVDEEVA, T.A., ANDREEVA, T.F. : Nitrogen nutrition and activities of CO_2-fixing enzymes and glyceraldehyde phosphate dehydrogenase in broad bean and maize. - Photosynthetica *7* : 140 - 145, 1973.

15466 - AVRATOVŠČUKOVÁ, N., ŘEŘÁBEK, J. : Porovnání fotosyntetické aktivity haploidních a diploidních rostlin *Nicotiana tabacum* L. a intenzity růstu tkáňových kultur z nich odvozených. [Comparison of photosynthetic activity of haploid and diploid plants of *Nicotiana tabacum* L. and intensity of growth of their tissues cultures.] - In : LANDA, Z., NOVÁK, F.J., OPATRNÝ, Z. (ed.) : Kolokvium Využití Kultur Rostlinných Explantátů *in vitro* v Genetice a Šlechtění. Pp. 241 - 247. Ústav exp. Bot. ČSAV, Praha - Olomouc 1973. [In Czech, ab : E.]

15467 - AVRON, M., BAMBERGER, E.S., ROTTENBERG, H., SCHULDINER, S. : The relation of energy linked proton movements to the mechanism of photophosphorylation in chloroplasts. - In : Ninth International Congress of Biochemistry, Stockholm, July 1-7, 1973. Abstract Book. P. 215. IUB, Stockholm 1973.

*15468 - BABADZHANOVA, M.A., GORENKOVA, A.G. : Vliyanie kinetina i leĭtsina na fiksatsiyu $C^{14}O_2$ fermentnymi preparatami, vydelennymi iz iskhodnykh i mutantnykh form *Arabidopsis thaliana*. [Effect of kinetin and leucine on the $^{14}CO_2$ fixation by enzyme preparations from wild and mutant forms of *Arabidopsis thaliana*.] - In : Geneticheskie Aspekty Fotosinteza. Tezisy Dokladov. Pp. 41 - 42. Donish, Dushanbe 1972. [In R.]

*15469 - BABADZHANOVA, M.A., KHAITOVA, L.T. : Fiksatsiya $C^{14}O_2$ v prisutstvii D-ribozo-

-5-fosfata i ATF fermentnymi preparatami, vydelennymi iz *Arabidopsis thaliana*
(L.) HEYNH. [Fixation of $^{14}CO_2$ in the presence of D-riboso-5-phosphate and
ATP by enzyme preparations isolated from *Arabidopsis thaliana* (L.) HEYNH.] -
Dokl. Akad. Nauk tadzh. SSR *15* (6) : 58 - 61, 1972. [In R.]

15470 - **BABAEV, T.B., AL'PEROVICH, L.I.** : Opticheskie kharakteristiki plenok β-karotina. [Optical characteristics of β-carotene layers.] - Zh. prikl. Spektroskop.
18 : 513 - 515, 1973. [In R.]

*15471 - **BABAEVA, T.N.** : O vliyanii tsitokininov na fotosinteticheskiɣ apparat mutantov gorokha. [Effect of cytokinins on the photosynthetic apparatus of pea mutants.] - In : Geneticheskie Aspekty Fotosinteza. Tezisy Dokladov. Pp. 96 -
- 97. Donish, Dushanbe 1972. [In R.]

15472 - **BABCOCK, G.T., SAUER, K.** : Electron paramagnetic resonance signal II in spinach chloroplasts. I. Kinetic analysis for untreated chloroplasts. - Biochim.
biophys. Acta *325* : 483 - 503, 1973.

15473 - **BABCOCK, G.T., SAUER, K.** : Electron paramagnetic resonance signal II in spinach chloroplasts. II. Alternative spectral forms and inhibitor effects on kinetics of signal II in flashing light. - Biochim. biophys. Acta *325* : 504 -
- 519, 1973.

*15474 - **BACCARI, V., CIAMPI, C., CORTI, R., FIRENZUOLI, A.M., GUERRITORE, A., MAGINI,
E., MASTRONUZZI, E., RAMPONI, G., VANNI, P., ZANOBINI, A.** : Ricerche sull'in-
compatibilità d'innesto nelle conifere. [Incompatibility of grafts in conifers.] - Ann. Acad. ital. Sci. forest. *17* : 35 - 100, 1968. [Chl; in Ital.,
ab : F.]

15475 - **BACCARINI MELANDRI, A., ZANNONI, D., MELANDRI, B.A.** : Energy transduction in
photosynthetic bacteria. VI. Respiratory sites of energy conservation in membranes from dark-grown cells of *Rhodopseudomonas capsulata*. - Biochim. biophys.
Acta *314* : 298 - 311, 1973.

15476 - **BACHMANN, M.D., ROBERTSON, D.S., BOWEN, C.C., ANDERSON, I.C.** : Chloroplast
ultrastructure in pigment-deficient mutants of *Zea mays* under reduced light.
- J. Ultrastruct. Res. *45* : 384 - 406, 1973.

15477 - **BACŁAWSKA-KRZEMIŃSKA, Z.** : Influence of light, water deficit and age of plant
on photosynthesis and air passage capacity in leaves of *Brassica oleracea* L.
var. *capitata alba* v. Ditmarska. - Hodowla Rośl., Aklimat. Nasienn. *17* : 303 -
- 328, 1973.

15478 - **BACŁAWSKA-KRZEMIŃSKA, Z., JARECKA, M., SKOŚKIEWICZ, K., WRÓBLEWSKA, H.** : Methods of air passage capacity, photosynthesis and water deficit measurement
applied at the department of plant physiology of the Warsaw University. - Hodowla Rośl., Aklimat. Nasienn. *17* : 297 - 302, 1973.

15479 - **BADOUR, S.S., FOO, S.K., WAYGOOD, E.R.** : Regulation of isocitrate metabolism
in *Chlamydomonas segnis*. - Plant Physiol. *51* (Suppl.) : 7, 1973.

15480 - **BAGAUTDINOVA, R.I., IVANOVA, N.I.** : Strukturnye izmeneniya fotosinteticheskogo apparata pri chastichnoɣ defoliatsii rasteniɣ kartofelya. [Changes of the
structure of photosynthetic apparatus after partial defoliation of potato
plants.] - In : Voprosy Regulyatsii Fotosinteza. Vol. 3. Pp. 125 - 130, 164.
Ural'. gos. Univ., Sverdlovsk 1973. [In R.]

15481 - **BAHL, J., LECHEVALLIER, D., MONÉGER, R.** : Sur les lipides de plastes de Blé
sélectionnés selon un protocole identique à partir de feuilles vertes, étiolées et verdissantes. - Compt. rend. Acad. Sci. Paris, Sér. D *276* : 969 - 972,
1973. [Chl.]

*15482 - **BAIA, V.** : Modificarea intensitătii fotosintezei la soia sub influenţa facto-
rilor de nutriţie minerală. [Photosynthesis intensity modification in soybeans
under the influence of mineral nutrition factor.] - Lucr. ştiinţ. Inst. agron.
Timişoara, Ser. Agron. *13* : 189 - 195, 1970. [In Roum., ab : E, R.]

15483 - **BAJWA, S.S., SASTRY, P.S.** : Enzymatic acylation of monogalactosyl monoglyceri-
de by leaf chloroplast acetone powder. - Indian J. Biochem. Biophys. *10* :
65 - 66, 1973.

*15484 - **BAKARDJIEVA, N., IVANOVA, Y.** : Interaction of gibberellic acid and certain
trace elements in the regulation of the growth and the pigment regime in hemp
(*Cannabis sativa* L.). - Dokl. Akad. sel'skokhoz. Nauk Bolg. *3* : 161 - 166,
1970.

15485 - **BAKER, A.F., HOLTON, R.W.** : Electrophoretic analysis of proteins and malic dehydrogenase isozymes in nine oscillatorian blue-green algae. - Phycologia
12 : 83 - 87, 1973. [Biliproteins.]

15486 - **BAKER, E.W., SMITH, G.D.** : Pleistocene changes in chlorophyll pigments. - In :
Advances in Organic Geochemistry 1973. Pp. 649 - 660. Éd. Technip, Paris 1973.

15487 - **BAKER, N.R., HARDWICK, K.** : Biochemical and physiological aspects of leaf development in cocoa (*Theobroma cacao*). I. Development of chlorophyll and photosynthetic activity. - New Phytol. *72* : 1315 - 1324, 1973.

15488 - **BAKER, R.A., WEAVER, E.C.** : A correlation of EPR spins with P_{700} in spinach
subchloroplast particles. - Photochem. Photobiol. *18* : 237 - 241, 1973.

15489 - **BAKKER-GRUNWALD, T., van DAM, K.** : The energy level associated with the light-
-triggered Mg^{2+}-dependent ATPase in spinach chloroplasts. - Biochim. biophys.
Acta *292* : 808 - 814, 1973.

15490 - **BAKRI, M.D.L.** : A photosynthetic study of olive necrotic 8147 mutant and normal maize (*Zea mays*). - Diss. Abstr. int. B *33* : 4667-B, 1973.

15491 - **BALASKO, J.A., SMITH, D.** : Carbohydrates in grasses : V. Incorporation of ^{14}C
into plant parts and nonstructural carbohydrates in timothy (*Phleum pratense*
L.) at three developmental stages. - Crop Sci. *13* : 19 - 22, 1973.

*15492 - **BALAUR, N.S., TSIULYANU, I.I.** : Izmenenie membrannoĭ sistemy kletok ětioliro-
vannykh prorostkov v rezul'tate svetoimpul'snogo oblucheniya semyan pshenitsy.
[Change in the membrane system of cells in etiolated shoots as a result of
irradiation of wheat seeds with light pulses.] - Izv. Akad. Nauk mold. SSR,
Ser. biol. khim. Nauk *1972* (1) : 11 - 16, 90, 1972. [Chloroplast; in R.]

15493 - **BALDING, F.R., CUNNINGHAM, G.L., PLUMMER, R.F.** : An inexpensive self contained
system for field measurements of gas exchange. - Photosynthetica *7* : 382 -
- 386, 1973.

15494 - **BALDRY, C.W., COOMBS, J.** : Regulation of photosynthetic carbon metabolism by
pH and Mg^{2+}. - Z. Pflanzenphysiol. *69* : 213 - 216, 1973.

15495 - **BALDY, C.** : Sur l'énergie active en photosynthèse. Son utilisation par des
graminées au cours de leur développement. Cas particulier de peuplements de
Blé. - Ann. agron. *24* : 1 - 31, 1973.

15496 - **BALDY, C.** : Contribution à l'étude de l'extinction de l'énergie d'origine solaire dans des peuplements de Blé tendre (*Triticum aestivum* L. em. THIELL.).
- Ann. agron. *24* : 507 - 532, 1973.

15497 - **BALDY, C., COMBRES, J.-C., BONHOMME, R.** : Utilisation de pyranomètres linéai-
res dans l'étude des éléments du climat lumineux dans la végétation. - In :
SLATYER, R.O. (ed.) : Plant Response to Climatic Factors. Pp. 45 - 49. Unesco,
Paris 1973.

15498 - **BALDY, C.M.** : Contribution à l'étude de la photosynthèse apparente du Blé.
Application de méthodes gravimétriques en conditions naturelles. - Oecol. Plant.
8 : 247 - 262, 1973.

15499 - **BALDY, C.M.** : Effets de la competition pour la lumière sur le tallage et le
developpement reproductif des variétés demi-naines de Blé tendre. - In :
Journées d'Étude "Demi-nanisme et Amélioration des Céréales à Paille", Groupe
de Travail "Céréales" de l'INRA. Pp. 1 - 15. CNRA, Versailles 1973.

15500 - **BALLESTER, A.** : Regularidad y heterogeneidad en los pigmentos fotosintetiza-
dores. [Regularity and heterogeneity of photosynthetic pigments.] - Invest.
Pesquera *37* : 245 - 293, 1973. [In Span., ab : E.]

15501 - **BALLESTER, A., PLANA, A.** : Análisis fluorimétrico contínuo de las clorofilas
fitoplanctónicas. [Continuous fluorimetric analysis of chlorophylls of phytoplankton.] - Afinidad *30* : 787 - 794, 1973. [In Span.]

15502 - **BALTSCHEFFSKY, M.** : Carotenoids, cytochromes and the energized state in *R. rubrum* chromatophores. - In : Ninth International Congress of Biochemistry, Stockholm, July 1-7, 1973. Abstract Book. P. 209. IUB, Stockholm 1973.

15503 - **BAMBERGER, E.S., ROTTENBERG, H., AVRON, M.** : Internal pH, ΔpH, and the kinetics of electron transport in chloroplasts. - Europe. J. Biochem. *34* : 557 - - 563, 1973.

15504 - **BANKIN, M.P.** : Kontrolirovanie khoda nakopleniya biomassy rasteniĭ po balansu uglekislotno-kislorodnogo gazoobmena. [Control of the course of plant biomass accumulation by means of the CO_2-O_2 gas exchange balance.] - In : Upravlenie Skorost'yu i Napravlennost'yu Biosinteza u Rasteniĭ. P. 132. Krasnoyarsk 1973. [In R.]

15505 - **BANSCHBACH, M.W.** : Evaluation of β-carotene binding to lamellar protein isolated from spinach (*Spinacia oleracea*) chloroplasts. - Diss. Abstr. int. B *33* : 4643-B - 4644-B, 1973.

15506 - **BANTHORPE, D.V., BAXENDALE, D.** : Model systems for the biological oxidation of monoterpene hydrocarbons. - Planta med. *23* : 239 - 250, 1973. [Chl.]

15507 - **BARANOV, A.A., KURMANGALIN, K.N., LIPINSKAYA, N.D.** : Regulyatsiya biosinteza khlorofilla v svyazi s vozrastom listovoĭ plastinki i razvitiem rasteniĭ. [Control of chlorophyll biosynthesis in relation to leaf area extension and plant development.] - In : Upravlenie Skorost'yu i Napravlennost'yu Biosinteza u Rasteniĭ. Pp. 73 - 74. Krasnoyarsk 1973. [In R.]

15508 - **BARASHKOV, B.I., CHIBISOV, A.K.** : Promezhutochnye produkty fotokhimicheskikh reaktsiĭ vodnykh rastvorov khlorofilla v prisutstvii *Tritona X-100*. [Intermediates of photochemical reactions of aqueous solutions of chlorophyll in the presence of *Triton X-100*.] - Biofizika *18* : 747 - 752, 1973. [In R, ab : E.]

15509 - **BARBER, J., SHIEH, Y.J.** : Effects of light on net Na^+ and K^+ transport in *Chlorella* and evidence for *in vivo* cyclic phosphorylation. - Plant Sci. Lett. *1* : 405 - 411, 1973.

15510 - **BARDUKADZE, D.A., CHANUKVADZE, S.A.** : K metodike izucheniya ploshchadi list'ev chaya, tsitrusovykh i lavra blagorodnogo. [Methods for measuring leaf area in tea, *Citrus* and *Laurus nobilis*.] - Subtrop. Kul't. *1973* (4) : 156 - 159, 1973. [In R.]

15511 - **BARNES, D.J., TAYLOR, D.L.** : *In situ* studies of calcification and photosynthetic carbon fixation in the coral *Montastrea annularis*. - Helgoländer wiss. Meeresuntersuchungen *24* : 284 - 291, 1973.

15512 - **BARNES, F.J., QURESHI, A.A., SEMMLER, E.J., PORTER, J.W.** : Prelycopersene pyrophosphate and lycopersene. Intermediates in carotene biosynthesis. - J. biol. Chem. *248* : 2768 - 2773, 1973.

15513 - **BARNES, L.W., KRIEG, D.R.** : Evidence for a trifluralin-potassium nitrate interaction affecting tomato seedling growth. - Crop Sci. *13* : 489 - 490, 1973. [Ps, Chl.]

15514 - **BAROOVA, S.R., HORVÁTH, I.** : Effect of light intensity on dry matter production and energy utilization in tomato plants. - Acta bot. Acad. Sci. hung. *18* : 273 - 280, 1973.

15515 - **BARR, R., BERG, S., KROGMANN, D.W., CRANE, F.L.** : Reversal of histone inhibition of chloroplast reactions by anions. - Plant Physiol. *51* (Suppl.) : 41, 1973.

15516 - **BARRS, H.D.** : Controlled environment studies of the effects of variable atmospheric water stress on photosynthesis, transpiration and water status of *Zea mays* L. and other species. - In : SLATYER, R.O. (ed.) : Plant Response to Climatic Factors. Pp. 249 - 258. Unesco, Paris 1973.

15517 - **BARSKY, E.L., SAMUILOV, V.D.** : Absorption changes of carotenoids and bacteriochlorophyll in energized chromatophores of *Rhodospirillum rubrum*. - Biochim. biophys. Acta *325* : 454 - 462, 1973.

15518 - **BARSKY, E.L., SAMUILOV, V.D.** : The bacteriochlorophyll absorption band shifts linked with the energy state of photosynthetic bacteria membranes. - J. Bioenerg. *4* : 391 - 395, 1973.

15519 - **BARTHOLOMEW, B.** : Drought response in the gas exchange of *Dudleya farinosa* *(Crassulaceae)* grown under natural conditions. - Photosynthetica *7* : 114 - 120, 1973.

15520 - **BARTKOV, B.I.** : O nelokal'nom raspredelenii assimilyatov v period plodonosheniya u rastenii̯ semei̯stva bobovykh. [Non-local distribution of photosynthates in the fruiting period in *Fabaceae*.] - In : Informatsionnye Materialy Akad. Nauk SSSR, Sibir. Otd., Sibir. Inst. Fiziol. Biokhim. Rast. (Irkutsk) *11* : 36 - 38, 1973. [In R.]

15521 - **BARTKOV, B.I., BARTKOVA, A.D.** : Ob ortostikhnoi̯ zonal'nosti postupleniya assimilyatov iz list'ev v plody gorokha. [Orthostichial zonality of photosynthate transport from leaves to pea fruits.] - Fiziol. Biokhim. kul't. Rast. *5* : 619 - 622, 1973. [In R, ab : E.]

15522 - **BASSHAM, J.A.** : The role of photosynthesis in green plants. - In : MILLER, L.P. (ed.) : Phytochemistry. Vol. I. The Process and Products of Photosynthesis. Pp. 38 - 74. Van Nostrand Reinhold Company, New York - Cincinnati - Toronto - London - Melbourne 1973.

15523 - **BASSHAM, J.A., KIRK, M.** : Sequence of formation of phosphoglycolate and glycolate in photosynthesizing *Chlorella pyrenoidosa*. - Plant Physiol. *52* : 407 - - 411, 1973.

*15524 - **BATALOV, R.B., USMANOV, P.D.** : Kolichestvennaya otsenka chastoty pigmentnykh mutatsii̯ na émbrionakh i prorostkakh v M_2 u *Arabidopsis thaliana*. [Estimation of frequency of pigment mutations in embryos and seedlings of M_2 of *Arabidopsis thaliana*.] - In : Geneticheskie Aspekty Fotosinteza. Tezisy Dokladov. Pp. 54 - 55. Donish, Dushanbe 1972. [In R.]

15525 - **BATTERSBY, A.R., GIBSON, K.H., McDONALD, E., MANDER, L.N., MORON, J., NIXON, L.N.** : Intermediates in porphyrin biosynthesis : studies with ^{14}C- and ^{13}C- -labelled pyrromethanes. - J. chem. Soc., chem. Commun. *1973* : 768 - 770, 1973.

15526 - **BATYUK, V.P.** : The effect of surface-active agents on the quantum requirement of photosynthesis determined by the thermodynamic method. - Biol. Plant. *15* : 161 - 165, 1973.

15527 - **BAUER, A., SCHLUNEGGER, U., ERISMANN, K.H.** : Untersuchungen zur Ammoniumassimilation in Aminosäuren und Proteine bei *Lemna minor* L. unter Photosynthesebedingungen. - Verhandl. schweiz. naturforsch. Ges. *1973* : 79 - 82, 1973.

15528 - **BAULD, J., BROCK, T.D.** : Ecological studies of *Chloroflexis,* a gliding photosynthetic bacterium. - Arch. Mikrobiol. *92* : 267 - 284, 1973. [Ps, Chl.]

15529 - **BAUMGARTNER, A.** : Estimation of the radiation and thermal micro-environment from meteorological and plant parameters. - In : SLATYER, R.O. (ed.) : Plant Response to Climatic Factors. Pp. 313 - 325. Unesco, Paris 1973.

15530 - **BAZZAZ, F.A.** : Photosynthesis of *Ambrosia artemisiifolia* L. plants grown in greenhouse and in the field. - Amer. Midland Naturalist *90* : 186 - 190, 1973.

15531 - **BAZZAZ, F.A., MEZGA, D.M.** : Primary productivity and microenvironment in an *Ambrosia*-dominated old field. - Amer. Midland Naturalist *90* : 70 - 78, 1973.

15532 - **BAZZAZ, M.B., GOVINDJEE** : Absorption and chlorophyll *a* fluorescence characteristics of Tris-treated and sonicated chloroplasts. - Plant Sci. Lett. *1* : 201 - 206, 1973.

15533 - **BAZZAZ, M.B., GOVINDJEE** : Photochemical properties of mesophyll and bundle sheath chloroplasts of maize. - Plant Physiol. *52* : 257 - 262, 1973.

15534 - **BEADLE, C.L., STEVENSON, K.R., NEUMANN, H.H., THURTELL, G.W., KING, K.M.** : Diffusive resistance, transpiration, and photosynthesis in single leaves of corn and sorghum in relation to leaf water potential. - Can. J. Plant Sci. *53*: 537 - 544, 1973.

15535 - **BEADLE, C.L., STEVENSON, K.R., THURTELL, G.W.** : Leaf temperature measurement and control in a gas-exchange cuvette. - Can. J. Plant Sci. *53* : 407 - 412, 1973.

15536 - **BEALE, S.I., CASTELFRANCO, P.A.** : ^{14}C incorporation from exogenous compounds into δ-aminolevulinic acid by greening cucumber cotyledons. - Biochem. biophys. Res. Commun. *52* : 143 - 149, 1973.

15537 - **BEALE, S.I., CASTELFRANCO, P.A.** : The formation of δ-aminolevulinic acid from labeled precursors by greening cucumber cotyledons. - Plant Physiol. *51* (Suppl.): 19, 1973.

15538 - **BEARDEN, A.J., MALKIN, R.** : Oxidation-reduction potential dependence of low--temperature photoreactions of chloroplast photosystem II. - Biochim. biophys. Acta *325* : 266 - 274, 1973.

15539 - **BEARDSELL, M.F., MITCHELL, K.J., THOMAS, R.G.** : Effects of water stress under contrasting environmental conditions on transpiration and photosynthesis in soybean. - J. exp. Bot. *24* : 579 - 586, 1973.

15540 - **BEARDSELL, M.F., MITCHELL, K.J., THOMAS, R.G.** : Transpiration and photosynthesis in soybean. Effects of temperature and vapour pressure deficit. - J. exp. Bot. *24* : 587 - 595, 1973.

15541 - **BECACOS-KONTOS, T.** : Environmental factors affecting production in Saronicos Gulf, Aegean Sea. - Bull. Inst. oceanogr. (Monaco) *71* (1423) : 1 - 16, 1973. [Chl.]

15542 - **BECACOS-KONTOS, T.** : Primary production investigations in the Saronicos Gulf, 1965 - 1967. - Rapp. Comm. int. Mer médit. *21* : 325 - 329, 1973.

15543 - **BECKER, J.F., GEACINTOV, N.E., van NOSTRAND, F., van METTER, R.** : Orientation of chlorophyll *in vivo*. Studies with magnetic field oriented *Chlorella*. - Biochem. biophys. Res. Commun. *51* : 597 - 602, 1973.

15544 - **BEDBROOK, J.R., MATTHEWS, R.E.F.** : Changes in the flow of early products of photosynthetic carbon fixation associated with the replication of TYMV. - Virology *53* : 84 - 91, 1973.

15545 - **BEDELL, G.W. II, GOVINDJEE** : Photophosphorylation in intact algae : Effects of inhibitors, intensity of light, electron acceptor and donors. - Plant Cell Physiol. *14* : 1081 - 1097, 1973.

15546 - **BEINFELD, M.C.** : A fast kinetic component of ESR signal II in photosynthesis. - Ann. N.Y. Acad. Sci. *222* : 871 - 883, 1973.

15547 - **BEKASOVA, O.D., EVSTIGNEEV, V.B.** : Vliyanie KCl na pryamoe i sensibilizirovannoe fikoëritrinom vosstanovlenie metilovogo krasnogo askorbinovoĭ kislotoĭ. [Effect of KCl on the direct and phycoerythrin-sensitized reduction of methyl red by ascorbic acid.] - Biofizika *18* : 243 - 250, 1973. [In R, ab : E.]

15548 - **BELAVSKAYA, A.P., SERAFIMOVICH, N.B.** : Produktsiya makrofitov nekotorykh ozer Pskovskoĭ oblasti. [Production of macrophytes of some lakes in the Pskov region.] - Rastit. Resursy *9* : 355 - 369, 1973. [In R.]

15549 - **BELL, L.N.** : Rastenie kak akkumulyator i preobrazovatel' solnechnoĭ ėnergii. [Plant as an accumulator and transformer of solar energy.] - Vestn. Akad. Nauk SSSR *1973* (2) : 33 - 41, 1973. [In R.]

15550 - **BELLY, R.T., TANSEY, M.R., BROCK, T.D.** : Algal excretion of ^{14}C-labeled compounds and microbial interactions in *Cyanidium caldarium* MATS. - J. Phycol. *9* : 123 - 127, 1973. [Ps.]

15551 - **BEN-AMOTZ, A., AVRON, M.** : NADP specific dihydroxyacetone reductase from *Dunaliella parva*. - FEBS Lett. *29* : 153 - 155, 1973.

15552 - **BEN-AZIZ, A., BRITTON, G., GOODWIN, T.W.** : Carotene epoxides in *Lycopersicon esculentum*. - Phytochemistry *12* : 2759 - 2764, 1973.

15553 - **BENCI, J.F., AASE, J.K., FERGUSON, A.H.** : Aerodynamic and energy balance comparisons between awned and nonawned barley. - Agron. J. *65* : 373 - 377, 1973.

15554 - **BENDER, M.M., ROUHANI, I., VINES, H.M., BLACK, C.C. Jr.** : ^{13}C/^{12}C ratio changes in crassulacean acid metabolism plants. - Plant Physiol. *52* : 427 - 430, 1973.

15555 - **BENDER, M.M., SMITH, D.** : Classification of starch- and fructosan-accumulating grasses as C-3 or C-4 species by carbon isotope analysis. - J. brit. Grassland Soc. *28* : 97 - 100, 1973.

15556 - **BENEDICT, C.R.** : The presence of ribulose 1,5-diphosphate carboxylase in the nonphotosynthetic endosperm of germinating castor beans. - Plant Physiol. *51* : 755 - 759, 1973.

15557 - BENEDICT, C.R., SMITH, R.H., KOHEL, R.J. : Incorporation of ^{14}C-photosynthate into developing cotton bolls, *Gossypium hirsutum* L. - Crop Sci. *13* : 88 - 91, 1973.

15558 - BENEMANN, J.R. : A model system for nitrogen fixation and hydrogen evolution by nonheterocystous blue-green algae. - Fed. Proc. *32* : 632 Abs, 1973. [Ps.]

15559 - BENEMANN, J.R., BERENSON, J.A., KAPLAN, N.O., KAMEN, M.D. : Hydrogen evolution by á chloroplast-ferredoxin-hydrogenase system. - Proc. nat. Acad. Sci. USA *70* : 2317 - 2320, 1973.

15560 - BENNETT, A., BOGORAD, L. : Complementary chromatic adaptation in a filamentous blue-green alga. - J. Cell Biol. *58* : 419 - 435, 1973.

15561 - BENNETT, J.H., HILL, A.C. : Inhibition of apparent photosynthesis by air pollutants. - J. environm. Qual. *2* : 526 - 530, 1973.

15562 - BENNOUN, P., LI, Y.-S. : New results on the mode of action of 3-(3,4-dichlorophenyl)-1,1-dimethylurea in spinach chloroplasts. - Biochim. biophys. Acta *292* : 162 - 168, 1973.

15563 - BENNUN, A. : Interacción del factor acoplante del cloroplasto con protones y agua. [Interaction of chloroplast coupling factor with protons of water.] - In : MEJÍA, R.H., MOGUILEVSKY, J.A. (ed.) : Recientes Adelantos en Biologia. Pp. 254 - 264. Buenos Aires 1971. [In Span.]

15564 - BENORTHAM, R.W., MATTSON, R.H., MITCHELL, H.L. : Pigmentation of poinsettias during short-day induction. - HortScience *8* : 132 - 133, 1973. [Chl, Car.]

15565 - BENSASSON, R., LAND, E.J. : Optical and kinetic properties of semireduced plastoquinone and ubiquinone : electron acceptors in photosynthesis. - Biochim. biophys. Acta *325* : 175 - 181, 1973.

15566 - BEREZIN, B.D., KOĬFMAN, O.I., ANDRIANOV, V.G. : Opredelenie konstant kislotnoĭ ionizatsii khlorofillovykh kislot metodom rastvorimosti. [Determination of the acid ionization constants of chlorophyllic acids by a solubility method.] - Zh. fiz. Khim. *47* : 1464 - 1466, 1973. [In R.]

15567 - BERG, A., SKRE, O., WIELGOLASKI, F.E., KJELVIK, S. : Leaf areas and angles, chlorophyll and reserve carbon in alpine and sub-alpine plant communities, Hardangervidda, Norway. - In : BLISS, L.C., WIELGOLASKI, F.E. (ed.) : Primary Production and Production Processes, Tundra Biome. Pp. 239 - 254. Tundra Biome Steering Comm., Edmonton 1973.

15568 - BERG, S., CIPOLLO, D., ARMSTRONG, B., KROGMANN, D.W. : Polycation inhibition of chloroplast Photosystem I. - Biochim. biophys. Acta *305* : 372 - 383, 1973.

15569 - BERGER, A. : Le potentiel hydrique et la résistance à la diffusion dans les stomates indicateurs de l'état hydrique de la plante. - In : SLATYER, R.O. (ed.) : Plant Response to Climatic Factors. Pp. 201 - 212. Unesco, Paris 1973.

15570 - BERGER, T.J., ORLANDO, J.A. : Purification and some properties of a protein factor required for light-dependent transhydrogenase in *Rhodopseudomonas spheroides*. - Arch. Biochem. Biophys. *159* : 25 - 31, 1973.

15571 - BERING, C.L., DILLEY, R.A., DODGE, S. : Chemical modifiers as probes of light--induced changes in chloroplast membranes. - Plant Physiol. *51* (Suppl.) : 67, 1973.

15572 - BERKALOFF, C., JUPIN, H. : Modifications du spectre d'émission de fluorescence et de l'ultrastructure du plaste de l'algue verte *Protosiphon botryoîdes* soumise à l'action de la streptomycine. - Protoplasma *80* : 41 - 55, 1973.

15573 - BERKOVÁ, E., DOUCHA, J., KUBÍN, Š., VENDLOVÁ, J., ZACHLEDER, V., ŠETLÍK, I. : The coupling of energy metabolism with synthetic and reproduction processes in cell cycles of *Scenedesmus quadricauda*. - Annu. Rep. algol. Lab. Třeboň *1970* : 31 - 51, 1973. [Ps, Chl.]

15574 - BERLAND, B., BONIN, D., COSTE, B., MAESTRINI, S., MINAS, H.J. : Influence des conditions hivernales sur les productions phyto- et zooplanctoniques en Méditerranée Nord-Occidentale. III. Caractérisation des eaux de surface au moyen de cultures d'algues. - Mar. Biol. *23* : 267 - 274, 1973. [Chl.]

15575 - **BERLAND, B.R., BONIN, D.J., MAESTRINI, S.Y., POINTIER, J.-P.** : Etude de la
fertilité des eaux marines au moyen de tests biologiques effectués avec des
cultures d'algues. II. Limitation nutritionnelle et viabilité de l'inoculum.
- Int. Rev. Ges. Hydrobiol. *58* : 203 - 220, 1973. [Chl.]

15576 - **BERLAND, B.R., BONIN, D.J., MAESTRINI, S.Y., POINTIER, J.P.** : Etude de la fertilité des eaux marines au moyen de tests biologiques effectués avec des cultures d'algues. III. Réponses de la diatomée *Skeletonema costatum* à différentes concentrations d'éléments nutritifs. - Int. Rev. ges. Hydrobiol. *58* :
401 - 416, 1973. [Chl.]

15577 - **BERLAND, B.R., BONIN, D.J., MAESTRINI, S.Y., POINTIER, J.P.** : Etude de la fertilité des eaux marines au moyen de tests biologiques effectués avec des cultures d'algues. IV. Etude d'eaux côtières méditerranéennes. - Int. Rev. ges.
Hydrobiol. *58* : 473 - 500, 1973. [Chl.]

15578 - **BERMAN, T.** : Modifications in filtration methods for the measurement of inorganic ^{14}C uptake by photosynthesizing algae. - J. Phycol. *9* : 327 - 330, 1973.

15579 - **BERNARD, J.M., BERNARD, F.A.** : Winter biomass in *Typha glauca* GODR. and *Spar-
ganium eurycarpum* ENGELM. - Bull. Torrey bot. Club *100* : 125 - 127, 1973.

15580 - **BERRY, J.A., BOWES, G.** : Removal of contaminant inorganic phosphate and phosphoglycolate from ribulose-1,5-diphosphate. - Carnegie Inst. Year Book *72* :
403 - 405, 1973.

15581 - **BERRY, J.A., BOWES, G.** : Oxygen uptake *in vitro* by RuDP carboxylase of *Chlamydomonas reinhardtii*. - Carnegie Inst. Year Book *72* : 405 - 407, 1973.

15582 - **BERZBORN, R.J., BISHOP, N.I.** : Isolation and properties of chloroplast particles of *Scenedesmus obliquus* D$_3$ with high photochemical activity. - Biochim.
biophys. Acta *292* : 700 - 714, 1973.

15583 - **BEWLEY, J.D., TUCKER, E.B.** : The effects of liquid nitrogen temperatures on
the metabolism of the moss *Tortula ruralis*. - Plant Physiol. *51* (Suppl.) : 64,
1973. [Ps.]

15584 - **BEZECNÝ, L.** : Vliv pořadí vývinu lat na výskyt chlorofylových mutací ovsa.
[The influence of the order in the development of panicle on the occurrence of
chlorophyll mutations in oats.] - Genet. Šlechtění *9* : 55 - 58, 1973. [In
Czech, ab : E, R.]

15585 - **BHAN, A.K., KAUL, M.L.H.** : Effect of ionizing radiations on dry matter production and pigment content in rice seedlings. - Sci. Culture *39* : 511 - 513,
1973.

15586 - **BIDWELL, R.G.S.** : A possible mechanism for the control of photoassimilate translocation. - Proc. Res. Inst. Pomol. Skierniewice *1973* (3) : 77 - 89, 1973.

15587 - **BIERHUIZEN, J.F.** : The effect of temperature on plant growth, development and
yield. - In : **SLATYER, R.O.** (ed.) : Plant Response to Climatic Factors. Pp.
89 - 98. Unesco, Paris 1973. [Ps.]

15588 - **BIGGINS, J.** : Kinetic behavior of cytochrome *f* in cyclic and noncyclic electron transport in *Porphyridium cruentum*. - Biochemistry *12* : 1165 - 1170, 1973.

*15589 - **BIKASIYAN, G.R., MUSTAFAEV, A., NEGMATOV, M.** : Khlorofil'nye mutanty *Gossypium
hirsutum* L., indutsirovannye *gamma*-luchami Co⁶⁰. [Chlorophyll mutants of *Gossypium hirsutum* L., induced by *gamma*-rays of Co⁶⁰.] - In : Geneticheskie Aspekty Fotosinteza. Tezisy Dokladov. P. 55. Donish, Dushanbe 1972. [In R.]

15590 - **BIL', K.Ya., STOĬLOV, M.A., KARPILOV, Yu.S.** : Osobennosti transporta produktov
fotosinteza u C-4-rasteniĭ. [Peculiarities of photosynthates transport in C-4
plants.] - In : Peredvizhenie Veshchestv u Rasteniĭ v Svyazi s Metabolizmom i
Biofizicheskimi Protsessami. Pp. 36 - 44. Gorkov. gos. Univ. M.I. Lobachevsko-
go, Gorkiĭ 1973. [In R.]

*15591 - **BINDLOSS, M.E., HOLDEN, A.V., BAILEY-WATTS, A.E., SMITH, I.R.** : Phytoplankton
production, chemical and physical conditions in Loch Leven. - In : **KAJAK, Z.,
HILLBRICHT-ILKOWSKA, A.** (ed.) : Productivity Problems of Freshwaters. Pp. 639 -
- 659. Pol. Sci. Publ., Warszawa 1972. [Ps, Chl.]

*15592 - **BINGHAM, E.T.** : Stomatal chloroplasts in alfalfa at four ploidy levels. - Crop Sci. *8* : 509 - 510, 1968.

15593 - **BIRD, I.F., CORNELIUS, M.J., DYER, T.A., KEYS, A.J.** : The purity of chloroplasts isolated in non-aqueous media. - J. exp. Bot. *24* : 211 - 215, 1973.

15594 - **BIRTH, G.S., ZACHARIAH, G.L.** : Spectrophotometer for biological applications. - Trans. ASAE *16* : 371 - 373, 1973. [Leaf reflectance and transmittance.]

15595 - **BISALPUTRA, T., BAILEY, A.** : The fine structure of the chloroplast envelope of a red alga, *Bangia fusco-purpurea*. - Protoplasma *76* : 443 - 454, 1973.

15596 - **BISCOE, P.V., LITTLETON, E.J., SCOTT, R.K.** : Stomatal control of gas exchange in barley awns. - Ann. appl. Biol. *75* : 285 - 297, 1973.

15597 - **BISCOE, P.V., UNSWORTH, M.H., PINCKNEY, H.R.** : The effects of low concentrations of sulphur dioxide on stomatal behaviour in *Vicia faba*. - New Phytol. *72* : 1299 - 1306, 1973.

15598 - **BISHOP, D.G.** : Inhibition of chloroplast photochemical reactions by polyene antibiotics. - Proc. aust. biochem. Soc. *6* : 47, 1973.

15599 - **BISHOP, D.G.** : Inhibition of photochemical activity in chloroplasts by the polyene antibiotic, filipin. - Arch. Biochem. Biophys. *154* : 520 - 526, 1973.

15600 - **BISHOP, D.G.** : Inhibition of photosynthetic electron transfer by amphotericin B. - Biochem. biophys. Res. Commun. *54* : 816 - 822, 1973.

15601 - **BISHOP, D.G., BAIN, J.M., SMILLIE, R.M.** : The effect of antibiotics on the ultrastructure and photochemical activity of a developing chloroplast. - J. exp. Bot. *24* : 361 - 375, 1973.

15602 - **BISHOP, N.I.** : Analysis of photosynthesis in green algae through mutation studies. - In : GIESE, A.C. (ed.) : Photophysiology. Current Topics in Photobiology and Photochemistry. Vol. 8. Pp. 65 - 96. Academic Press, New York - London 1973.

15603 - **BITTERMAN, M., DYKYJOVÁ, D.** : Optimal shape of greenhouse roofs deduced from the solar shape of tree crowns and other plant surfaces. - In : The Sun in the Service of Mankind. (Int. Congr.) Pp. V.22-1 - V.22-10. Unesco, Paris 1973.

15604 - **BJÖRKMAN, O.** : Comparative studies on photosynthesis in higher plants. - In : GIESE, A.C. (ed.) : Photophysiology. Current Topics in Photobiology and Photochemistry. Vol. 8. Pp. 1 - 63. Academic Press, New York - London 1973.

15605 - **BJÖRKMAN, O., BERRY, J.** : High-efficiency photosynthesis. - Sci. Amer. *229* : 80 - 87, 91 - 93, 132, 1973.

15606 - **BJÖRKMAN, O., NOBS, M., BERRY, J., MOONEY, H., NICHOLSON, F., CATANZARO, B.** : Physiological adaptation to diverse environments : Approaches and facilities to study plant responses to contrasting thermal and water regimes. - Carnegie Inst. Year Book *72* : 393 - 403, 1973. [Ps.]

15607 - **BJÖRKMAN, O., TROUGHTON, J., NOBS, M.** : Photosynthesis in relation to leaf structure. - Brookhaven Symp. Biol. *25* - Basic Mechanisms in Plant Morphogenesis : 206 - 226, 1973.

*15608 - **BKHAGAT, V.P.** : Raspredelenie RDF i FÉP-karboksilazy u indutsirovannykh mutantov yachmenya i ikh svyaz' s nitratami. [Distribution of RUDP and PEP carboxylases in induced mutants of barley and their relation to nitrate.] - In : Geneticheskie Aspekty Fotosinteza. Tezisy Dokladov. P. 42. Donish, Dushanbe 1972. [In R.]

15609 - **BLACK, C.C., CAMPBELL, W.H., CHEN, T.M., DITTRICH, P.** : The monocotyledons : their evolution and comparative biology. III. Pathways of carbon metabolism related to net carbon dioxide assimilation by monocotyledons. - Quart. Rev. Biol. *48* : 299 - 313, 1973.

15610 - **BLACK, C.C. Jr.** : Photosynthetic carbon fixation in relation to net CO_2 uptake. - Annu. Rev. Plant Physiol. *24* : 253 - 286, 1973.

15611 - **BLACKBURN, M.** : Regressions between biological oceanographic measurements in the eastern tropical Pacific and their significance to ecological efficiency. - Limnol. Oceanogr. *18* : 552 - 563, 1973. [Chl.]

15612 - **BLAGONRAVOVA, L.N., KHOLCHENKOV, V.A.** : Vliyanie insektitsidov na porazhenie list'ev yabloni miniruyushcheĭ mol'yu, soderzhanie v nikh khlorofilla i na kachestvo plodov s"emnoĭ zrelosti. [Effect of insecticides on the affection of apple leaves by *Stigmella malella*, their chlorophyll content and quality of fruit of harvest maturity.] - Khim. sel'. Khoz. *11* (10) : 37 - 39, 1973. [In R.]

15613 - **BLAIR, G.E., ELLIS, R.J.** : Protein synthesis in chloroplasts. I. Light-driven synthesis of the large subunit of fraction I protein by isolated pea chloroplasts. - Biochim. biophys. Acta *319* : 223 - 234, 1973.

15614 - **BLASCO, D.** : Estudio de las variaciones de la relación fluorescencia *in vivo*/ /clorofila *a*, y su aplicación en oceanografía. Influencia de la limitación de diferentes nutrientes, efecto del día y noche, y dependencia de la especie estudiada. [Variations of the ratio fluorescence *in vivo*/chlorophyll *a* and its application in oceanography. Effect of limitation by various nutrients, effect of day and night, and dependence on the species studies.] - Invest. Pesq. *37* : 533 - 556, 1973. [In Span., ab : E.]

15615   **BLIGNY, R., BISCH, A.-M., GARREC, J.-P., FOURCY, A.** : Observations morphologiques et structurales des effets du fluor sur les cires épicuticulaires et sur les chloroplastes des aiguilles de sapin (*Abies alba* MILL.). - J. Microscop. (Paris) *17* : 207 - 214, 1973.

15616 - **BLISS, L.C., KERIK, J.** : Primary production of plant communities of the Truelove Lowland, Devon Island, Canada - rock outcrops. - In : BLISS, L.C., WIELGOLASKI, F.E. (ed.) : Primary Production and Production Processes, Tundra Biome. Pp. 27 - 36. Tundra Biome Steering Comm., Edmonton 1973.

B15617 - **BLISS, L.C., WIELGOLASKI, F.E.** (ed.) : Primary Production and Production Processes, Tundra Biome. - Tundra Biome Steering Comm., Edmonton 1973.

*15618 - **BLIXT, S.** : Linkage studies in *Pisum*. VIII. Gene-symbolization of 15 *chlorotica*-mutants. - Agri Horti. Genet. *26* : 82 - 87, 1968. [Chl.]

*15619 - **BLIXT, S.** : Linkage studies in *Pisum*. IX. Linkage of the gene *chi5* in chromosome II. - Agri Horti. Genet. *26* : 88 - 99, 1968. [Chl.]

15620 - **BOASSON, R., GIBBS, S.P.** : Chloroplast replication in synchronously dividing *Euglena gracilis*. - Planta *115* : 125 - 134, 1973.

15621 - **BOBROVSKIĬ, A.P., KHOLMOGOROV, V.E.** : O prirode pervichnogo fotokhimicheskogo akta vzaimodeĭstviya khlorofilla s khinonami. [Nature of the primary photochemical act of the chlorophyll-quinone interaction.] - Dokl. Akad. Nauk SSSR *208* : 1472 - 1475, 1973. [In R.]

15622 - **BOBROVSKIĬ, A.P., KHOLMOGOROV, V.E.** : Spektry pogloshcheniya i EPR produktov odnoėlektronnogo fotookisleniya bakteriokhlorofilla. [Absorption and EPR spectra of products of one-electron photooxidation of bacteriochlorophyll.] - Zh. prikl. Spektroskop. *18* : 706 - 713, 1973. [In R.]

15623 - **BOCHEV, B., CHICHEV, P.** : Fiziologo-biokhimichni prouchvaniya na nyakoi mezhduvidovi khibridi mezhdu *T. aestivum* i *T. durum* pri razlichen khranitelen rezhim. I. Prouchvane na intenzivnostta na fotosintezata. [Physiological and biochemical investigations of some interspecies hybrids between *T. aestivum* and *T. durum* in different nutritional regimes. I. Photosynthetic rate.] - Izv. Inst. Fiziol. Rast. "M. Popov" *18* : 235 - 242, 1973. [In Bulg., ab : E, R.]

15624 - **BOCHEV, B., CHICHEV, P.** : Fiziologo-biokhimichni prouchvaniya na nyakoi izmeneni i aneuploidni formi ot *T. durum* DESF. [Physiological and biochemical investigations of some changes and aneuploidic forms of *T. durum* DESF.] - Izv. Inst. Fiziol. Rast. "M. Popov" *18* : 243 - 251, 1973. [In Bulg., ab : E, R.]

15625 - **BOE, A.A., LEE, T.S., TAPIO, D.D., BANKO, T.J.** : Effect of SADH on radish. - HortScience *8* : 497 - 498, 1973. [Ps.]

15626 - **BOGACHEVA, I.I., GOLUBKOVA, B.M., KISLYAKOVA, T.E.** : Evolyutsionnyĭ podkhod v izuchenii fotosinteticheskogo apparata selaginelly. [Evolutionary approach to the studies of photosynthetic apparatus of *Selaginella*.] - In : Upravlenie Skorost'yu i Napravlennost'yu Biosinteza u Rasteniĭ. P. 6. Krasnoyarsk 1973. [In R.]

*15627 - BOGACHEVA, M.I., KISLYAKOVA, T.E., GOLUBKOVA, B.M. : Evolyutsionnyĭ podkhod
v izuchenii struktury i funktsii fotosinteticheskogo apparata razlichnykh vidov. [Evolutionary approach to the study of structure and function of the photosynthetic apparatus in different species.] - In : Geneticheskie Aspekty Fotosinteza. Tezisy Dokladov. Pp. 12 - 13. Donish, Dushanbe 1972. [In R.]

12628 - BOGDANOVIĆ, M. : Chlorophyll formation in the dark. I. Chlorophyll in pine
seedlings. - Physiol. Plant. 29 : 17 - 18, 1973.

15629 - BOGDANOVIĆ, M. : Chlorophyll formation in the dark. II. Chlorophyll in wheat
leaves transplanted to pine megagametophytes. - Physiol. Plant. 29 : 19 - 21,
1973.

15630 - BÖGER, P., LIEN, S.S., SAN PIETRO, A. : Fluorescence studies on ferredoxin-
-NADP reductase. - Z. Naturforsch. 28C : 505 - 510, 1973.

15631 - BÖHME, H., CRAMER, W.A. : Uncoupler-dependent decrease in midpoint potential
of the chloroplast cytochrome b_6. - Biochim. biophys. Acta 325 : 275 - 283,
1973.

15632 - BÖHME, H., CRAMER, W.A. : Studies on the chloroplast cytochrome b_6. - In :
Ninth International Congress of Biochemistry, Stockholm, July 1 - 7, 1973. Abstract Book. P. 217. IUB, Stockholm 1973.

15633 - BOĬCHENKO, E.A. : Vosstanavlivayushchaya sila soedineniĭ zheleza s flavinnuk-
leotidami rasteniĭ. [Reducing power of compounds of iron with flavin nucleotides in plants.] - Fiziol. Rast. 20 : 488 - 492, 1973. [Ps; in R, ab : E.]

15634 - BOĬCHEV, A., RANGELOV, B. : Vliyanie na nyakoi kherbitsidi v"rkhu khlorofilno-
to i v"glekhidratnoto s"d"rzhanie na lozite i zapasenostta na pochvata s vlaga
i makroelementi. [Effect of some herbicides on chlorophyll and carbohydrate
contents in grapevine and supply of water and macroelements in the soil.] -
Gradinar. lozar. Nauka 10 (5) : 127 - 134, 1973. [In Bulg., ab : F, R.]

15635 - BOLDUC, R. : Déformations ultrastructurales chez les membranes du chloroplaste
causées par le froid chez le Blé d'hiver. - Plant Physiol. 51 (Suppl.) : 27,
1973.

15636 - BOLLE-JONES, E.W., SANEI, F., PAHLAVANI, A. : Incidence and control of iron
deficiency chlorosis in fruit trees in Iran. - Exp. Agr. 9 : 241 - 247, 1973.

15637 - BOLTON, J.R., COST, K. : Flash photolysis-electron spin resonance : a kinetic
study of endogenous light-induced free radicals in reaction center preparations
from Rhodopseudomonas spheroides. - Photochem. Photobiol. 18 : 417 - 421, 1973.

15638 - BONHOMME, R. : Analyse de la surface des taches de soleil, de l'indice foliaire
et de l'inclinaison moyenne des feuilles à l'aide de photographies hémisphéri-
ques. - In : SLATYER, R.O. (ed.) : Plant Response to Climatic Factors. Pp.
369 - 376. Unesco, Paris 1973.

15639 - BONNET, L., CAPBLANCQ, J. : Phytoplancton et productivité primaire d'un lac
d'altitude dans les Pyrénées. Analyse factorielle des correspondances appliquée
aux relevés effectués dans le lac de Port-Bielh. - Ann. Limnol. 9 : 183 - 192,
1973.

*15640 - BOOS, G.V., SEMICHEV, V.N. : Fotosinteticheskaya aktivnost' list'ev tomatov
v svyazi s yavleniem geterozisa. [Photosynthetic activity of tomato leaves in
relation to the phenomenon of heterosis.] - In : Geneticheskie Aspekty Fotosinteza. Tezisy Dokladov. Pp. 98 - 99. Donish, Dushanbe 1972. [In R.]

15641 - BORG, D.C., FAJER, J., FORMAN, A., FELTON, R.H., DOLPHIN, D. : Pi-radical ions
from the oxidation and reduction of chlorins, chlorophyll, and bacteriochlorophyll. - In : IV International Biophysics Congress. Symposial Papers. Vol. 2.
Pp. 528 - 538. Pushchino 1973.

15642 - BORGNA, P., CALDERARA, G., VICARINI, L. : Attività inibente la reazione di Hill
e fitotossicità di m-alchiltioanilidi. [Inhibitory activity of Hill reaction
and phototoxicity of m-alkylthionilides.] - Farmaco, Ed. sci. 28 : 791 - 799,
1973. [In Ital., ab : E.]

15643 - BORISOV, A.Yu. : Pervichnye protsessy pri fotosinteze. [Primary processes of
photosynthesis.] - In : Sovremennye Problemy Fotosinteza. Pp. 161 - 174. Izd.
mosk. Univ., Moskva 1973. [In R.]

15644 - **BORISOV, A.Yu., GODIK, V.I.** : Excitation energy transfer in photosynthesis. Biochim. biophys. Acta *301* : 227 - 248, 1973.

15645 - **BORISOV, A.Yu., IL'INA, M.D.** : Kvantovyǐ vykhod fotookisleniya P_{700} v "legkikh" digitoninovykh fragmentakh khloroplastov gorokha. [Quantum yield of P_{700} photooxidation in "light" digitonin fragments of pea chloroplasts.] - Mol. Biol. (Moskva) *7* : 212 - 219, 1973. [In R, ab : E.]

15646 - **BORISOV, A.Yu., IL'INA, M.D.** : The fluorescence lifetime and energy migration mechanism in Photosystem I of plants. - Biochim. biophys. Acta *305* : 364 - 371, 1973.

15647 - **BORISOV, A.Yu., IL'INA, M.D.** : The quantum yield of reaction center photooxidation in subchloroplast fragments enriched with photosystem I. - Biochim. biophys. Acta *325* : 240 - 246, 1973.

15648 - **BORISOV, A.Yu., IVANOVSKIǏ, R.N., KONDRAT'EVA, E.N.** : Sravnitel'noe issledovanie lyuminestsentsii zelenykh i purpurnykh fotosinteziruyushchikh bakteriǐ v oblasti 440 mmk. [Comparative studies of luminescence of green and purple phototrophic bacteria in the region around 440 nm.] - Biokhimiya *38* : 59 - 62, 1973. [In R, ab : E.]

15649 - **BÖRNER, R.** : Zur Isolierung von Membranfraktionen aus Protoplasten nach deren präparativer Gewinnung aus Blättern von *Bryophyllum daigremontianum*. - Biol. Zentralbl. *92* : 583 - 594, 1973.

15650 - **BÖRNER, T.** : Struktur und Funktion der genetischen Information in den Plastiden. VI. Zur Funktion von Plastiden-DNA, Kern-DNA, plastidaler und cytoplasmatischer Proteinsynthese beim Aufbau der Chloroplasten - eine tabellarische Übersicht. - Biol. Zentralbl. *92* : 545 - 561, 1973.

15651 - **BORZENKOVA, R.A.** : Gormonal'naya regulyatsiya fotosinteza. [Hormonal regulation of photosynthesis.] - In : Voprosy Regulyatsii Fotosinteza. Vol. 3. Pp. 45 - 57, 161. Ural'. gos. Univ., Sverdlovsk 1973. [In R.]

15652 - **BORZENKOVA, R.A., MOKRONOSOV, A.T.** : Ėndogennye faktory, opredelyayushchie transport assimilyatov v klubni kartofelya. [Endogenous factors determining photosynthate transport into potato tubers.] - Tr. biol.-pochv. Inst., dal'-nevost. nauch. Tsentr Akad. Nauk SSSR *20* (123) : 148 - 152, 1973. [In R, ab : E.]

15653 - **BOSCHETTI, A., BOGDANOV, S.** : Different effects of streptomycin on the ribosomes from sensitive and resistant mutants of *Chlamydomonas reinhardi*. - Eur. J. Biochem. *35* : 482 - 488, 1973. [Chloroplast ribosomes.]

15654 - **BOSCHETTI, A., WALZ, A.** : Streptomycininduzierte, reversible Vergilbung bei *Chlamydomonas reinhardii*. - Arch. Mikrobiol. *89* : 1 - 14, 1973.

15655 - **BOSIAN, G.** : Die Bedeutung der Luftfeuchtigkeit für die CO_2-Assimilation und den Stoffwechsel der höheren Pflanzen (*Vitis vinifera*) : Problem und Praxis künstlicher, klimatisierender Beregnung. - Ber. deut. bot. Ges. *86* : 447 - - 458, 1973.

15656 - **BOUGES-BOCQUET, B.** : Limiting steps in Photosystem II and water decomposition in *Chlorella* and spinach chloroplasts. - Biochim. biophys. Acta *292* : 772 - - 785, 1973.

15657 - **BOUGES-BOCQUET, B.** : Electron transfer between the two photosystems in spinach chloroplasts. - Biochim. biophys. Acta *314* : 250 - 256, 1973.

15658 - **BOUGES-BOCQUET, B., BENNOUN, P., TABOURY, J.** : Deactivation of oxygen precursors in presence of 3(3,4-dichlorophenyl)-1,1-dimethylurea and phenylurethane. - Biochim. biophys. Acta *325* : 247 - 254, 1973.

15659 - **BOULDIN, D.R., LATHWELL, D.J., GOYETTE, E.A., LAUER, D.A.** : Changes in water chemistry in marshes over a 12-year period following establishment. - N.Y. Fish Game J. *20* : 129 - 146, 1973. [Ps.]

15660 - **BOUNIAS, M.** : Equipment pigmentaire des cotylédons d'*Arabidopsis* après irradiation "*gamma*" des graines. - *Arabidopsis* Inform. Serv. *10* : 26 - 27, 1973.

15661 - **BOUNIAS, M.** : Incorporation de Leucine-^{14}C dans les pigments photosynthétiques, les sucres et aminoacides libres de plants d'*Arabidopsis* : Témoins et mutants *viridis*. - *Arabidopsis* Inform. Serv. *10* : 29 - 30, 1973.

15662 - **BOURCELIER, C., DAUSSANT, J.** : La fraction I protéique du limbe de Blé : purification et caractérisation immunochimique préliminaire. - Compt. rend. Acad. Sci. Paris, Sér. D *276* : 2525 - 2528, 1973.

*15663 - **BOYD, C.E.** : Sources of CO_2 for nuisance blooms of algae. - Weed Sci. *20* : 492 - 497, 1972. [Ps.]

15664 - **BOYD, C.E.** : Summer algal communities and primary productivity in fish ponds. - Hydrobiologia *41* : 357 - 390, 1973. [Chl.]

*15665 - **BOYD, C.E., VICKERS, D.H.** : Relationships between production, nutrient accumulation, and chlorophyll synthesis in an *Eleocharis quadrangulata* population. - Can. J. Bot. *49* : 883 - 888, 1971.

15666 - **BOYER, J.S., POTTER, J.R.** : Chloroplast response to low leaf water potentials. I. Role of turgor. - Plant Physiol. *51* : 989 - 992, 1973.

15667 - **BOYLEN, C.W., BROCK, T.D.** : Effects of thermal additions from the Yellowstone geyser basins on the benthic algae of the Firehole River. - Ecology *54* : 1282 - - 1291, 1973. [Chl.]

15668 - **BRADBEER, J.W.** : The synthesis of chloroplast enzymes. - In : **MILBORROW, B.V.** (ed.) : Biosynthesis and its Control in Plants. Pp. 279 - 302. Academic Press, London - New York 1973.

15669 - **BRADEEN, D.A., WINGET, G.D., GOULD, J.M., ORT, D.R.** : Site-specific inhibition of photophosphorylation in isolated spinach chloroplasts by mercuric chloride. - Plant Physiol. *52* : 680 - 682, 1973.

*15670 - **BRAĬON, O.V.** : Deyaki fiziologichni osoblyvosti zelenykh plastyd derevnoï chastyny pagoniv roslyn. [Physiological peculiarities of green plastids from woody parts of plant seedlings.] - Visnyk Kyïv. Univ., Ser. Biol. *13* : 60 - 66, 1971. [In Ukr., ab : G.]

B15671 - **BRAĬON, O.V.** : Fluorestsentna Mikroskopiya Roslynnykh Tkanyn i Klityn. [Fluorescent Microscopy of Plant Tissues and Cells.] - Vyshcha Shkola, Kyïv 1973. [Chl, Car; in Ukr.]

15672 - **BRAND, J., SAN PIETRO, A.** : Polylysine-enhanced effectiveness of plastocyanin in photosystem I. - Biochim. biophys. Acta *325* : 255 - 265, 1973.

15673 - **BRANDLE, J.R., SCHNARE, P.D., HINCKLEY, T.M., BROWN, G.N.** : Changes in polysomes of black locust seedlings during dehydration-rehydration cycles. - Physiol. Plant. *29* : 406 - 409, 1973. [Ps.]

15674 - **BRANDON, P.C., van BOEKEL-MOL, T.N.** : Properties of purified malic enzyme in relation to Crassulacean acid metabolism. - Europe. J. Biochem. *35* : 62 - 69, 1973.

15675 - **BRANDON, P.C., ELGERSMA, O.** : Effects of α-benzyl-α-bromo-malodinitrile on the primary electron acceptor of Photosystem II in spinach chloroplasts. - Biochim. biophys. Acta *292* : 753 - 762, 1973.

15676 - **BRANDT, A.B., LANSE, E., AVEROVA, L.** : Svetovoĭ rezhim poseva kak spetsificheskiĭ faktor formirovaniya urozhaya v usloviyakh zharkikh stran. [Light regime as a specific factor for crop formation in tropical countries.] - Fiziol. Rast. *20* : 967 - 971, 1973. [In R, ab : E.]

15677 - **BRANDT, A.B., SHARIPOV, K.A., KISELEVA, M.I.** : Dinamika nakopleniya pigmentov sinkhronnoĭ kul'turoĭ khlorelly i fotosinteticheskaya produktivnost' razlichnykh oblasteĭ spektra. [Dynamics of pigment accumulation in a synchronous *Chlorella* culture and photosynthetic productivity of various spectral regions.] - In : Upravlenie Biosintezom Mikroorganizmov. Pp. 102 - 104. Krasnoyarsk 1973. [In R.]

15678 - **BRANGEON, J.** : Compared ontogeny of the two types of chloroplasts of *Zea mays*. - J. Microscop. *16* : 233 - 242, 1973.

15679 - **BRANGEON, J.** : Effect of irradiance on granal configurations of *Zea mays* bundle sheath chloroplasts. - Photosynthetica *7* : 365 - 372, 1973.

15680 - **BRANTSEVICH, L.G., CHEKMACHEVA, V.V.** : Vliyanie azotobaktera na pigmentnyĭ kompleks v prorostkakh kukuruzy. [Effect of azotobacter on pigment complex in maize seedlings.] - Fiziol. Biokhim. kul't. Rast. *5* : 392 - 396, 1973. [In R, ab : E.]

15681 - **BRÅTEN, T.** : Autoradiographic evidence for the rapid disintegration of one chloroplast in the zygote of the green alga *Ulva mutabilis*. - J. Cell Sci. *12* : 385 - 389, 1973.

15682 - **BRATKOWSKA, I., NIEWIADOMSKI, H.** : The relation between autoxidation of triglycerides and chlorophylls transformation. - Zesz. probl. Postępow Nauk roln. *136* : 71 - 75, 1973.

15683 - **BRAVDO, B., CANVIN, D.T.** : The use of ^{14}C as a tracer for photorespiration. - Plant Physiol. *51* (Suppl.) : 42, 1973.

15684 - **BRAZEE, R.D., FOX, R.D.** : Analyzing atmospheric turbulence in plant canopies. - Agr. Sci. Rev. *11* : 7 - 23, 1973.

15685 - **BREEN, P.J., MURAOKA, T.** : Effect of indolbutyric acid on distribution of ^{14}C--photosynthate in softwood cuttings of "Marianna 2624" plum. - J. amer. Soc. hort. Sci. *98* : 436 - 439, 1973.

15686 - **BRETON, J., BECKER, J.F., GEACINTOV, N.E.** : Fluorescence depolarization study of randomly oriented and magneto-oriented spinach chloroplasts in suspension. - Biochem. biophys. Res. Commun. *54* : 1403 - 1409, 1973.

15687 - **BRETON, J., MICHEL-VILLAZ, M., PAILLOTIN, G.** : Orientation of pigments and structural proteins in the photosynthetic membrane of spinach chloroplasts : A liner dichroism study. - Biochim. biophys. Acta *314* : 42 - 56, 1973.

15688 - **BRETT, W.J., SINGER, A.C.** : Chlorophyll concentration in leaves of *Juniperus virginiana* L., measured over a 2-year period. - Amer. Midland Naturalist *90* : 194 - 200, 1973.

15689 - **BREVEDAN, E.R., HODGES, H.F.** : Effects of moisture deficits on ^{14}C translocation in corn (*Zea mays* L.). - Plant Physiol. *52* : 436 - 439, 1973.

15690 - **BREZEANU, A., PAUCĂ-COMĂNESCU, M., TĂCINĂ, F.** : The influence of light on production and on some structure elements of the forest herbaceous layer. - Rev. roum. Biol., Sér. Bot. *18* (2) : 83 - 95, 1973.

*15691 - **BREZEANU, A., TIŢU, H.** : Morphological and ultrastructural modifications induced by (^{60}Co) *gamma* radiations in *Festuca pratensis* HUDS. - Rev. roum. Biol., Sér. Bot. *17* : 219 - 226, 1972. [Chloroplast.]

15692 - **BŘEZINA, V.** : Analyse der Chloroplastenbewegung bei *Mougeotia* nach Starklichtreiz. - Arch. Hydrobiol. Suppl. *41* : 333 - 340, 1973.

15693 - **BRIANTAIS, J.-M., VERNOTTE, C., MOYA, I.** : Intersystem exciton transfer in isolated chloroplasts. - Biochim. biophys. Acta *325* : 530 - 538, 1973.

*15694 - **BRIN, G.P., ALIEV, Z.Sh.** : Deĭstvie rastvoriteleĭ na fotokhimicheskuyu aktivnost' i spektral'nye svoĭstva khloroplastov. [Effect of solvents on the photochemical activity and spectral properties of chloroplasts.] - In : Geneticheskie Aspekty Fotosinteza. Tezisy Dokladov. P. 83. Donish, Dushanbe 1972. [In R.]

15695 - **BRITTAIN, E.G., CAMERON, R.J.** : Photosynthesis of leaves of some *Eucalyptus* species. - New Zealand J. Bot. *11* : 153 - 162, 1973.

15696 - **BRITTON, G., GOODWIN, T.W.** : Chlorophyll, carotenoid pigments and sterols. - In : BUTLER, G.W., BAILEY, R.W. (ed.) : Chemistry and Biochemistry of Herbage. Vol. 1. Pp. 477 - 510. Academic Press, London - New York 1973.

15697 - **BRITZ, S.J., SELIGER, H.H.** : Endogenous and photoperiodic diurnal rhythms of *in vivo* light absorption and scattering in the green alga *Ulva lactuca* L. - Biol. Bull. *144* : 12 - 18, 1973. [Chl.]

*15698 - **BROCK, T.D.** : Microbial adaptation to extremes of temperature and pH. - In : BERNSTEIN, I.A. (ed.) : Biochemical Responses to Environmental Stress. Pp. 32 - - 37. Plenum Press, New York - london 1971. [Ps, Chl.]

15699 - **BROCKMANN, H. Jr., KNOBLOCH, G., SCHWEER, I., TROWITZSCH, W.** : Die Alkoholkomponente des Bacteriochlorophyll *a* aus *Rhodospirillum rubrum*. - Arch. Mikrobiol. *90* : 161 - 164, 1973.

15700 - **BRODY, M., WHITE, J.E.** : Environmental regulation of enzymes in the microbodies and mitochondria of dark-grown, greening, and light-grown *Euglena Gracilis*. - Dev. Biol. *al* : 48 - 361, 1973. [Ps.]

15701 - BRODY, S.S. : Surface properties of plastocyanin at an air-water interface. - Z. Naturforsch. *28c* : 397 - 400, 1973.

15702 - BROOKER, M.P., EDWARDS, R.W. : Effects of the herbicide paraquat on the ecology of a reservoir : I. Botanical and chemical aspects. - Freshwater Biol. *3* : 157 - 175, 1973. [Ps, Chl.]

15703 - BROOKER, M.P., EDWARDS, R.W. : Effects of the herbicide paraquat on the ecology of a reservoir. II. Community metabolism. - Freshwater Biol. *3* : 383 - 389, 1973. [Ps.]

15704 - BROOKING, I.R., TAYLOR, A.O. : Plants under climatic stress. V. Chilling and light effects on radiocarbon exchange between photosynthetic intermediates of *Sorghum*. - Plant Physiol. *52* : 180 - 182, 1973.

15705 - BROOKS, C., GANTT, E. : Comparison of phycoerythrins (542, 566 nm) from cryptophycean algae. - Arch. Mikrobiol. *88* : 193 - 204, 1973.

15706 - BROUWER, R., KLEINENDORST, A., LOCHER, J.T. : Growth responses of maize plants to temperature. - In : SLATYER, R.O. (ed.) : Plant Response to Climatic Factors. Pp. 169 - 174. Unesco, Paris 1973. [Ps.]

15707 - BROVCHENKO, M.I. : Transport organicheskikh veshchestv v listovoĭ plastinke sakharnoĭ svekly. [Transport of organic substances in the leaf blade of sugar beet plants.] - Tr. biol.-pochv. Inst., dal'nevost. nauch. Tsentr Akad. Nauk SSSR *20* (123) : 53 - 58, 1973. [In R, ab : E.]

15708 - BROWN, J.S. : Separation of photosynthetic systems I and II. - In : GIESE, A.C. (ed.) : Photophysiology. Current Topics in Photobiology and Photochemistry. Vol. 8. Pp. 97 - 112. Academic Press, New York - London 1973.

15709 - BROWN, J.S. : The effect of manganese on DCIP reduction. - Carnegie Inst. Year Book *72* : 359 - 361, 1973.

15710 - BROWN, J.S., GASANOV, R.A., FRENCH, C.S. : A comparative study of the forms of chlorophyll and photochemical activity of system 1 and system 2 fractions from spinach and *Dunaliella*. - Carnegie Inst. Year Book *72* : 351 - 359, 1973.

15711 - BROWN, R.H., ETHREDGE, W.J., KING, J.W. : Influence of succinic acid 2,2-dimethylhydrazide on yield and morphological characteristics of starr peanuts (*Arachis hypogaea* L.). - Crop Sci. *13* : 507 - 510, 1973. [Growth analysis.]

15712 - BROWN, W.V., SMITH, B.N. : *Evolvulus alsinoides* is not a Kranz species. - Bull. Torrey bot. Club *100* : 348 - 349, 1973.

15713 - BRUN, L.J., KANEMASU, E.T., POWERS, W.L. : Estimating transpiration resistance. - Agron. J. *65* : 326 - 328, 1973. [Growth analysis.]

15714 - BRYLINSKY, M., MANN, K.H. : An analysis of factors governing productivity in lakes and reservoirs. - Limnol. Oceanogr. *18* : 1 - 14, 1973.

15715 - BRZOSKA, W. : Dry matter production and energy utilization of high mountain plants in the Austrian Alps. - Oecol. Plant. *8* : 63 - 70, 1973.

15716 - BRZOSKA, W. : D. Stoffproduktion und Energiehaushalt von Nivalpflanzen. - In : ELLENBERG, H. (ed.) : Ökosystemforschung. Pp. 225 - 233. Springer-Verlag, Berlin - Heidelberg - New York 1973.

15717 - BUCHANAN, B.B. : Ferredoxin and carbon assimilation. - In : LOVENBERG, W. (ed.) : Iron-Sulfur Proteins. Vol. 1. Biological Properties. Pp. 129 - 150. Academic Press, New York - London 1973.

15718 - BUCHANAN, B.B., MAGYAROSY, A.C., SCHÜRMANN, P. : Effect of virus infection on photosynthesis by chloroplasts. - Fed. Proc. *32* : 522 Abs, 1973.

15719 - BUCHANAN, B.B., SCHÜRMANN, P. : Regulation of ribulose 1,5-diphosphate carboxylase in the photosynthetic assimilation of carbon dioxide. - J. biol. Chem. *248* : 4956 - 4964, 1973.

15720 - BUKHBINDER, A.A. : Dinamika nakopleniya pigmentov v list'yakh gerani, vyrashchennoĭ na allyuvial'nykh pochvakh Kolkhidskoĭ nizmennosti. [Dynamics of pigment accumulation in leaves of geranium grown on alluvial soils of the Kolkhid lowland.] - Subtrop. Kul't. *1973* (3) : 107 - 111, 1973. [In R.]

15721 - **BULJAN, M., HURE, J., PUCHER-PETKOVIĆ, T.** : Hidrografske i produkcione prilike u Malostonskom zaljevu. [Hydrographic and productivity conditions in the Bay of Mali Ston.] - Acta adriat. *15* (2) : 1 - 60, 1973. [Primary production.]

15722 - **BULYCHEV, A.A.** : Spektr deĭstviya bystrykh fotoindutsirovannykh izmeneniĭ transmembrannogo potentsiala khloroplasta. [Spectrum of activity of fast photoinduced changes of chloroplast transmembrane potential.] - Vestn. mosk. Univ., Ser. 6 - Biol. Pochvoved. *28* (1) : 98 - 99, 1973. [In R, ab : E.]

15723 - **BULYCHEV, A.A., ANDRIANOV, V.K., KURELLA, G.A., LITVIN, F.F.** : Bystrye fotoélektricheskie protsessy na membranakh khloroplasta i ikh svyaz' s dvumya fotokhimicheskimi reaktsiyami fotosinteza. 1. Analiz kinetiki fotoindutsirovannykh izmeneniĭ potentsiala khloroplasta. [Rapid photoelectric processes on chloroplast membranes and their relation to the two photochemical reactions of photosynthesis. 1. Analysis of kinetics of photoinduced changes in chloroplast potential.] - In : Biofizika Membran. Pp. 104 - 109. Kaunas 1973. [In R.]

15724 - **BULYCHEV, A.A., ANDRIANOV, V.K., KURELLA, G.A., LITVIN, F.F.** : Bystrye fotoélektricheskie protsessy na membranakh khloroplasta i ikh svyaz' s dvumya fotokhimicheskimi reaktsiyami fotosinteza. 2. Vliyanie smeshcheniya membrannogo potentsiala na fotoélektricheskuyu reaktsiyu khloroplasta. [Rapid photoelectric processes on chloroplast membranes and their relation to the two photochemical reactions of photosynthesis. 2. Effect of the shift of membrane potential on the photoelectric reaction of chloroplast.] - In : Biofizika Membran. Pp. 110 - - 113. Kaunas 1973. [In R.]

15725 - **BUNT, J.S., MOUNT, Z.** : *In situ* studies on the tropical benthos. VI. Oxygen exchange associated with the sediments near hydro lab. - Tech. Rep. Univ. Miami, Rosenstiel School mar. atmos. Sci. *73060* : 1 - 7, 1973.

15726 - **BURBA, M., ELSTNER, E.F.** : Photosynthetische Aktivitäten isolierter Chloroplasten der Zuckerrübe. I. Allgemeine Grundlagen und Methodik. - Z. Zuckerindustrie *23* : 609 - 615, 1973.

15727 - **BURDEN, E.M., HORTON, A.A.** : Site of inhibition by 2-amino-1,1,3-tricyanopropene of photosynthetic oxygen evolution by spinach leaf chloroplasts. - Biochem. J. *134* : 663 - 665, 1973.

15728 - **BURIAN, K.** : A. *Phragmites communis* TRIN. im Röhricht des Neusiedler Sees. Wachstum, Produktion und Wasserverbrauch. - In : **ELLENBERG, H.** (ed.) : Ökosystemforschung. Pp. 61 - 78. Springer-Verlag, Berlin - Heidelberg - New York 1973. [Ps.]

15729 - **BURK, J.H., DICK-PEDDIE, W.A.** : Comparative production of *Larrea divaricata* CAV. on three geomorphic surfaces in southern New Mexico. - Ecology *54* : 1094- - 1102, 1973.

15730 - **BURNELL, J.N., ANDERSON, J.W.** : Adenosine 5'-sulphatophosphate kinase activity in spinach leaf tissue. - Biochem. J. *134* : 565 - 579, 1973. [In chloroplasts.]

15731 - **BURRIS, J.S., EDJE, O.T., WAHAB, A.H.** : Effects of seed size on seedling performance in soybeans : II. Seedling growth and photosynthesis and field performance. - Crop Sci. *13* : 207 - 210, 1973.

*15732 - **BURTSEVA, R.A.** : Vliyanie vnekornevoĭ podkormki superfosfatom na raspredelenie radiougleroda sredi produktov fotosinteza (vodno-spirtovoĭ fraktsii) u soi. [Effect of extra-root nutrition by superphosphate on the distribution of ^{14}C among photosynthates (water-alcoholic fraction) in soybean.] - In : Tezisy Dokladov Vsesoyuznoĭ Konferentsii po Ispol'zovaniyu Radiatsionnoĭ Tekhniki v Sel'skom Khozyaĭstve. Vol. III. Pp. 39 - 40. Kishinev 1972. [In R.]

15733 - **BURTSEVA, R.A., KURILENKO, L.A.** : Ob ispol'zovanii produktov fotosinteza u soi pri vnekormovom pitanii. [Use of photosynthates in soybean plants under extra- -root nutrition.] - Tr. biol.-pochv. Inst., dal'nevost. nauch. Tsentr Akad. Nauk SSSR *20* (123) : 191 - 194, 1973. [In R, ab : E.]

15734 - **BURTSEVA, R.A., VORONKOVA, N.M.** : Raspredelenie mechenykh assimilyatov u soi pri porazhenii rasteniĭ virusnoĭ infektsieĭ. [Distribution of labelled assimilates in virus affected soybean plants.] - Tr. biol.-pochv. Inst., dal'nevost. nauch. Tsentr Akad. Nauk SSSR *13* (Biokhimicheskie Issledovaniya na Sovetskom Dal'nem Vostoke) : 25 - 28, 1973. [In R, ab : E.]

15735 - BURZLAFF, D.F., HAM, G.L., KEHR, W.R. : *In situ* estimation of alfalfa forage yields. - Agron. J. *65* : 644 - 646, 1973.

15736 - BUTLER, W.L. : Primary photochemistry of photosystem II of photosynthesis. - Accounts chem. Res. *8* : 177 - 184, 1973.

15737 - BUTLER, W.L. : Primary photochemical reaction of photosystem II in chloroplasts. - In : Ninth International Congress of Biochemistry. Stockholm,July 1 - 7, 1973. Abstract Book. P. 206. IUB, Stockholm 1973.

15738 - BUTLER, W.L., VISSER, J.W.M., SIMONS, H.L. : The kinetics of light-induced changes of $C-550$, cytochrome b_{559} and fluorescence yield in chloroplasts at low temperature. - Biochim. biophys. Acta *292* : 140 - 151, 1973.

15739 - BUTLER, W.L., VISSER, J.W.M., SIMONS, H.L. : The back reaction in the primary electron transfer couple of Photosystem II of photosynthesis. - Biochim. biophys. Acta *325* : 539 - 545, 1973.

15740 - BUTTERFASS, T. : Control of plastid division by means of nuclear DNA amount. - Protoplasma *76* : 167 - 195, 1973.

15741 - BYKOV, O.D. : Primary carbon metabolic pool and the use of new photosynthates in respiration during photosynthesis. - Photosynthetica *7* : 232 - 237, 1973.

15742 - BYKOV, O.D. : Metody izmereniya konstanty skorosti fotosinteza i skorosti fotodykhaniya po dannym infrakrasnogo gazovogo analiza. [Methods of calculation of photosynthetic rate constant and the rate of photorespiration based on the infra-red gas analysis data.] - In : Metody Kompleksnogo Izucheniya Fotosinteza. Vol. 2. Pp. 15 - 32. VASKHNIL, Leningrad 1973. [In R.]

15743 - BYKOV, O.D., ÉSYUNINA, A.I. : Uproshchennyĭ ékspress-metod analiza organicheskikh kislot rasteniĭ s pomoshch'yu khromatografii na bumage. [Simple rapid method for analysis of organic acids of plants by means of paper chromatography.] - In : Metody Kompleksnogo Izucheniya Fotosinteza. Vol. 2. Pp. 246 - 251. VASKHNIL, Leningrad 1973. [In R.]

*15744 - BYKOV, O.D., GALKIN, V.I., ZHITLOVA, N.A., KOSHKIN, V.A. : Vliyanie temperatury na izmenenie intensivnosti fotosinteza i dykhaniya diploidov i poliploidov kartofelya. [Effect of temperature on the changes of photosynthetic rate and respiration in diploids and polyploids of *Solanum*.] - In : Geneticheskie Aspekty Fotosinteza. Tezisy Dokladov. Pp. 99 - 100. Donish, Dushanbe 1972. [In R.]

15745 - BYKOV, O.D., IVANOV, O.V. : Ustanovka dlya issledovaniya kinetiki fotosinteticheskogo gazoobmena po $C^{12}O_2$ i $C^{14}O_2$. [Apparatus for investigation of photosynthetic gas exchange kinetics with $^{12}CO_2$ and $^{14}CO_2$.] - In : Metody Kompleksnogo Izucheniya Fotosinteza. Vol. 2. Pp. 33 - 54. VASKHNIL, Leningrad 1973. [In R.]

15746 - BYKOV, O.D., KOSHKIN, V.A. : Usovershenstvovannaya ustanovka dlya massovykh opredeleniĭ potentsial'noĭ intensivnosti fotosinteza radiometricheskim metodom. [Improved apparatus for mass determination of potential photosynthetic rate by radiometric method.] - In : Metody Kompleksnogo Izucheniya Fotosinteza. Vol. 2. Pp. 55 - 62. VASKHNIL, Leningrad 1973. [In R.]

15747 - BYKOV, O.D., LOKUTSIEVSKAYA, L.K., SAMOĬLOVA, L.A. : Modelirovanie fotosinteticheskogo gazoobmena : II. Izmenenie kineticheskikh parametrov vydeleniya CO_2 bikarbonat-karbonatnymi rastvorami. [Models of photosynthetic gas exchange : II. Measuring of kinetic parameters of CO_2 release by bicarbonate-carbonate solutions.] - In : Metody Kompleksnogo Izucheniya Fotosinteza. Vol. 2. Pp. 85 - 101. VASKHNIL, Leningrad 1973. [In R.]

15748 - BYKOV, O.D., LOKUTSIEVSKAYA, L.K., SAMOĬLOVA, L.A., IVANOV, O.V. : Modelirovanie fotosinteticheskogo gazoobmena : I. Izmerenie parametrov sorbtsii CO_2 shchelochnymi rastvorami. [Models of photosynthetic gas exchange : I. Measuring of parameters of CO_2 uptake by alkaline solutions.] - In : Metody Kompleksnogo Izucheniya Fotosinteza. Vol. 2. Pp. 63 - 84. VASKHNIL, Leningrad 1973. [In R.]

15749 - BYRNE, G.F. : An approach to growth curve analysis. - Agr. Meteorol. *11* : 161 - 168, 1973.

*15750 - **BYRNE, G.F., ROSE, C.W., BEGG, J.E., TORSSELL, B.W.R., McPHERSON, H.G.** : Instrumentation for crop-environment measurement in a tropical savannah climate. - Div. Land Res. tech. Paper. No. 32. Pp. 1 - 19. CSIRO, Australia 1971. [Ps.]

15751 - **CALABRESE, G., FELICINI, G.P.** :Ricerche sui pigmenti delle alghe rosse. III : Influenza della luce sui pigmenti di *Petroglossum nicaeense* (DUBY) SCHOTTER in coltura. [Pigments of red algae. III : Effect of illuminance on pigment formation in *Petroglossum nicaeense* (DUBY) SCHOTTER cultured *in vitro*.] - G. bot. ital. *107* : 1 - 8, 1973. [In Ital., ab. : E.]

15752 - **CALABRESE, G., FELICINI, G.P.** : Research on red algal pigments. 5. The effect of the intensity of white and green light on the rate of photosynthesis and its relationship to pigment components in *Gracilaria compressa* (C.AG.)GREV. (*Rhodophyceae, Gigartinales*). - Phycologia *12* : 195 - 199, 1973.

15753 - **CALDER, D.M.** : The effect of temperature on growth and dry weight distribution of population of *Poa annua* L. - In : SLATYER, R.O. (ed.) : Plant Response to Climatic Factors. Pp. 145 - 152. Unesco, Paris 1973. [Growth analysis.]

15754 - **CALDER, F.W., CANHAM, W.D., FENSOM, D.S.** : Some effects of Alar-85 on the physiology of alfalfa and Ladino clover. - Can. J. Plant Sci. *53* : 269 - 278, 1973. [Chl.]

15755 - **CALDER, J.A., PARKER, P.L.** : Geochemical implications of induced changes in C^{13} fractionation by blue-green algae. - Geochim. cosmochim. Acta *37* : 133 - - 140, 1973.

*15756 - **CALÈ, M.T., FIGLIOLIA, A.** : Studi sul potere assimilante dei cloroplasti in rapporto alla fertilizzazione e al regime idrico. [Effect of mineral nutrition and water relations on chloroplast assimilatory power.] - Ann. Ist. sper. Nutr. Piante (Rome) *2* (2) : 15 - 34, 1971. [In Ital., ab : E.]

15757 - **CALJON, A.** : Ekosystematische studie van het fytoplankton van het Molsbroek bij Lokkeren (O.Vl.). [Ecosystematic study of phytoplankton in Molsbroek near Lokkeren (O.Vl.).] - Biol. Jaarb. *41* : 71 - 115, 1973. [Chl; in Flemish, ab : E.]

15758 - **CALLAGHAN, T.V.** : Studies on the factors affecting the primary production of bi-polar *Phleum alpinum*. - In : BLISS, L.C., WIELGOLASKI, F.E. (ed.) : Primary Production and Production Processes, Tundra Biome. Pp. 153 - 167. Tundra Biome Steering Comm., Edmonton 1973.

15759 - **CALLOW, M.E., WOOLHOUSE, H.W.** : Changes in nucleic-acid metabolism in regreening leaves of *Perilla*. - J. exp. Bot. *24* : 285 - 294, 1973.

15760 - **CALVAYRAC, R.M., CLAISSE, M.L.** : Etude spectrale à basse temperature chez *Euglena gracilis* Z en culture synchrone sur milieu lactate : Variation de teneur en cytochrome 556. - Planta *112* : 17 - 24, 1973.

15761 - **CAMMACK, R.** : "Super-reduction" of *Chromatium* high-potential iron-sulphur protein in the presence of dimethyl sulphoxide. - Biochem. biophys. Res. Commun. *54* : 548 - 554, 1973.

15762 - **CAMPBELL, L.G.** : Studies on the biosynthesis of the chloroplast enzyme ribulosediphosphate carboxylase. - Diss. Abstr. Int. B *34* : 52-B, 1973.

15763 - **CAMPBELL, W.P., GRIFFITHS, D.A.** : Pathogenicity of *Verticillium dahliae* to potato in Victoria, Australia. - Plant Disease Reporter *57* : 735 - 738, 1973. [Chl.]

15764 - **CAPBLANCQ, J.** : Phytobenthos et productivité primaire d'un lac de haute montagne dans les Pyrénées centrales. - Ann. Limnol. *9* : 193 - 230, 1973. [Chl.]

15765 - **CAPLIN, S.M.** : Translocation of ^{14}C metabolites in carrot root. - Amer. J. Bot. *60* : 703 - 707, 1973.

15766 - **CARDE, J.-P.** : Le tissu de transfert (=cellules de Strasburger) dans les aiguilles de Pin maritime (*Pinus pinaster* AIT.) I. Étude histologique et infrastructurale du tissu adulte. - J. Microscop. *17* : 65 - 88, 1973. [Chl, chloroplast.]

15767 - **CARDENAS, R., GAUSMAN, H.W.** : Relation of light reflectance of six barley lines with chlorophyll assays and optical film densities. - Agron. J. *65* : 518 - - 519, 1973.

15768 - **CARMELI, C., RACKER, E.** : Partial resolution of the enzymes catalyzing photo-phosphorylation. XIV. Reconstitution of chlorophyll-deficient vesicles catalyzing phosphate-adenosine triphosphate exchange. - J. biol. Chem. *248* : 8281 - 8287, 1973.

15769 - **CARPENTER, D.J., JITTS, H.R.** : A remote operating submarine irradiance meter. - Deep-Sea Res. *20* : 859 - 865, 1973.

15770 - **CARR, N.G.** : Metabolic control and autotrophic physiology. - In : CARR, N.G., WHITTON, B.A. (ed.) : The Biology of Blue-Green Algae. Bot. Monogr. 9. Pp. 39 - 65. Blackwell sci. Publ., Oxford - London - Edinburgh - Melbourne 1973. [Ps.]

15771 - **CARTLEDGE, O., CONNOR, D.J.** : Photosynthetic efficiency of tropical and temperate grass canopies. - Photosynthetica *7* : 109 - 113, 1973.

15772 - **CASE, G.D., PARSON, W.W.** : Redistribution of electric charge accompanying photosynthetic electron transport in *Chromatium*. - Biochim. biophys. Acta *292* : 677 - 684, 1973.

15773 - **CASE, G.D., PARSON, W.W.** : Shifts of bacteriochlorophyll and carotenoid absorption bands linked to cytochrome *c*-555 photooxidation in *Chromatium*. - Biochim. biophys. Acta *325* : 441 - 453, 1973.

15774 - **CASERTA, G., CERVIGNI, T.** : A piezoelectric transducer model for phosphorylation in photosynthetic membranes. - J. theor. Biol. *41* : 127 - 142, 1973.

15775 - **CASTELFRANCO, P.A., RICH, P.M., GINZTON, L.** : Chlorophyll synthesis from exogenous δ-aminolevulinic acid in greening cucumber cotyledons. - Plant Physiol. *51* (Suppl.) : 19, 1973.

15776 - **CASTENHOLZ, R.W.** : Ecology of blue-green algae in hot springs. - In : CARR, N.G., WHITTON, B.A. (ed.) : The Biology of Blue-Green Algae. Bot. Monogr. 9. Pp. 379 - 414. Blackwell sci. Publ., Oxford - London - Edinburgh - Melbourne 1973.

15777 - **CASTENHOLZ, R.W.** : The possible photosynthetic use of sulfide by the filamentous phototrophic bacteria of hot springs. - Limnol. Oceanogr. *18* : 863 - 876, 1973.

15778 - **CASWELL, H., REED, F., STEPHENSON, S.N., WERNER, P.A.** : Photosynthetic pathways and selective herbivory : a hypothesis. - Amer. Naturalist *107* : 465 - 480, 1973.

15779 - **ČATSKÝ, J., CHARTIER, P., DJAVANCHIR, A.** : Assimilation nette, utilisation de l'eau et microclimat d'un champ de maïs. IV. - Évolution diurne de la résistance stomatique et du déficit de saturation des feuilles; conséquences sur la fixation du CO_2. - Ann. agron. *24* : 287 - 305, 1973.

15780 - **CAVALEIRO, J.A.S., KENNER, G.W., SMITH, K.M.** : Pyrroles and related compounds. Part XXIII. Protoporphyrin-1. - J. chem. Soc., Perkin Trans. I *21* : 2478 - - 2485, 1973.

15781 - **CECHOVA, I., DAVIS, E.M.** : Trend surface analysis and seasonal distribution patterns of primary nutrients and chlorophyll in unstratified Gulf Coast estuaries. - Water Resources Res. *9* : 1543 - 1554, 1973.

15782 - **CERFF, R.** : Glyceraldehyde 3-phosphate dehydrogenases and glyoxylate reductase. I. Their regulation under continuous red and far red light in the cotyledons of *Sinapis alba* L. - Plant Physiol. *51* : 76 - 81, 1973.

15783 - **CHAĬKA, M.T., SAVCHENKO, G.E.** : Metabolizm pigmentov v protsesse razvitiya zelenogo lista. [Pigment metabolism in the course of development of a green leaf.] - In : Formirovanie Pigmentnogo Apparata Fotosinteza. Pp. 105 - 129. Nauka i Tekhnika, Minsk 1973. [In R.]

15784 - **CHALUPA, V.** : Růst lesních dřevin při zvýšeném obsahu kysličníku uhličitého v ovzduší. [Growth of forest trees in an atmosphere with increased CO_2 content.] - Lesnická Práce *52* : 548 - 551, 1973. [Ps ; in Czech.]

15785 - **CHAN, P.H.** : The role of chloroplast and nuclear DNAs in coding for the large and small subunits of fraction I proteins isolated from *Nicotiana* species and reciprocal, interspecific hybrids. - Diss. Abstr. Int. B *33* : 4127, 1973.

15786 - **CHANG, C.S., HUANG, B.K.** : Plant growth simulation based on net carbon dioxide consumption. - Trans. ASAE *16* : 724 - 727, 1973.

15787 - **CHANG, C.W.** : Carbonic anhydrase of cotton. - Plant Physiol. *51* (Suppl.) : 41, 1973.

15788 - **CHANG, T.-P.** : An interpretation : The "nonblackbody effect" in photosynthesis. - Taiwania *18* : 9 - 12, 1973.

15789 - **CHAPMAN, D.J.** : Biliproteins and bile pigments. - In : **CARR, N.G., WHITTON, B.A.** (ed.) : The Biology of Blue-Green Algae. Pp. 162 - 185. Blackwell sci. Publ., Oxford - London - Edinburgh - Melbourne 1973.

15790 - **CHARLES-EDWARDS, D.A., THORNLEY, J.H.M.** : Light interception by an isolated plant. A simple model. - Ann. Bot. *37* : 919 - 928, 1973.

*15791 - **CHARTIER, P.** : Solar radiation and crop photosynthesis. - In : Coll. Franco- -Israélien sur le Bilan des Rayonnements et l'Agriculture. Pp. 1 - 17. Bet Dagan 1971.

15792 - **CHARTIER, P.** : Rectificatif à propos de l'article "Étude du microclimat lumineux dans la végétation" paru dans les Annales Agronomiques 1966, 17 (5), 571 - 602. - Ann. agron. *24* : 395 - 396, 1973.

15793 - **CHARTIER, P., BECKER, M., BONHOMME, R., BONY, J.P.** : Effets physiologiques et caractérisation du rayonnement solaire dans le cadre d'une méthode d'aménagement sylvicole en forêt dense africaine. - Bois Forêts trop. *152* : 19 - 35, 1973.

15794 - **CHARTIER, P., BONHOMME, R., VARLET-GRANCHER, C.** : Captation de l'energie solaire par une culture de maïs. - In : The Sun in the Service of Mankind. Pp. V. 2-1 - V.2-11. Unesco, Paris 1973.

15795 - **CHATTERTON, N.J.** : Product inhibition of photosynthesis in alfalfa leaves as related to specific leaf weight. - Crop Sci. *13* : 284 - 285, 1973.

15796 - **CHATTERTON, N.J., LEE, D.R.** : Leaf chamber to measure photosynthesis and transpiration of intact grass leaf sections. - Crop Sci. *13* : 576 - 577, 1973.

15797 - **CHAUVET, J.-P., JOURNEAUX, R., VIOVY, R.** : Photo-oxydation de la chlorophylle *a* dans le binaire *Triton X 100*-eau. - Compt. rend. Acad. Sci. Paris, Sér. C *277* : 527 - 530, 1973.

15798 - **CHAWDHRY, M.A., SAGAR, G.R.** : An autoradiographic study of the distribution of ^{14}C labelled assimilates at different stages of development of *Oxalis latifolia* H.B.K. and *O. pes-caprae* L. - Weed Res. *13* : 430 - 437, 1973.

15799 - **CHECCUCCI, A.** : Photomotion methodology in flagellates. - In : **CHECCUCCI, A., WEALE, R.A.** (ed.) : Primary Molecular Events in Photobiology. Pp. 217 - 244. Elsevier, Amsterdam - London - New York 1973. [Ps, Chl.]

B15800 - **CHECCUCCI, A., WEALE, R.A.** (ed.) : Primary Molecular Events in Photobiology. - Elsevier, Amsterdam - London - New York 1973.

*15801 - **CHELOVSKAYA, L.N., NIKOLAEV, A.G.** : Sravnitel'naya kharakteristika sortov morkovi razlichnogo proiskhozhdeniya po soderzhaniyu pigmentov. [Pigment content in different cultivars of carrots.] - Tr. Khim. prirodnykh Soedinenii *8* : 99 - - 105, 1969. [In R.]

15802 - **CHEN, T.M., CAMPBELL, W.H., DITTRICH, P., BLACK, C.C.** : Distribution of carboxylation enzymes in isolated mesophyll cells and bundle sheath strands of C_4 plants. - Biochem. biophys. Res. Commun. *51* : 461 - 467, 1973.

15803 - **CHEN, T.M., DITTRICH, P., CAMPBELL, W.H., BLACK, C.C. Jr.** : Resolution of a mature nutsedge leaf into upper, and lower epidermis, mesophyll cells, and bundle sheath strands and their enzyme complements. - Plant Physiol. *51* (Suppl.) : 6, 1973.

15804 - **CHENIAE, G.M., MARTIN, I.F.** : Absence of oxygen-evolving capacity in dark-grown *Chlorella* : the photoactivation of oxygen-evolving centers. - Photochem. Photobiol. *17* : 441 - 459, 1973.

15805 - **CHERMNYKH, L.N., CHUGUNOVA, N.G., KOSOBRYUKHOV, A.A.** : O regulyatsii fotosinteticheskoi aktivnosti rastenii faktorami vneshney sredy. [Control of plant

photosynthetic activity by environmental factors.] - In : Upravlenie Skorost'-
yu i Napravlennost'yu Biosinteza u Rastenii. Pp. 103 - 104. Krasnoyarsk 1973.
[In R.]

15806 - CHERNAVINA, I.A., ZHOGOVA, E.P., NIKIFOROVA, T.A. : Plastotsianin i tsitokhrom
f v ēlektrontransportnoi tsepi fotosinteza khloroplastov rastenii ovsa. [Plas-
tocyanine and cytochrome f in electron transport chain of photosynthesis in
oat chloroplasts.] - Fiziol. Rast. 20 : 988 - 994, 1973. [In R, ab : E.]

*15807 - CHERNYAD'EV, I.I., USPENSKAYA, V.Ė., KONDRAT'EVA, E.N., DOMAN, N.G. : Ob assi-
milyatsii ugleroda iskhodnym i mutantnym shtammami *Rhodopseudomonas palustris*.
[Assimilation of carbon by wild and mutant strains of *Rhodopseudomonas palustris*.] - In : Geneticheskie Aspekty Fotosinteza. Tezisy Dokladov. P. 48. Donish, Dushanbe 1972. [In R.]

15808 - CHERNYSHEV, V.D. : O spektral'nom diffuznom otrazhenii sveta khvoei drevesnykh
vidov Dal'nego Vostoka. [Spectral diffusive reflectance of light by needles of
coniferous trees of Far East.] - Fiziol. Rast. 20 : 321 - 325, 1973. [In R,
ab : E.]

15809 - CHERNYSHEV, V.D. : Fotosintez i dykhanie podrosta v shirokolistvenno-khvoinykh
lesakh yuzhnogo Primor'ya. [Photosynthesis and respiratory activity of the regrowth in broad-leaved-coniferous forests in the south of Primor'e territory.]
- Tr. biol.-pochv. Inst. Vladivostok, nov. Ser. 16 (119 - Fiziologiya i Ėkologiya Drevesnykh Rastenii Primor'ya) : 7 - 22, 1973. [In R, ab : E.]

15810 - CHERNYSHEV, V.D. : Sostav i kolichestvo pigmentov khvoi podrosta v razlichnykh
fitotsenoticheskikh usloviyakh. [Composition and number of needle pigments of
regrowth under different phytocenotic conditions.] - Tr. biol.-pochv. Inst.
Vladivostok, nov. Ser. 16 (119 - Fiziologiya i Ėkologiya Drevesnykh Rastenii
Primor'ya) : 23 - 33, 1973. [In R, ab : E.]

15811 - CHERNYSHEV, V.D. : Pokazatel' zateneniya pochvy drevostoem. [Index of soil
shading by a forest stand.] - Tr. biol.-pochv. Inst. Vladivostok, nov. Ser. 16
(119 - Fiziologiya i Ėkologiya Drevesnykh Rastenii Primor'ya) : 164 - 168,
1973. [Chl; in R, ab : E.]

15812 - CHERNYSHEV, V.D. : K metodike izmereniya temperatury vozdukha v mnogoyarusnykh
fitotsenozakh. [Technique of measuring air temperature in multilevel phytocoe-
noses.] - Tr. biol.-pochv. Inst. Vladivostok, nov. Ser. 16 (119 - Fiziologiya
i Ėkologiya Drevesnykh Rastenii Primor'ya) : 169 - 174, 1973. [In R, ab : E.]

15813 - CHERNYSHEV, V.D., CHIZHIKOV, A.A. : O primenenii samopishushchikh mikroamper-
millivol'tmetrov dlya izucheniya rezhima i balansa solnechnoi ēnergii v lesu.
[Self-recording microampermillivoltmeter utilization for the study of solar
energy regime and balance in a forest.] - Tr. biol.-pochv. Inst. Vladivostok,
nov. Ser. 16 (119 - Fiziologiya i Ėkologiya Drevesnykh Rastenii Primor'ya) :
175 - 177, 1973. [In R, ab : E.]

15814 - CHERNYSHEV, V.D., PETROV, P.G. : Ob ēkologo-fiziologicheskikh izmeneniyakh
v kedrovom drevostoe Primor'ya posle rubok ukhoda. [Ecological and physiological changes in Primor'e cedar stand after maintenance cuttings.] - Tr. biol.-
-pochv. Inst. Vladivostok, nov. Ser. 16 (119 - Fiziologiya i Ėkologiya Drevesnykh Rastenii Primor'ya) : 53 - 60, 1973. [Chl; in R, ab : E.]

*15815 - CHESNOKOV, V.A., STEPANOVA, A.M. : Proiskhozhdenie khloroplastov. [Genesis of
chloroplasts.] - In : Geneticheskie Aspekty Fotosinteza. Tezisy Dokladov. Pp.
17 - 18. Donish, Dushanbe 1972. [In R.]

15816 - CHETVERIKOVA, N.I. : Sostav vtorichnykh produktov fotosinteza v list'yakh i
sozrevayushchikh bobakh soi. [Composition of secondary products of photosynthesis in leaves and ripening seeds of soybean.] - Tr. biol.-pochv. Inst., dal'-
nevost. nauch. Tsentr Akad. Nauk SSSR 13 (116 - Biokhimicheskie Issledovaniya
na Sovetskom Dal'nem Vostoke) : 29 - 48, 1973. [In R, ab : E.]

15817 - CHETVERIKOVA, N.I., KOSMAKOVA, V.E., ZVEREVA, E.G., ZHEMCHUGOVA, V.P., CHUB,
A.I., DZIZENKO, G.K. : Vliyanie pereuvlazhneniya pochvy na sostav produktov
fotosinteza u soi. [Effect of surplus water content in soil on the composition
of photosynthates in soybean.] - Uch. Zap. dal'nevost. gos. Univ. 61 (Ustoichivost' Rastenii k Pereuvlazhneniyu Pochvy na Dal'nem Vostoke i Deistvie v Ētikh
Usloviyakh Mikroēlementov) : 35 - 42, 1973. [In R.]

15818 - **CHETVERIKOVA, N.I., ZHEMCHUGOVA, V.P.** : O prevrashchenii vtorichnykh produktov fotosinteza v sozrevayushchikh semenakh podsolnechnika. [Transformation of secondary products of photosynthesis in ripening seeds of sunflower.] - Tr. biol.-pochv. Inst., dal'nevost. nauch. Tsentr Akad. Nauk SSSR *13* (116 - Biokhimicheskie issledovaniya na Sovetskom Dal'nem Vostoke) : 49 - 60, 1973. [In R, ab : E.]

15819 - **CHETVERIKOVA, N.I., ZHEMCHUGOVA, V.P.** : Vliyanie virusnoĭ infektsii na sostav vtorichnykh produktov fotosinteza u soi. [Effect of virus infection on the composition of secondary products of photosynthesis in soybean plants.] - Tr. biol.-pochv. Inst., dal'nevost. nauch. Tsentr Akad. Nauk SSSR *14* (Virusnye Bolezni Rasteniĭ Dal'nego Vostoka) : 34 - 45, 1973. [In R, ab : E.]

15820 - **CHETVERIKOVA, N.I., ZHEMCHUGOVA, V.P.** : Dinamika postupleniya produktov fotosinteza i ispol'zovanie ikh v sinteze zapasnykh veshchestv semyan podsolnechnika. [Dynamics of photosynthates transport into ripening seeds and their utilization in the synthesis of reserve substances in the sunflower plants.] - Tr. biol.-pochv. Inst., dal'nevost. nauch. Tsentr Akad. Nauk SSSR *20* (123) : 276 - 280, 1973. [In R, ab : E.]

15821 - **CHETVERIKOVA, N.I., ZHEMCHUGOVA, V.P., CHUB, A.I.** : Puti ispol'zovaniya sobstvennykh i pritekayushchikh assimilyatov v molodykh list'yakh soi. [Ways of utilizing "own" and "foreign" photosynthates in young leaves of soybean plants.] - Tr. biol.-pochv. Inst., dal'nevost. nauch. Tsentr Akad. Nauk SSSR *20* (123) : 71 - 75, 1973. [In R, ab : E.]

15822 - **CHEUNG, Y.-N.S., NOBEL, P.S.** : Amino acid uptake by pea leaf fragments : Specificity, energy sources, and mechanism. - Plant Physiol. *52* : 633 - 637, 1973. [Relation to Ps.]

15823 - **CHEVALLIER, D.** : Rôle du manganèse dans la germination des spores de *Funaria hygrometrica*. I - Etude de l'absorption du manganèse et de sa localisation cellulaire. - Physiol. vég. *11* : 461 - 473, 1973. [Ps, Chl.]

15824 - **CHEVALLIER, D.** : Rôle du manganèse dans la germination des spores de *Funaria hygrometrica*. II - Précisions sur l'état du manganèse absorbé au moyen de la résonance paramagnétique electronique. - Physiol. vég. *11* : 475 - 486, 1973. [Ps, Chl.]

15825 - **CHIKOV, V.I., LOZOVAYA, V.V.** : Vliyanie insektitsidov na postfotosinteticheskiĭ metabolizm C^{14}-assimilyatov v list'yakh bobov. [Effect of insecticides on postphotosynthetic metabolism of ^{14}C-assimilates in broadbean leaves.] - Fiziol. Rast. *20* : 317 - 320, 1973. [In R, ab : E.]

15826 - **CHIMIKLIS, P.E., KARLANDER, E.P.** : Light and calcium interactions in *Chlorella* inhibited by sodium chloride. - Plant Physiol. *51* : 48 - 56, 1973. [Ps, Chl.]

15827 - **CHOLLET, R.** : Oxygen effects on dark $^{14}CO_2$ fixation and glycolate metabolism in isolated maize bundle sheath strands. - Plant Physiol. *51* (Suppl.) : 7, 1973.

15828 - **CHOLLET, R.** : The effect of oxygen on $^{14}CO_2$ fixation in mesophyll cells isolated from *Digitaria sanguinalis* (L.) SCOP. leaves. - Biochem. biophys. Res. Commun. *55* : 850 - 856, 1973.

15829 - **CHOLLET, R., OGREN, W.L.** : Photosynthetic carbon metabolism in isolated maize bundle sheath strands. - Plant Physiol. *51* : 787 - 792, 1973.

*15830 - **CHOUDHARY, V.B.** : Seasonal variation in standing crop and net above-ground production in *Dichanthium annulatum* grassland at Varanasi. - In : International Symposium on Tropical Ecology with an Emphasis on Organic Production. Pp. 51 - - 57. 1972.

15831 - **CHRISTY, A.L.** : Translocation kinetics in relation to source-leaf photosynthesis and carbohydrate concentrations in sugar beet. - Diss. Abstr. int. B *33* : 3507-B - 3508-B, 1973.

15832 - **CHRISTY, A.L., SWANSON, C.A.** : Translocation kinetics in relation to source--leaf photosynthesis and carbohydrate concentrations. - Plant Physiol. *51* (Suppl.) : 62, 1973.

*15833 - **CHROMETZKA, P.** : Über die Carotinoide der Blütenblätter von *Oenothera* I. Quali-

tative und quantitative Untersuchungen bei normalgelben, sulfurea- und vetau-
rea-Blüten. - Theor. appl. Genet. *41* : 205 - 207, 1971.

15834 - CHU, D.K., BASSHAM, J.A. : Activation and inhibition of ribulose 1,5-diphospha-
te carboxylase by 6-phosphogluconate. - Plant Physiol. *52* : 373 - 379, 1973.

15835 - CHUA, N.-H., BLOBEL, G., SIEKEVITZ, P. : Isolation of cytoplasmic and chloroplast ribosomes and their dissociation into active subunits from *Chlamydomonas
reinhardtii*. - J. Cell Biol. *57* : 798 - 814, 1973.

15836 - CHUA, N.-H., BLOBEL, G., SIEKEVITZ, P., PALADE, G.E. : Attachment of chloroplast polysomes to thylakoid membranes in *Chlamydomonas reinhardtii*. - Proc.
nat. Acad. Sci. USA *70* : 1554 - 1558, 1973.

15837 - CHUCHALIN, A.I. : Produktivnost' poseva redisa v svyazi s intensivnost'yu potoka FAR i gustotoy poseva. [Productivity of radish crop in relation to PhAR
flux and canopy density.] - In : Informatsionnye Materialy Akad. Nauk SSSR,
Sibir. Otd., Sibir. Inst. Fiziol. Biokhim. Rast., Irkutsk *11* : 35 - 36, 1973.
[In R.]

15838 - CHUCHALIN, A.I., SHILENKO, M.P., TIKHOMIROV, A.A. : Fotosinteticheskiy potentsial, biologicheskaya produktivnost' i koeffitsient ispol'zovaniya FAR posevom v nepreryvnoy kul'ture pshenitsy. [Photosynthetic potential, biological
productivity and coefficient of PhAR utilization by a canopy in a continuous
wheat culture.] - In : Upravlenie Skorost'yu i Napravlennost'yu Biosinteza
u Rasteniy. Pp. 176 - 177. Krasnoyarsk 1973. [In R.]

*15839 - CHUNAEV, A.S., KVITKO, K.V. : Supressiya priznaka "svetochuvstvitel'nosti"
u *Chlamydomonas reinhardi*. [Suppression of phenomenon "light-sensitivity"in
Chlamydomonas reinhardi.] - In : Geneticheskie Aspekty Fotosinteza. Tezisy
Dokladov. Pp. 77 - 78. Donish, Dushanbe 1972. [In R.]

15840 - CLARK, J.B., LISTER, G.R. : A comparative study of the photosynthetic action
spectra for a deciduous and four coniferous trees. - Plant Physiol. *51* (Suppl.):
20, 1973.

15841 - CLARK, R.N., HILER, E.A. : Plant measurements as indicators of crop water deficit. - Crop Sci. *13* : 466 - 469, 1973. [Productivity.]

15842 - CLASBY, R.C., HORNER, R., ALEXANDER, V. : An *in situ* method for measuring primary productivity of arctic sea ice algae. - J. Fish. Res. Board Can. *30* :
835 - 838, 1973.

15843 - CLAYTON, R.K. : Primary processes in bacterial photosynthesis. - Annu. Rev.
Biophys. Bioeng. *2* : 131 - 156, 1973.

15844 - CLAYTON, R.K. : The position and prospects of solar energy conversion. - Anais
Acad. bras. Ciênc. *45* (Suppl. - Novas Tendencias em Fotobiologia) : 103 - 109,
1973. [Ps.]

15845 - CLEARE, M., PERCIVAL, E. : Carbohydrates of the freshwater alga *Tribonema aequa-
le*. II. Preliminary photosynthetic studies with ^{14}C. - Brit. phycol. J. *8* :
181 - 184, 1973.

15846 - CLEVE, K. van : Energy and biomass relationships in alder (*Alnus*) ecosystems
developing on the Tanana River floodplain near Fairbanks, Alaska. - Arctic
alpine Res. *5* : 253 - 260, 1973.

15847 - CLIFFORD, P.E., MARSHALL, C., SAGAR, G.R. : An examination of the value of
^{14}C-urea as a source of ^{14}CO$_2$ for studies of assimilate distribution. - Ann.
Bot. *36* : 37 - 44, 1973.

15848 - CLIFFORD, P.E., MARSHALL, C., SAGAR, G.R. : The reciprocal transfer of radiocarbon between a developing tiller and its parent shoot in vegetative plants
of *Lolium multiflorum* LAM. - Ann. Bot. *37* : 777 - 785, 1973. [Ps.]

15849 - CLYMO, R.S. : The growth of *Sphagnum* : some effects of environment. - J. Ecol.
61 : 849 - 869, 1973. [Growth analysis.]

15850 - COBB, A.H., WELLBURN, A.R. : Developmental changes in the levels of SDS-extract-
able polypeptides during plastid morphogenesis. - Planta *114* : 131 - 142, 1973.

15851 - COCK, J.H., YOSHIDA, S. : Photosynthesis, crop growth, and respiration of a
tall and short rice varieties. - Soil Sci. Plant Nutr. *19* : 53 - 59, 1973.

15852 - **COCKS, P.S.** : The influence of temperature and density on the growth of communities of subterranean clover (*Trifolium subterraneum* L. cv. Mount Barker). -
Aust. J. agr. Res. *24* : 479 - 495, 1973. [Growth analysis.]

15853 - **CODD, G.A., SCHMID, G.H., KOWALLIK, W.** : Further enzymic studies and electron
microscopy of the microbodies of a mutant of *Chlorella vulgaris*. - Arch. Mikrobiol. *92* : 21 - 38, 1973. [Chloroplast.]

15854 - **CODD, G.A., STEWART, W.D.P.** : Pathways of glycollate metabolism in the blue-
-green alga *Anabaena cylindrica*. - Arch. Mikrobiol. *94* : 11 - 28, 1973.

15855 - **CODD, G.A., STEWART, W.D.P.** : The photometabolism of glycollate in the nitro-
gen-fixing blue-green alga, *Anabaena cylindrica*. - In : Symposium on Prokaryotic Photosynthetic Microorganisms. Pp. 101 - 102. Univ. Freiburg, Freiburg 1973.

15856 - **COGDELL, R.J., PRINCE, R.C., CROFTS, A.R.** : Light induced H^+ uptake catalysed
by photochemical reaction centres from *Rhodopseudomonas spheroides* R26. - FEBS
Lett. *35* : 204 - 208, 1973.

15857 - **COHEN, S.S.** : Mitochondria and chloroplasts revisited. - Amer. Scientist *61* :
437 - 445, 1973.

15858 - **COHEN, S.S.** : Recent studies on the origins of cellular organelles. - In :
SRB, A.M. (ed.) : Genes, Enzymes and Populations. Pp. 27 - 52. Plenum Publ.
Corp., New York 1973. [Chloroplast.]

15859 - **COLBOW, K.** : Energy transfer in photosynthesis. - Biochim. biophys. Acta *314* :
320 - 327, 1973.

15860 - **COLBOW, K.** : Chlorophyll in phospholipid vesicles. - Biochim. biophys. Acta
318 : 4 - 9, 1973.

15861 - **COLLINS, N.J.** : Productivity of selected bryophyte communities in the maritime
Antarctic. - In : **BLISS, L.C., WIELGOLASKI, F.E.** (ed.) : Primary Production
and Production Processes, Tundra Biome. Pp. 177 - 183. Tundra Biome Steering
Comm., Edmonton 1973.

15862 - **CONNELLY, J.L., JONES, O.T.G., SAUNDERS, V.A., YATES, D.W.** : Kinetic and thermodynamic properties of membrane-bound cytochromes of aerobically and photosynthetically grown *Rhodopseudomonas spheroides*. - Biochim. biophys. Acta *292*:
644 - 653, 1973.

15863 - **CONNOLLY, J.S., GORMAN, D.S., SEELY, G.R.** : Laser flash photolysis studies of
chlorin and porphyrin systems. I. Energetics of the triplet state of bacteriochlorophyll. - Ann. New York Acad. Sci. *206* : 649 - 669, 1973.

15864 - **COOK, J.R.** : Unbalanced growth and replication of chloroplast populations in
Euglena gracilis. - J. gen. Microbiol. *75* : 51 - 60, 1973. [Chl.]

15865 - **COOK, J.R., LI, T.C.C.** : Influence of culture pH on chloroplast structure in
Euglena gracilis. - J. Protozool. *20* : 652 - 653, 1973.

15866 - **COOKE, R.C.** : The use of activated charcoal for the removal of oxygen from gas
systems. - Limnol. Oceanogr. *18* : 150 - 152, 1973.

15867 - **COOMBS, J., BALDRY, C.W., BROWN, J.E.** : The C-4 pathway in *Pennisetum purpureum*. III. Structure and photosynthesis. - Planta *110* : 121 - 129, 1973.

15868 - **COOMBS, J., BALDRY, C.W., BUCKE, C.** : The C-4 pathway in *Pennisetum purpureum*.
I. The allosteric nature of PEP carboxylase. - Planta *110* : 95 - 107, 1973.

15869 - **COOMBS, J., BALDRY, C.W., BUCKE, C.** : The C-4 pathway in *Pennisetum purpureum*.
II. Malate dehydrogenase and malic enzyme. - Planta *110* : 109 - 120, 1973.

15870 - **COOPER, D.C.** : Enhancement of net primary productivity by herbivore grazing in
aquatic laboratory microcosms. - Limnol. Oceanogr. *18* : 31 - 37, 1973.

15871 - **COOPER, D.C., COPELAND, B.J.** : Responses of continuous-series estuarine micro-
ecosystems to point-source input variations. - Ecol. Monogr. *43* : 213 - 236,
1973. [Primary production.]

15872 - **CORLEY, R.H.V.** : Effects of plant density on growth and yield of oil palm. -
Exp. Agr. *9* : 169 - 180, 1973. [Growth analysis.]

15873 - **CORLEY, R.H.V., HARDON, J.J., OOI, S.C.** : Some evidence for genetically con-

trolled variation in photosynthetic rate of oil palm seedlings. - Euphytica
22 : 48 - 55, 1973.

15874 - CORNIC, G. : Etude de l'inhibition de la respiration par la lumière chez la
Moutarde blanche (*Sinapis alba* L.). - Physiol. vég. *11* : 663 - 679, 1973.

*15875 - CORTI, R., MAGINI, E., CIAMPI, C., BACCARI, V., GUERRITORE, A., RAMPONI, G.,
FIRENZUOLI, A.M., VANNI, P., MASTRONUZZI, E., ZANOBINI, A. : Note sur l'incom-
patibilité de greffe chez les conifères. - Silvae Genet. *17* : 121 - 130, 1968.
[Chl.]

15876 - COST, K., BOLTON, J.R. : Flash photolysis-electron spin resonance studies of
the interaction of phenazine methosulfate with reaction-center preparations
from the R-26 mutant of *Rhodopseudomonas spheroides*. - Photochem. Photobiol.
18 : 423 - 428, 1973.

15877 - COSTES, C., BAZIER, R., EVRARD, C. : Recherches sur la stabilité des chloroplastes sous la lumière artificielle : effets de l'éclairement sur les pigments et sur les lipides. - Physiol. vég. *11* : 301 - 326, 1973.

15878 - COSTES, C., GAUDILLÈRE, J.-P. : Sur un modèle de l'assimilation photosynthétique brute du CO_2 par la feuille. - Compt. rend. Acad. Sci. Paris, Sér. D *277* :
2821 - 2824, 1973.

15879 - COUILLAULT, J., BONNEMAIN, J.-L. : Transport et distribution des assimilats chez
Trapa natans L. - Physiol. vég. *11* : 45 - 53, 1973.

*15880 - COX, M.T., JACKSON, A.H., KENNER, G.W. : Pyrroles and related compounds. Part
XIX. Synthesis of phylloporphyrins related to *Chlorobium* chlorophyll (660). -
J. chem. Soc. C *1971* : 1974 - 1981, 1971.

15881 - COX, R.P., BENDALL, D.S. : The functions of plastoquinone and β-carotene in
photosystem II of chloroplasts. - In : Ninth International Congress of Biochemistry, Stockholm,1 - 7 July 1973. Abstract Book. P. 217. IUB, Stockholm
1973.

15882 - CRAIGIE, J.S. : Storage products. - In : STEWART, W.D.P. (ed.) : Algal Physiology and Biochemistry. Bot. Monogr. Vol. Io. Pp. 206 - 235. Blackwell sci.
Publ., Oxford - London - Edinburgh - Melbourne 1973.

15883 - CRAN, D.G., POSSINGHAM, J.V. : The fine structure of avocado plastids. - Ann.
Bot. *37* : 993 - 997, 1973.

15884 - CREED, D., HALES, B.J., PORTER, G. : Photochemistry of the plastoquinones. -
Proc. roy. Soc. London A *334* : 505 - 521, 1973.

15885 - CRESPI, H.L., NORRIS, J.R. : ESR studies with ^{2}H-flavin and ^{2}H-flavoprotein. -
In : Proc. 1st Int. Conf. Stable Isotop. Chem., Biol., Med. CONF-730525. Pp.
171 - 176. Nat. tech. Inform. Service, US Dep. Commerce, Springfield, Va. 1973.
[Ps.]

15886 - CROOKSTON, R.K., MOSS, D.N. : A variation of C_4 leaf anatomy in *Arundinella
hirta (Gramineae)*. - Plant Physiol. *52* : 397 - 402, 1973.

15887 - CROTTY, W.J., LEDBETTER, M.C. : Membrane continuities involving chloroplasts
and other organelles in plant cells. - Science *182* : 839 - 841, 1973.

15888 - CRUZ, A.A. de la, NAQVI, S.M. : Mirex incorporation in the environment : uptake
in aquatic organisms and effects on the rates of photosynthesis and respiration. - Arch. environm. Contamination Toxicol. *1* : 255 - 264, 1973.

15889 - CULBERT-RUNQUIST, J.A., HADSELL, R.M., LOACH, P.A. : Dependency on environmental redox potential of photophosphorylation in *Rhodopseudomonas spheroides*. -
Biochemistry *12* : 3508 - 3514, 1973.

15890 - CURRIE, P.O., MORRIS, M.J., NEAL, D.L. : Uses and capabilities of electronic
capacitance instruments for estimating standing herbage. Part 2. Sown ranges.
- J. brit. Grassland Soc. *28* : 155 - 160, 1973.

15891 - CURRIE, R.I., FISHER, A.E., HARGREAVES, P.M. : Arabian Sea upwelling. - In :
ZEITSCHEL, B. (ed.) : The Biology of the Indian Ocean. Pp. 37 - 52. Springer-
-Verlag, Berlin - Heidelberg - New York 1973. [Chl.]

15892 - CURTIS, V.A., SIEDOW, J.N., SAN PIETRO, A. : Studies on photosystem I. II. Involvement of ferredoxin in cyclic electron flow. - Arch. Biochem. Biophys. *158* : 898 - 902, 1973.

15893 - **CZAPLEWSKI, R.L., PARKER, M.** : Use of a BOD oxygen probe for estimating primary productivity. - Limnol. Oceanogr. *18* : 152 - 154, 1973.

B15894 - **CZARNOWSKI, M.** : Ekofizjologiczne Studia nad Szacunkową Metodą Oceny Produkcji Fotosyntetycznej Liści Wybranych Gatunków Roślin. [Ecophysiological Studies on the Rating Method for Evaluating the Photosynthetic Production of Leaves of Selected Plant Species.] - Pol. Akad. Nauk, Zakł. Fizjol. Rośl., Pracow. Fotosyntezy, Kraków 1973. [in Pol.]

15895 - **CZECZUGA, B.** : The role of bacteria of the genus *Chlorobium* purifying water reservoirs from H_2S/cellular and extracellular production. - In : GENOVESE, S. (ed.) : Atti del 5º Colloquio Internazionale di Oceanografia Medica. Pp. 487 - - 494. Editrice Libraria Bonanzinga, Messina 1973. [Ps.]

15896 - **CZECZUGA, B., GRADZKI, F.** : Relationship between extracellular and cellular production in the sulphuric green bactreium *Chlorobium limicola* NADS. (*Chlorobacteriaceae*) as compared to primary production of phytoplankton. - Hydrobiologia *42* : 85 - 95, 1973.

15897 - **DAHLÉN, J.A.H.** : Chlorophyll content monitoring of Swedish rapeseed and its significance in oil quality. - J. amer. Oil Chemists' Soc. *50* : 312A, 314A - - 317A, 327A, 1973.

15898 - **DALEY, R.J.** : Experimental characterization of lacustrine chlorophyll diagenesis. II. Bacterial, viral and herbivore grazing effects. - Arch. Hydrobiol. *72* : 409 - 439, 1973.

15899 - **DALEY, R.J., BROWN, S.R.** : Experimental characterization of lacustrine chlorophyll diagenesis. I. Physiological and environmental effects. - Arch. Hydrobiol. *72* : 277 - 304, 1973.

15900 - **DALEY, R.J., BROWN, S.R.** : Chlorophyll, nitrogen, and photosynthetic patterns during growth and senescence of two blue-green algae. - J. Phycol. *9* : 395 - - 401, 1973.

15901 - **DALEY, R.J., GRAY, C.B.J., BROWN, S.R.** : A quantitative, semiroutine method for determining algal and sedimentary chlorophyll derivatives. - J. Fish. Res. Board Can. *30* : 345 - 356, 1973.

15902 - **DALEY, R.J., GRAY, C.B.J., BROWN, S.R.** : Reversed-phase thin-layer chromatography of chlorophyll derivatives. - J. Chromatogr. *76* : 175 - 183, 1973.

15903 - **DAMAGNEZ, J.** : Les bilans hydriques et énergétiques et l'étude des facteurs du milieu. - In : Soil-Moisture and Irrigation Studies II. Pp. 155 - 168. Int. at. Energy Agency, Vienna 1973. [Ps.]

15904 - **DAMISCH, W.** : Beiträge zur Ertragsphysiologie des Getreides. 2. Mitt. Die Attraktions-Produktionsbeziehungen in der Kornfüllungsperiode bei verschiedenen Winterweizengenotypen. - Arch. Züchtungsforsch. *3* : 285 - 296, 1973.

15905 - **DANDONNEAU, Y.** : Étude du phytoplancton sur le plateau continental de Côte d' Ivoire. III. Facteurs dynamiques et variations spatiotemporelles. - Cah. ORSTOM, Sér. Océanogr. *11* : 431 - 454, 1973. [Chl.]

15906 - **DANTSEVICH, Yu.D.** : O deĭstvii desikantov na fotosintez podsolnechnika. [Effect of desiccants on sunflower photosynthesis.] - Fiziol. Biokhim. kul't. Rast. *5* : 540 - 542, 1973. [in R, ab : E.]

15907 - **DANTUMA, G.** : Rates of photosynthesis in leaves of wheat and barley varieties. - Neth. J. agr. Sci. *21* : 188 - 198, 1973.

15908 - **DARBINYAN, N.O.** : Vozniknovenie khlorotichnykh rasteniĭ v pervom pokolenii gibridov ozimoĭ myagkoĭ pshenitsy. [Appearance of chlorotic plants in the first generation of winter wheat hybrids.] - Genetika *9* (6) : 5 - 10, 1973. [in R, ab : E.]

15909 - **DAS, G.** : Aspects of metabolic development in an illuminated synchronous culture of *Scenedesmus obtusiusculus*. - Can. J. Bot. *51* : 113 - 120, 1973. [Ps, Chl, Car.]

15910 - **DAS, G.** : Influence of calcium on the metabolism of chlorophyll, carotene, nucleic acid, and protein in *Scenedesmus*. - Can. J. Bot. *51* : 121 - 125, 1973.

15911 - **DAS, G.** : Possible mechanism of gibberellin-induced chlorosis in lettuce seedlings. - Can. J. Bot. *51* : 175 - 184, 1973.

15912 - **DAS, G.** : Gibberellin-induced changes in ultrastructure of lettuce cotyledons. - Ann. Bot. *37* : 999 - 1004, 1973. [Chloroplast.]

15913 - **DAS, V.S.R., RAGHAVENDRA, A.S.** : A screening of the dicotyledonous weed flora for the occurrence of C_4 dicarboxylic acid pathway of photosynthesis. - Proc. Indian Acad. Sci., Sect. B *77* : 93 - 100, 1973.

15914 - **DAUNICHT, H.-J., LENZ, F.** : Das Verhalten von Paprikapflanzen mit unterschiedlichem Fruchtbehang bei Behandlung mit 3 CO_2-Konzentrationen. - Gartenbauwissenschaft *38* : 533 - 546, 1973.

15915 - **DAVIES, D.D., PATIL, K.D.** : Control of CO_2 fixation by cell-free extracts of *Chlamydomonas reinhardtii*. - Plant Physiol. *51* : 1142 - 1144, 1973.

15916 - **DAVIES, R.C., GORCHEIN, A., NEUBERGER, A., SANDY, J.D., TAIT, G.H.** : Biosynthesis of bacteriochlorophyll. - Nature *245* : 15 - 19, 1973.

15917 - **DAVIES, R.C., NEUBERGER, A.** : Polypyrroles formed from porphobilinogen and amines by uroporphyrinogen synthetase of *Rhodopseudomonas spheroides*. - Biochem. J. *133* : 471 - 492, 1973.

15918 - **DAVIES, W.J., KOZLOWSKI, T.T.** : Short- and long-term effects of film-forming antitranspirants. - Amer. J. Bot. *60* (4 Suppl.) : 24, 1973. [Ps.]

15919 - **DAWSON, F.H.** : Notes on the production of stream bryophytes in the High Pyrenees (France). - Ann. Limnol. *9* : 231 - 240, 1973.

15920 - **DEBELYĬ, G.A., PTASHENCHUK, V.N.** : Retsessivnye khlorofil'nye mutatsii v semennom potomstve rastenyĭ M_1 yarovoĭ viki. [Recessive chlorophyll mutations in the seed progeny of M_1 plants of spring vetch.] - Genetika *9* (8) : 45 - 49, 1973. [In R, ab : E.]

15921 - **DEKAPRELEVICH, L.L., NGUEN KHYU NGIA** : Chastota pestrolistnykh form i ikh sootnoshenie s morfologicheskimi izmeneniyami v M_1 u ozimoĭ pshenitsy posle vozdeĭstviya khimicheskimi mutagenami. [Appearance of variegated forms and their correlation with the morphological alterations in the M_1 of winter wheat after treatment with chemical mutagens.] - Genetika *9* (12) : 135 - 137, 1973. [In R, ab : E.]

15922 - **DELRIEU, M.-J.** : Explication de l'amortissement de la séquence de jets d'oxygène dégagés à la suite d'éclairs saturants. - Compt. rend. Acad. Sci. Paris, Sér. D *277* : 2809 - 2812, 1973.

*15923 - **DE LUCA, P., MUSACCHIO, A., TADDEI, R.** : Diverso comportamento in eterotrofia delle due forme di "*Cyanidium caldarium*"dei Campi Flegrei (Napoli). [Various heterotrophic behaviour of forms of *Cyanidium caldarium* of Campi Flegrei (Naples).] - Delpinoa, nuova Ser. *12-13* : 19 - 27, 1970-1971 (1972). [Chl, Car; in Ital., ab : E.]

*15924 - **DEMCHENKO, S.I., AVETISOV, V.A., BUTENKO, R.G.** : Analiz prirody khlorofil'nykh khimer u *Arabidopsis* v M_1 s pomoshch'yu kul'tury tkani. [Analysis of chlorophyll chimaerae origin in *Arabidopsis* in M_1 by means of tissue cultures.] - In : Geneticheskie Aspekty Fotosinteza. Tezisy Dokladov. Pp. 58 - 59. Donish, Dushanbe 1972. [In R.]

15925 - **DEMCHENKO, S.I., AVETISOV, V.A., BUTENKO, P.G.** : An enquiry into chlorophyll chimerae in *Arabidopsis* in M_1 by means of tissue culture. - *Arabidopsis* Inform. Service *10* : 17 - 18, 1973.

*15926 - **DEMCHENKO, S.I., SEROVA, R.Ya.** : Vzaimosvyaz' khlorofil'nykh khimer v M_1 s mutatsionnym protsessom u *Arabidopsis*. [Relationship between the chlorophyll chimaerae in M_1 and the mutation process in *Arabidopsis thaliana*.] - In : Geneticheskie Aspekty Fotosinteza. Tezisy Dokladov. Pp. 59 - 60. Donish, Dushanbe 1972. [In R.]

15927 - **DEMIDOV, E.D., BELL, L.N.** : Vliyanie sinego i krasnogo sveta na fotofosforilirovanie v izolirovannykh khloroplastakh gorokha. [Effect of blue and red light on photophosphorylation in isolated chloroplasts of pea.] - Fiziol. Rast. *20* : 292 - 299, 1973. [In R, ab : E.]

15928 - DEMINA, O.M., KHARLAMOVA, É.I., YANGALYCHEVA, L.Kh. : Statsionarnye issledova-niya lugovoĭ rastitel'nosti v nizov'yakh reki Chu. [Long-term studies of meadow vegetation in the lower Chu River.] - Bot. Zh. *58* : 806 - 814, 1973. [Productivity; in R, ab : E.]

15929 - DENIS, P., GUICHERD, R., DAMAGNEZ, J., MERMIER, M. : Mesure du rayonnement net: mise au point d'un pyrradiomètre differentiel à circulation d'eau. - In ⋅ The Sun in the Service of Mankind. 10 pp. Unesco, Paris 1973.

15930 - DERIEUX, M., KERREST, R., MONTALANT, Y. : Étude de la variabilité génétique de la surface foliaire et de l'activité photosynthétique chez quelques hybrides de maïs. - Ann. Amélior. Plantes *23* : 95 - 107, 1973.

15931 - DERIEUX, M., KERREST, R., MONTALANT, Y. : Influence du gène opaque 2 sur l' activité photosynthétique d'un hybride de maïs. - Ann. Amélior. Plantes *23* : 109 - 113, 1973.

15932 - DESCOMPS, S., DEROCHE, M.-E. : Action de l'éclairement continu sur l'appareil photosynthétique de la Tomate. - Physiol. vég. *11* : 615 - 631, 1973.

15933 - DESJARDINS, R.L., SINCLAIR, T.R., LEMON, E.R. : Light fluctuations in corn. - Agron. J. *65* : 904 - 908, 1973.

15934 - DETERS, D., NELSON, N., WHARTON, D.C. : Subunit structure and function of coupling factor 1 (CF_1) from spinach chloroplasts. - Fed. Proc. *32* : 516 Abs, 1973.

15935 - DETLING, J.K., KLIKOFF, L.G. : Physiological response to moisture stress as a factor in halophyte distribution. - Amer. Midland Naturalist *90* : 307 - 318, 1973. [Ps.]

*15936 - DEVAĬ, M., RAĬKI, S. : Fotosinteticheskie protsessy v ozimoĭ pshenitse, prois-khodyashchie pri temperaturakh okolo 0 °C, i ikh rol' v mikroévolyutsii. [Photosynthetic processes in winter wheat, proceeding at temperatures around 0 °C, and their importance in microevolution.] - In : Geneticheskie Aspekty Fotosinteza. Tezisy Dokladov. Pp. 87 - 88. Donish, Dushanbe 1972. [In R.]

15937 - DEVAUX, J. : Contribution a l'etude des populations phytoplanctoniques du Lac de Tazenat (Puy-de-Dome). - Ann. Sta. biol. Besse-en-Chandesse *1973* (7) : 1 - - 101, 1973. [Ps.]

15938 - DEWAR, M.A., BARBER, J. : Cation regulation in *Anacystis nidulans*. - Planta *113* : 143 - 155, 1973. [Ps, Chl.]

15939 - DIAKOFF, S., SCHEIBE, J. : Action spectra for chromatic adaptation in *Tolypothrix tenuis*. - Plant Physiol. *51* : 382 - 385, 1973. [Biliproteins.]

15940 - DICKMAN, M. : Changes in phytoplankton following nitrate and phosphate additions to large enclosures in Marion Lake, British Columbia. - Schweiz. Z. Hydrol.*35* : 114 - 120, 1973. [Primary productivity.]

15941 - DIERICKX, P.J., VENDRIG, J.C. : The influence of the synthetic growth regulator α-chloro-β-(3-chlor-o-tolyl)propionitril on some phytochrome-mediated processes. - Physiol. Plant. *28* : 374 - 377, 1973. [Car.]

15942 - DIJKMANS, H. : Aggregation of chlorophylls *in vitro*. Absorption spectroscopy of chlorophyll *a* in water-methanol solutions. - Europe. J. Biochem. *32* : 233 - - 236, 1973.

15943 - DIJKMANS, H., AGHION, J. : Aggregation of chlorophylls *in vitro*. Role of the magnesium atom and of the phytyl chain in the chlorophyll *a* molecule. - Europe. J. Biochem. *32* : 237 - 241, 1973.

15944 - DILLEY, A.C., HELMOND, I. : The estimation of net radiation and potential evapotranspiration using atmometer measurements. - Agr. Meteorol. *12* : 1 - 11, 1973.

*15945 - DIMITROV, Kh. : Prouchvane v"rkhu asimilatsiyata, dishaneto, kompensatsionnite temperaturi i transpiratsiyata na nyakoi gorsko-d"rvesni vidove. [Photosynthesis, respiration, compensation temperatures and transpiration of certain forest

arboreal species.] - Fiziol. Rast. (Sofia) *1* : 113 - 122, 1970. [In Bulg.,
ab : E, R.]

15946 - **DINA, S.J., KLIKOFF, L.G.** :.Carbon dioxide exchange by several streamside and
scrub oak community species of Red Butte Canyon, Utah. - Amer. Midland Naturalist *89* : 70 - 80, 1973.

15947 - **DINANT, M., AGHION, J.** : Agregation des chlorophylles *in vitro* - II. Photodécoloration de la chlorophylle *a* adsorbée sur des particules de lipoproteines
extraites du lait. - Photochem. Photobiol. *17* : 25 - 30, 1973.

15948 - **DINER, B., MAUZERALL, D.** : Photosynthetic oxygen production and the size of
the photosynthetic unit in a cell-free preparation from *Cyanidium caldarium*. -
Biochim. biophys. Acta *292* : 285 - 290, 1973.

15949 - **DINER, B., MAUZERALL, D.** : Feedback controlling oxygen production in a cross-
-reaction between two photosystems in photosynthesis. - Biochim. biophys. Acta
305 : 329 - 352, 1973.

15950 - **DINER, B., MAUZERALL, D.** : The turnover times of photosynthesis and redox properties of the pool of electron carriers between the photosystems. - Biochim.
biophys. Acta *305* : 353 - 363, 1973.

15951 - **DIRKS, W., RICHTER, G.** : Die Wirkung des Cytostaticum "Proresid" auf das Wachstum und die Blaulicht-induzierte Chloroplastendifferenzierung isolierter und
normaler Keimlingswurzeln von *Pisum sativum*. - Planta *112* : 101 - 120, 1973.

15952 - **DITTMER, H.J.** : Clipping effects on Bermuda grass biomass. - Ecology *54* : 217 -
- 219, 1973.

15953 - **DITTRICH, P., CAMPBELL, W.H., BLACK, C.C. Jr.** : Phosphoenolpyruvate carboxyki-
nase in plants exhibiting crassulacean acid metabolism. - Plant Physiol. *52* :
357 - 361, 1973.

15954 - **DITTRICH, P., CAMPBELL, W.H., CHEN, T.M., BLACK, C.C. Jr.** : PEP-carboxykinase,
the decarboxylating enzyme in certain CAM-plants. - Plant Physiol. *51* (Suppl.1:
7, 1973.

15955 - **DITTRICH, P., SALIN, M.L., BLACK, C.C.** : Conversion of carbon 4 of malate into
products of the pentose cycle by isolated bundle sheath strands of *Digitaria
sanguinalis* (L.)SCOP. leaves. - Biochem. biophys. Res. Commun. *55* : 104 - 110,
1973.

15956 - **DOBREVA, S., STOEV, K., SLAVCHEVA, T., GADEVSKA, A.** : Kamera za tseli lozi pri
izsledvane na sumarnata fotosinteza ili dishaneto. [Chamber for whole shrubs
of grapevine for studying the sum of photosynthesis or respiration.] - Gradinar
lozar. Nauka *10* (2) : 79 - 87, 1973. [In Bulg., ab : F, R.]

15957 - **DÖHLER, G.** : Einfluß kurzer Dunkelperioden auf CO_2-Aufnahme und Phosphoenolpy-
ruvat-Carboxylierung während der Photosynthese-Induktion bei *Chlorella vulgaris*. - Arch. Mikrobiol. *90* : 333 - 341, 1973.

15958 - **DÖHLER, G.** : Neue Ergebnisse über die Induktionseffekte der photosynthetischen
CO_2-Aufnahme bei *Anacystis* und *Chlorella*. - Ber. deut. bot. Ges. *86* : 371 -
- 379, 1973.

15959 - **DÖHLER, G.** : Regulation der photosynthetischen CO_2-Aufnahme bei der Grünalge
Chlorella. - Hoppe-Seyler's Z. physiol. Chem. *354* : 1181, 1973. [Ps.]

15960 - **DÖHLER, G.** : Wirkung intermittierender Belichtung auf die ^{14}C-markierten Pro-
dukte während der Photosynthese-Induktion von *Chlorella vulgaris*. - Z. Pflanzenphysiol. *69* : 142 - 151, 1973.

15961 - **DÖHLER, G., PRZYBYLLA, K.-R.** : Einfluß der Temperatur auf die Lichtatmung der
Blaualge *Anacystis nidulans*. - Planta *110* : 153 - 158, 1973.

15962 - **DOKULIL, M.** : Planktonic primary production within the *Phragmites* community of
Lake Neusiedlersee (Austria). - Pol. Arch. Hydrobiol. *20* : 175 - 180, 1973.

15963 - **DOKULIL, M.** : E. Zur Steuerung der planktischen Primärproduktion durch die
Schwebstoffe. - In : **ELLENBERG, H.** (ed.) : Ökosystemforschung. Pp. 109 - 110.
Springer-Verlag, Berlin - Heidelberg - New York 1973.

15964 - **DOMAN, N.G.** : Regulyatsiya karboksiliruyushchikh fermentov i optimizatsiya

urozhaya. [Control of carboxylation enzymes and optimization of yield.] - In :
Upravlenie Skorost'yu i Napravlennost'yu Biosinteza u Rasteniĭ. Pp. 15 - 16.
Krasnoyarsk 1973. [In R.]

15965 - **DOMAŃSKA, H., KŁOSIŃSKA-RYCERSKA, B.** : Kilka uwag dotyczących mechanizmu dzia-
łania herbicydów w powiązaniu z ich wpływem na przebieg fotosyntezy i przemia-
ny białkowe. [Mechanism of herbicide action with respect to their effect on
the course of photosynthesis and protein transformations.] - Postępy Nauk rol.
20 (3) : 37 - 44, 1973. [In Pol.]

15966 - **DONALDSON, C., BLACKMAN, G.E.** : A further analysis of hybrid vigour in *Zea
mays* during the vegetative phase. - Ann. Bot. *37* : 905 - 917, 1973. [Growth
analysis.]

15967 - **DONGMANN, G., NÜRNBERG, H.W., WAGENER, K.** : Die $H_2{}^{18}O$-Anreicherung in den
Blättern transpirierender Pflanzen und ihre Bedeutung für die stationäre ^{18}O-
-Überhöhung in der Atmosphäre. - Ber. Kernforschungsanlage Jülich *974* : 1 -
- 95, 1973. [Ps.]

15968 - **DONZE, M.** : Pigment systems and electron transport in *Anabaena* and its hetero-
cysts. - In : Abstr. Symp. Prokaryotic Photosynthetic Organisms, September
19 - 23, 1973. Pp. 42 - 43. Freiburg i.Br. 1973.

B15969 - **DOROKHOV, B.L.** : Izmenenie Fotosinteticheskoĭ Deyatel'nosti u Materinskikh
Rasteniĭ pri Razlichnom Opylenii i u Ikh Potomstva. [Change in Photosynthetic
Activity of Maternal Plants with Different Pollination and of Their Progeny.]
- Shtiintsa, Kishinev 1973. [In R.]

15970 - **DOROKHOV, B.L.** : Kratkie itogi izucheniya v AN MSSR vliyaniya mineral'nogo pitaniya na fotosinteticheskuyu deyatel'nost' ozimoĭ pshenitsy, fasoli, gorokha.
[Effects of mineral nutrition on photosynthetic activity of winter wheat,
bean and pea studied in the Academy of Sciences of the Moldavian SSR.] - Fiziol. Biokhim. kul't. Rast. *5* : 181 - 186, 1973. [In R, ab : E.]

15971 - **DOUCE, R., BENSON, A.A.** : Components of the chloroplast envelope. - In : Ninth
International Congress of Biochemistry, Stockholm, 1 - 7 July, 1973. Abstract
Book. P. 286. IUB, Stockholm 1973.

15972 - **DOUCE, R., HOLTZ, R.B., BENSON, A.A.** : Isolation and properties of the envelope of spinach chloroplasts. - J. biol. Chem. *248* : 7215 - 7222, 1973.

15973 - **DOUCHA, J., BERKOVÁ, E., ZACHLEDER, V., ŠETLÍK, I.** : Synchronous populations
of *Scenedesmus quadricauda* in which cellular division occurs in light. - Annu.
Rep. algol. Lab. Třeboň *1970* : 52 - 68, 1973. [Ps, Chl.]

15974 - **DOUGLAS, G.W., RAMSDEN, J.** : Representation of leaf orientation data by equal-
-area projection techniques. - Can. J. Bot. *51* : 1081 - 1088, 1973.

15975 - **DOŬNAR, U.S.** : Zalezhnasts' velichyni ŭradzhayu zernya yachmenyu ad chystaĭ
praduktsyĭnastsi fotasintézu. [Dependence of barley grain yield on the net
productivity of photosynthesis.] - Vestsi Akad.Navuk belarus.SSR, Ser. biyal.
Navuk *1973* (6) : 16 - 19, 129, 1973. [In Belorus., ab : R.]

15976 - **DOWNES, R.W., CONNOR, D.J.** : Effect of growth environment on gas exchange characteristics of brigalow (*Acacia harpophylla* F. MUELL.). - Photosynthetica *7* :
34 - 40, 1973.

15977 - **DOWNTON, W.J.S., HAWKER, J.S.** : Sucrose and starch metabolism in mesophyll and
bundle sheath cells of maize. - Proc. aust. biochem. Soc. *6* : 30, 1973.

15978 - **DOYLE, G.J.** : Primary production estimates of native blanket bog and meadow
vegetation growing on reclaimed peat at Glenamoy, Ireland. - In : BLISS, L.C.,
WIELGOLASKI, F.E. (ed.) : Primary Production and Production Processes, Tundra
Biome. Pp. 141 - 151. Tundra Biome Steering Comm., Edmonton 1973.

15979 - **DRAKE, B.G., RASCHKE, K.** : Chilling causes reductions in photosynthesis and
stomatal conductance in leaves of *Xanthium strumarium* L. - Plant Physiol. *51*
(Suppl.) : 25, 1973.

15980 - **DRAPEAU, A.J., LE VAN THANH** : Application du laser aux études de mécanisme photosynthétique. Partie I. Étude des plantes. Partie II. Étude des algues. -
Année biol. *12* (5 - 6) : 193 - 208, 1973.

15981 - **DRAPEAU, A.J., LE VAN THANH** : Application du laser aux études du mécanisme photosynthétique. Partie III. Études des bactéries. - Année biol. *12* (11 - 12): 525 - 533, 1973.

15982 - **DRAXLER, G.** : D. Gaswechselmessungen an *Utricularia vulgaris*. - In : ELLENBERG, H. (ed.) : Ökosystemforschung. Pp. 103 - 107. Springer-Verlag, Berlin - Heidelberg - New York 1973.

15983 - **DRIESSCHE, R. van den** : Different effects of nitrate and ammonium forms of nitrogen on growth and photosynthesis of slash pine seedlings. - Aust. Forestry *36* : 125 - 137, 1973.

15984 - **DROKOVA, I.G., POPOVA, R.C.** : Porivnyal'na kharakterystyka karotynonosnosti deyakykh shtamiv *Dunaliella salina* TEOD. v umovakh masovoĭ kul'tury. [Comparative characteristics of carotene content of some strains of *Dunaliella salina* TEOD. in mass culture.] - Ukr. bot. Zh. *30* : 329 - 331, 1973. [In Ukr., ab : E.]

15985 - **DROKOVA, I.G., POPOVA, R.C.** : Pro vplyv spektral'nogo skladu svitla na vodorist' *Dunaliella salina* TEOD. [Effect of spectral composition of light on the alga *Dunaliella salina* TEOD.] - Ukr. bot. Zh. *31* : 121 - 124, 134, 1973. [Car; in Ukr., ab : E, R.]

15986 - **DRUYAN, M.E., NORRIS, J.R., KATZ, J.J.** : Electron spin resonance of [^{25}Mg] chlorophyll *a*. - J. amer. chem. Soc. *95* : 1682 - 1683, 1973.

15987 - **DUBOIS, J.** : Action de la lumière sur la croissance et la teneur en pigments plastidiaux des tissus isolés de carotte. - Bull. Soc. bot. France *120* : 3 - - 26, 1973.

15988 - **DUCREY, M.** : Appreciation du rayonnement solaire dans et sous le couvert forestier. - In : The Sun in the Service of Mankind. Pp. V-11-1 - V.11-13. Unesco, Paris 1973.

15989 - **DUDZIAK, B., KRUPA, Z., BASZYŃSKI, T.** : Effect of chloramphenicol on the synthesis of plastid benzoquinones and pigments in greening cells of *Euglena gracilis*. - Ann. UMCS (Lublin), Sect. C *28* (3) : 23 - 30, 1973.

15990 - **DUFFUS, C.M., ROSIE, R.** : Some enzyme activities associated with the chlorophyll containing layers of the immature barley pericarp. - Planta *114* : 219 - - 226, 1973.

15991 - **DUJARDIN, E.** : Pigment-lipoprotein complexes in the lyophilized etiolated leaf. - Photosynthetica *7* : 121 - 131, 1973.

15992 - **DUNAEVA, S.E., BYKOV, O.D.** : Raspredelenie krakhmala i ul'trastruktura khloroplastov rasteniĭ, razlichayushchikhsya pervichnymi produktami fotosinteza. [Distribution of starch and the ultrastructure of chloroplasts in plants differing in the primary products of photosynthesis.] - Tr. biol.-pochv. Inst., dal'nevost. nauch. Tsentr Akad. Nauk SSSR *20* (123) : 44 - 47, 1973. [In R, ab : E.]

15993 - **DUNCAN, M.J.** : *In situ* studies of growth and pigmentation of the phaeophycean *Nereocystis luetkeana*. - Holgoländer wiss. Meeresuntersuchungen *24* : 510 - 525, 1973.

15994 - **DUNCAN, W.G., SHAVER, D.L., WILLIAMS, W.A.** : Insolation and temperature effects on maize growth and yield. - Crop Sci. *13* : 187 - 191, 1973. [Ps efficiency]

15995 - **DUNN, G.M., KETEL, E.J., ROUTLEY, D.G., COUTURE, R.M.** : Effects of temperature and photoperiod on a virescent mutant in smooth bromegrass. - Crop Sci. *13* : 69 - 72, 1973. [Chl.]

15996 - **DUNSTAN, W.M.** : A comparison of the photosynthesis - light intensity relationship in phylogenetically different marine microalgae. - J. exp. mar. Biol. Ecol. *13* : 181 - 187, 1973.

15997 - **DUNSTONE, R.L., GIFFORD, R.M., EVANS, L.T.** : Photosynthetic characteristics of modern and primitive wheat species in relation to ontogeny and adaptation to light. - Aust. J. biol. Sci. *26* : 295 - 307, 1973.

*15998 - **DURZAN, D.J., MIA, A.J., RAMAIAH, P.K.** : The metabolism and subcellular organization of the jack pine embryo (*Pinus banksiana*) during germination. - Can. J. Bot. *49* : 927 - 938, 1971. [Chl.]

15999 - **DUTTON, P.L., LEIGH, J.S.** : Electron spin resonance characterization of *Chromatium* D hemes, non-heme irons and the components involved in primary photochemistry. - Biochim. biophys. Acta *314* : 178 - 190, 1973.

16000 - **DUTTON, P.L., LEIGH, J.S. Jr., REED, D.W.** : Primary events in the photosynthetic reaction centre from *Rhodopseudomonas spheroides* strain R26 : triplet and oxidized states of bacteriochlorophyll and the identification of the primary electron acceptor. - Biochim. biophys. Acta *292* : 654 - 664, 1973.

16001 - **DUTTON, P.L., LEIGH, J.S., WRAIGHT, C.A.** : Direct measurement of the midpoint potential of the primary electron acceptor in *Rhodopseudomonas spheroides in situ* and in the isolated state : some relationships with pH and *o*-phenanthroline. - FEBS Lett. *36* : 169 - 173, 1973.

16002 - **DUYSEN, M.E., GALITZ, D.S.** : Chloroplast pigment formation in water-stressed wheat leaves. - Plant Physiol. *51* (Suppl.) : 19, 1973.

16003 - **DUYSENS, L.N.M.** : Function of cytochrome *C422* (*C555*) in the *Athiorhodacea Chromatium*. - In : Abstr. Symp. Prokaryotic Photosynthetic Organisms, September 19 - 23, 1973. Pp. 38 - 39. Freiburg i.Br. 1973.

16004 - **DVOJKOVIČ-PENAVA, Z.** : Promjene fotosintetske aktivnosti plastida tijekom njihovih pretvorbi. [Changes in photosynthetic activity of plastids during their transformation.] - Acta bot. croat. *32* : 63 - 68, 1973. [In Croat., ab : E.]

16005 - **DVORNIN, A.V.** : Vliyanie plotnosti posadki privivok v gidroponnye ustanovki na nekotorye fiziologicheskie protsessy sazhentsev. [Effect of density of grafting in a hydroponic equipment on some physiological processes of seedlings.] - Tr. kishinev. sel'skokhoz. Inst. M.V. Frunze *118* (Vinogradarstvo) : 19 - 23, 103, 1973. [Ps, Chl; in R.]

16006 - **D'YACHENKO, A.P.** : Kachestvennyǐ sostav produktov fotosinteza i geterotrofnoǐ fiksatsii CO_2 u mkhov. [Qualitative composition of photosynthates and heterotrophic CO_2 fixation in mosses.] - In : Voprosy Regulyatsii Fotosinteza. Vol. 3. Pp. 153 - 158, 166. Ural'. gos. Univ., Sverdlovsk 1973. [In R.]

16007 - **DYE, A.J.** : Carbon dioxide exchange of blue grama swards in the field. - Diss. Abstr. int. B *33* : 3508-P, 1973.

16008 - **DYKYJOVÁ, D.** : Specific differences in vertical structures and radiation profiles in the helophyte stands. (A survey of comparative measurements.) - In : HEJNÝ, S. (ed.) : Ecosystem Study on Wetland Biome in Czechoslovakia, Czechosl. IBP/PT-PP Report No. 3. Pp. 121 - 131. Třeboň 1973.

16009 - **DYKYJOVÁ, D., HEJNÝ, S., KVĚT, J.** : Proposal for international comparative investigations of production by stands of reed (*Phragmites communis*). - Folia geobot. phytotaxon. *8* : 435 - 442, 1973.

16010 - **DYKYJOVÁ, D., HRADECKÁ, D.** : Productivity of reed-bed stands in relation to the ecotype, microclimate and trophic conditions of the habitat. - Pol. Arch. Hydrobiol. *20* : 111 - 119, 1973.

*16011 - **DZHABAROV, Kh.** : Geneticheskiǐ analiz mutantov khlopchatnika *Gossypium hirsutum* L. [Genetic analysis of mutants of *Gossypium hirsutum* L.] - In : Geneticheskie Aspekty Fotosinteza. Tezisy Dokladov. Pp. 60 - 61. Donish, Dushanbe 1972. [Chl; in R.]

*16012 - **DZHUMAEV, M.** : Sortovoe raznoobrazie i produktivnost' khlopchatnika. [Cotton cultivars differences and productivity.] - In : Geneticheskie Aspekty Fotosinteza. Tezisy Dokladov. Pp. 100 - 101. Donish, Dushanbe 1972. [In R.]

16013 - **EAGLES, C.F.** : Effect of light intensity on growth of natural populations of *Dactylis glomerata* L. - Ann. Bot. *37* : 253 - 262, 1973. [Growth analysis.]

16014 - **EARLEY, J.E.** : Oxygen evolution : A molecular model for the photosynthetic process, based on an inorganic example. - Inorg. nucl. chem. Lett. *9* : 487 - 490, 1973.

16015 - **EASTIN, J.D., HULTQUIST, J.H., SULLIVAN, C.Y.** : Physiologic maturity in grain sorghum. - Crop Sci. *13* : 175 - 178, 1973. [^{14}C assimilate accumulation.]

16016 - **ECKARDT, F.E.** : L'enceinte climatisée en tant qu'outil permettant de relier les recherches de laboratoire et de terrain. - In : SLATYER, R.O. (ed.) :

Plant Response to Climatic Factors. Pp. 295 - 310. Unesco, Paris 1973.

16017 - **ECKARDT, F.E.** : Plant strategy, CO_2-exchange and primary production. - Oecol. Plant. *8* : 309 - 312, 1973.

16018 - **ECKARDT, F.E., METHY, M., SAUVEZON, R.** : Interception et utilisation de l'énergie solaire par différents types de végétation dans la région méditerranéenne. - In : The Sun in the Service of Mankind. Pp. V.8-1 - V.8-10. Unesco, Paris 1973.

16019 - **EDWARDS, G.E., KANAI, R., KU, S.B., GUTIERREZ, M.** : Distribution of chlorophylls, enzymes, and some photosynthetic activities of mesophyll protoplasts and bundle sheath cells from C_4 plants. - Plant Physiol. *51* (Suppl.) : 6, 1973.

16020 - **EDWARDS, G.E., MOHAMED, A.K.** : Reduction in carbonic anhydrase activity in zinc deficient leaves of *Phaseolus vulgaris* L. - Crop Sci. *13* : 351 - 354, 1973. [Chl.]

16021 - **EDWARDS, J.A.** : Vascular plant production in the maritime Antarctic. - In : BLISS, L.C., WIELGOLASKI, F.E. (ed.) : Primary Production and Production Processes, Tundra Biome. Pp. 169 - 175. Tundra Biome Steering Comm., Edmonton 1973.

16022 - **EFIMTSEV, E.I.** : Funktsional'naya svyaz' pigmentnoĭ sistemy otdel'noĭ rastitel'noĭ kletki *Nitella flexilis* s fotoindutsirovannymi izmeneniyami okislitel'no--vosstanovitel'nogo potentsiala i potentsiala pokoya. [Functional relation of the pigment system of individual plant cells of *Nitella flexilis* to the photoinduced changes of redox potential and resting potential.] - Tr. mosk. Obshch. Ispyt. Prirody *49* (Problemy Biofotokhimii) : 155 - 160, 1973. [In R, ab : E.]

16023 - **EFIMTSEV, E.I., IGNATOV, N.V.** : Spektral'nyĭ analiz biologicheskikh sistem, kak metod opredeleniya ustoĭchivosti organizmov k izmeneniyu vneshnikh usloviĭ. [Spectral analysis of biological systems as a method of determining resistance of organisms to the changes of external conditions.] - In : Biofizicheskie Aspekty Zagryazneniya Biosfery. Pp. 63 - 64. Nauka, Moskva 1973. [Chl; in R.]

16024 - **EGAN, J.M. Jr., SCHIFF, J.A.** : The light requirements for chlorophyll and alkaline DNase formation during light-induced chloroplast development in *Euglena gracilis*, KLEBS var. *bacillaris* PRINGSHEIM. - Plant Physiol. *51* (Suppl.) : 23, 1973.

16025 - **EGGER, K., SITTE, P.** : Plastiden und Photosynthese. - In : HIRSCH, G.C., RUSKA, H., SITTE, P. (ed.) : Grundlagen der Cytologie. Pp. 345 - 387. VEB G. Fischer, Jena 1973.

16026 - **EGIERSZDORFF, S., TOMASZEWSKI, M.** : Auxin dependent accumulation of photosynthates in cambium and wood formation in Scots pine. - Proc. Res. Inst. Pomol., Skierniewice, Ser. E *1973* : 181 - 189, 1973.

16027 - **EGLI, D.E., LEGGETT, J.E.** : Dry matter accumulation patterns in determinate and indeterminate soybeans. - Crop Sci. *13* : 220 - 222, 1973.

16028 - **EGOROVA, G.D., MASHENKOV, V.A., SOLOV'EV, K.N., YUSHKEVICH, N.A.** : Vliyanie tsiklopentanonogo kol'tsa na spektral'no-lyuminestsentnye svoĭstva porfirinov. [Effect of cyclopentanone ring on spectral-luminescent properties of porphyrins.] - Biofizika *18* : 40 - 47, 1973. [In R, ab : E.]

16029 - **EGUNJOBI, J.K.** : Studies on the primary productivity of a regulary burnt tropical savanna. - Ann. Univ. Abidjan, Ser. E *6* (2) : 152 - 169, 1973.

16030 - **EHEART, M.S., ODLAND, D.** : Use of ammonium compounds for chlorophyll retention in frozen green vegetables. - J. Food Sci. *38* : 202 - 205, 1973.

16031 - **EICKENBUSCH, J.D., BECK, E.** : Evidence for involvement of 2 types of reaction in glycolate formation during photosynthesis in isolated spinach chloroplasts. - FEBS Lett. *31* : 225 - 228, 1973.

16032 - **EICKMEIER, W.G., ADAMS, M.S.** : Net photosynthesis and respiration of *Cladonia eomocyna* (ACH.) NYL. from the Rocky Mountains and comparison with three eastern alpine lichens. - Amer. Midland Naturalist *89* : 58 - 69, 1973.

16033 - **EĬNOR, L.O.** : Fotookislenie askorbinovoĭ kisloty khloroplastami i "zelenymi

preparatami". [Photooxidation of ascorbic acid by chloroplasts and "green preparations".] - Fiziol. Biokhim. kul't. Rast. 5 : 495 - 500, 1973. [In R, ab :
E.]

B16034 - ĔĬNOR, L.O. : Rekonstruirovanie Ėnergeticheskikh Mekhanizmov Fotosinteza. [Re-
construction of Energetic Mechanisms of Photosynthesis.] - Naukova Dumka, Kiev
1973. [In R.]

16035 - EISENBACH, M., CARMELI, C. : Sites along the electron-transport chain controlled by the energy-conversion system in chloroplasts. - Europe. J. Biochem. 37 :
361 - 366, 1973.

*16036 - ELEY, J.H. : Effect of carbon dioxide concentration on pigmentation in the
blue-green alga Anacystis nidulans. - Plant Cell Physiol. 12 : 311 - 316, 1971.

16037 - EL-GHAWAS, M.I., KHALIL, H.A. : The production and evaluation of a Triticum-
-Agropyron hybrid. - Egypt. J. Bot. 16 : 483 - 499, 1973. [Chl, Car.]

16038 - ELIZAROVA, V.A. : Soderzhanie khlorofilla v edinitse biomassy fitoplanktona
Rybinskogo vodokhranilishcha. [Chlorophyll content in unit of phytoplankton
biomass of the Rybinskoe reservoir.] - In : Krugovorot Veshchestva i Ėnergii
v Ozerakh i Vodokhranilishchakh. Vol. I. Pp. 127 - 128. Listvenichnoe na Baĭ-
kale 1973. [In R.]

16039 - ELIZAROVA, V.A. : Sostav i soderzhanie rastitel'nykh pigmentov v vodakh Rybinskogo vodokhranilishcha. [Plant pigments content and composition in waters of
the Rybinskoe reservoir.] - Gidrobiol. Zh. 9 (2) : 23 - 33, 1973. [Chl, Car;
in R, ab : E.]

B16040 - ELLENBERG, H. (ed.) : Ökosystemforschung. - Springer-Verlag, Berlin - Heidelberg - New York 1973. [Ps.]

16041 - ELLIS, J., KANAMORI, S. : An evaluation of the Miller method for dissolved oxygen analysis. - Limnol. Oceanogr. 18 : 1002 - 1005, 1973.

16042 - ELLSWORTH, R.K., HSING, A.S. : The reduction of vinyl side-chains of Mg-proto-
porphyrin IX monomethyl ester in vitro. - Biochim. biophys. Acta 313 : 119 -
- 129, 1973.

16043 - ELLSWORTH, R.K., LAWRENCE, G.D. : Synthesis of magnesium-protoporphyrin IX in
vitro. - Photosynthetica 7 : 73 - 86, 1973.

16044 - ELLSWORTH, R.K., NOWAK, C.A. : A method for the biological preparation and
thin-layer chromatographic purification of ^{14}C-protochlorophyllide a. - Anal.
Biochem. 51 : 656 - 662, 1973.

16045 - ELLSWORTH, R.K., NOWAK, C.A. : The inability of crude homogenates of etiolated
wheat seedlings containing protochlorophyllase to convert ^{14}C-protochlorophyll-
ide to ^{14}C-protochlorophyll. - Photosynthetica 7 : 246 - 251, 1973.

16046 - ELMORE, C.D. : Contributions of the capsule wall and bracts to the developing
cotton fruit. - Crop Sci. 13 : 751 - 752, 1973. [Ps, Chl.]

16047 - EL-SAYED, S.Z., JITTS, H.R. : Phytoplankton production in the southeastern
Indian Ocean. - In : ZEITSCHEL, B. (ed.) : The Biology of the Indian Ocean.
Pp. 131 - 142. Springer-Verlag, Berlin - Heidelberg - New York 1973. [Chl.]

16048 - ELSTNER, E.F., HEUPEL, A. : On the decarboxylation of α-keto acids by isolated
chloroplasts. - Biochim. biophys. Acta 325 : 182 - 188, 1973.

16049 - ELSTNER, E.F., KRAMER, R. : Role of the superoxide free radical ion in photosynthetic ascorbate oxidation and ascorbate-mediated photophosphorylation. -
Biochim. biophys. Acta 314 : 340 - 353, 1973.

16050 - EMEL'YANOV, L.G., SIDOROVA, T.V. : Ovodnennost', soderzhanie pigmentov v list'-
yakh i produktivnost' rasteniĭ ovsa pri razlichnoĭ vlazhnosti torfyanoĭ pochvy.
[Water and pigment contents in leaves and productivity of oat plants at various moisture of peat soil.] - In : VECHER, A.S. (ed.) : Fotosintez i Ustoĭchivost' Rasteniĭ. Pp. 76 - 81. Nauka i Tekhnika, Minsk 1973. [Chl; in R.]

16051 - EMERSON, S., BROECKER, W., SCHINDLER, D.W. : Gas-exchange rates in a small lake
as determined by the radon method. - J. Fish. Res. Board Can. 30 : 1475 - 1484,
1973.

16052 - **EMMETT, J.M., WALKER, D.A.** : Thermal uncoupling in chloroplasts. Inhibition of photophosphorylation without depression of light-induced pH change. - Arch. Biochem. Biophys. *157* : 106 - 113, 1973.

16053 - **ENGSTRÖM, G.** : Phytoplankton distribution in the archipelago measured as chlorophyll *a*. Nutrients and their influence on the algae in the Stockholm Archipelago during 1970. No. 4. - Oikos *1973* (Suppl. 15) : 176 - 178, 1973.

16054 - **ENOS, P., VAN VALEN, L.** : Photosynthesis and atmospheric oxygen. - Science *180* : 515 - 516, 1973.

16055 - **ENYI, B.A.C.** : A spacing/time of planting trial with cowpea (*Vigna unguiculata* (L) WALP). - Ghana J. Sci. *13* : 78 - 85, 1973. [Growth analysis.]

16056 - **ENYI, B.A.C.** : Growth rates of three cassava varieties (*Manihot esculenta* CRANTZ) under varying population densities. - J. agr. Sci. *81* : 15 - 28, 1973.

16057 - **ENYI, B.A.C.** : Effect of plant population on growth and yield of soya bean (*Glycine max*). - J. agr. Sci. (Cambridge) *81* : 131 - 138, 1973. [Growth analysis.]

16058 - **EPEL, B.L., NEUMANN, J.** : The mechanism of the oxidation of ascorbate and Mn^{2+} by chloroplasts. The role of the radical superoxide. - Biochim. biophys. Acta *325* : 520 - 529, 1973.

*16059 - **EPPLEY, R.W.** : Temperature and phytoplankton growth in the sea. - Fish Bull. *70* : 1063 - 1085, 1972. [Ps, Chl.]

16060 - **EPSTEIN, E., BRAVDO, B.** : Effects of three nematicides on the physiology of rose infected with *Meloidogyne hapla*. - Phytopathology *63* : 1411 - 1414, 1973. [Ps, Chl.]

16061 - **ERIXON, K., RENGER, G.** : Electron donation to PS II in intact and tris-washed chloroplasts by tetraphenylboron. - In : Ninth International Congress of Biochemistry. Stockholm, 1 - 7 July, 1973. Abstract Book. P. 241. IUB, Stockholm 1973.

16062 - **ERMAKOV, E.I.** : Vliyanie korneobitaemykh sred na produktivnost' rasteniĭ v kontroliruemykh usloviyakh. [Effect of root media on the productivity of plants in controlled conditions.] - In : Upravlenie Skorost'yu i Napravlennost'yu Biosinteza u Rasteniĭ. Pp. 53 - 54. Krasnoyarsk 1973. [In R.]

16063 - **ERMOLIN, I.E.** : Vliyanie ėkologicheskikh usloviĭ na spektr pogloshcheniya solnechnoĭ ėnergii list'yami podrosta duba. [The influence of ecological factors on the absorption spectrum of solar energy of oak leaves (*Quercus robur* L.).] - Vestn. mosk. Univ., Ser. VI - Biol., Pochvoved. *28* (4) : 51 - 56, 1973. [In R, ab : E.]

16064 - **EROKHIN, Yu.E., MOSKALENKO, A.A.** : Kharakteristika belkov (chislo tsepeĭ i molekulyarnye vesa) pigment-lipoproteinovykh kompleksov *Chromatium*. [Characteristics of proteins (number of chains and molecular weight) of pigment-lipoprotein complexes of *Chromatium*.] - Dokl. Akad. Nauk SSSR *212* : 495 - 497, 1973. [In R.]

16065 - **ERTL, M., TOMAJKA, J.** : Primary production of the periphyton in the littoral of the Danube. - Hydrobiologia *42* : 429 - 444, 1973. [Chl.]

16066 - **ESTES, G.O., KOCH, D.W., BRUETSCH, T.F.** : Influence of potassium nutrition on net CO_2 uptake and growth in maize (*Zea mays* L.). - Agron. J. *65* : 972 - 975, 1973.

16067 - **ETIENNE, A.-L.** : Utilisation des méthodes d'écoulement en photosynthèse. - Advan. Rad. Res., Phys. Chem. *1* : 339 - 346, 1973.

16068 - **ETTL, H.**: Morphological changes of the chloroplast during asexual reproduction in the genus *Chlamydomonas*. - Annu. Rep. algol. Lab. Třeboň *1970* : 19 - 25, 1973.

16069 - **ETTL, H., BŘEZINA, V., MARVAN, P.** : Methodical notes on assessment of productivity in littoral algae. - In : KVĚT, J. (ed.) : Littoral of the Nesyt Fishpond. Studie ČSAV *15*. Pp. 111 - 115. Academia, Praha 1973.

16070 - **ETTL, H., GREEN, J.C.** : *Chlamydomonas reginae* sp. nov. (*Chlorophyceae*), a new marine flagellate with unusual chloroplast differentiation. - J. mar. biol. Ass. UK *53* : 975 - 985, 1973.

16071 - EVANS, E.H., CROFTS, A.R. : The relationship between delayed fluorescence and the H$^+$ gradient in chloroplasts. - Biochim. biophys. Acta *292* : 130 - 139, 1973.

16072 - EVANS, E.L., ALLEN, M.M. : Phycobilisomes in *Anacystis nidulans*. - J. Bacteriol. *113* : 403 - 408, 1973. [Chl, biliproteins.]

16073 - EVANS, L.T. : The effect of light on plant growth, development and yield. - In: SLATYER, R.O. (ed.) : Plant Response to Climatic Factors. Pp. 21 - 35. Unesco, Paris 1973.

16074 - EVANS, L.V., CALLOW, J.A., CALLOW, M.E. : Structural and physiological studies on the parasitic red alga *Holmsella*. - New Phytol. *72* : 393 - 402, 1973. [Ps.]

16075 - EVANS, M.C.W., REEVES, S.G., TELFER, A. : The detection of a bound ferredoxin in the photosynthetic lamellae of blue-green algae and other oxygen evolving photosynthetic organisms. - Biochem. biophys. Res. Commun. *51* : 593 - 596, 1973.

16076　　EVSTIGNEEV, V.B. : O mekhanizme i regulyatsii fotosensibiliziruyushchego deĭstviya khlorofilla. [Mechanism and control of the photosensibilizing action of chlorophyll.] - In : Sovremennye Problemy Fotosinteza. Pp. 109 - 125. Izd. mosk. Univ., Moskva 1973. [In R.]

15077 - EVSTIGNEEV, V.B. : On evolution of the photosynthetic pigments. - Space Life Sci. *4* : 448 - 454, 1973.

16078 - EVSTIGNEEV, V.B., BEKASOVA, O.D. : Fotokhimicheskaya aktivnost' fikotsianina i fikoeritrina. [Photochemical activity of phycocyanin and phycoerythrin.] - Izv. Akad. Nauk SSSR, Ser. biol. *1973* : 344 - 356, 1973. [In R, ab : E.]

16079 - EVSTIGNEEV, V.B., KAZAKOVA, A.A., KISELEV, B.A. : O vliyanii kislotnosti sredy na fotopotentsial i izmenenie pH pri vzaimodeĭstvii khlorofilla s gidrokhinonom v raznykh rastvoritelyakh. [Effect of medium acidity on photopotential and pH change when chlorophyll interacts with hydroquinon in different solvents.] - Biofizika *18* : 53 - 58, 1973. [In R, ab : E.]

16080 - EVSTIGNEEV, V.B., OLOVYANISHNIKOVA, G.D. : Fotokhimicheskie svoĭstva analogov khlorofilla - Zn-feofitinov. [Photochemical properties of chlorophyll analogues - Zn-pheophytins.] - Mol. Biol. (Moskva) *7* : 195 - 202, 1973. [In R, ab : E.]

16081 - EZE, J.M.O. : The vegetative growth of *Helianthus annuus* and *Phaseolus vulgaris* as affected by seasonal factors in Freetown, Sierra Leone. - Ann. Bot. *37* : 315 - 329, 1973. [Growth analysis.]

16082 - FABIAN, I. : Untersuchungen über den Xanthophyllcyclus bei an P-mangelnden Maispflanzen. - Rev. roum. Biol., Sér. Bot. *18* : 163 - 169, 1973.

16083 - FABIAN-GALAN, G. : The mode of ^{14}C labelling the carotenoids of the sunflower normal and phosphorus-lacking leaves. - Rev. roum. Biol., Sér. Bot. *18* : 153 - - 161, 1973.

*16084 - FADEEL, A.A., AL-SANI, N. : Studies on the changes of the inner micro-structure during the chloroplast development in chlorophyllous roots of wheat. - Bot. Notiser *125* : 477 - 482, 1972.

16085 - FAGERBERG, W.R., DAWES, C.J. : An electron microscopic study of the sporophytic and gametophytic plants of *Padina vickersiae* HOYT. - J. Phycol. *9* : 199 - 204, 1973. [Chloroplast.]

16086 - FAHMY, R., MOMTAZ, A., SALEH, H.M., FOUAD, S.F. : Physiological studies on certain cultivars of kenaf (*Hibiscus cannabinus*). - Egypt. J. Bot. *16* : 329 - - 335, 1973. [Chl, Car.]

16087 - FAHMY, R., SELIM, O.A., EL-SHARAWIE, M.O., NAGUIB, E. : Influence of boron and manganese upon growth and development of flax (*Linum usittatissimum*). - Egypt. J. Bot. *16* : 59 - 70, 1973. [Chl, Car.]

16088 - FAIR, P., TEW, J., CRESSWELL, C.F. : Enzyme activities associated with carbon dioxide exchange in illuminated leaves of *Hordeum vulgare* L. I. Effects of

light period, leaf age, and position, on carbon dioxide compensation point
(Γ). - Ann. Bot. *37* : 831 - 844, 1973.

16089 - **FAIR, P., TEW, J., CRESSWELL, C.F.** : Enzyme activities associated with carbon
dioxide exchange in illuminated leaves of *Hordeum vulgare* L. II. Effects of
external concentrations of carbon dioxide and oxygen. - Ann. Bot. *37* : 1035 -
- 1039, 1973.

16090 - **FAJER, J., BORG, D.C., FORMAN, A., DOLPHIN, D., FELTON, R.H.** : Anion radical
of bacteriochlorophyll. - J. amer. chem. Soc. *95* : 2739 - 2741, 1973.

16091 - **FALKOWSKI, M., KUKUŁKA, I., KOZŁOWSKI, S.** : Zależność między nawożeniem azoto-
wym a zawartością chlorofilu (*a* + *b*) w trawach pastwiskowych. [Relation between nitrogen fertilization and chlorophyll (*a* + *b*) content in pasture grasses.]
- Rocz. Nauk roln., Ser. F *78* : 7 - 16, 1973. [In Pol., ab : E, R.]

*16092 - **FALUDI-DANIEL', A.** : Sintez pigmentov i fotosinteticheskaya aktivnost' karoti-
noid-nedostatochnykh mutantov kukuruzy. [Pigment synthesis and photosynthetic
activity in carotenoid-deficient mutants of maize.] - In : Geneticheskie Aspekty Fotosinteza. Tezisy Dokladov. Pp. 28 - 29. Donish, Dushanbe 1972. [In R.]

16093 - **FALUDI-DÁNIEL, Á., DEMETER, S., GARAY, A.S.** : Circular dichroism spectra of
granal and agranal chloroplasts of maize. - Plant Physiol. *52* : 54 - 56, 1973.

*16094 - **FAM TKHAN' KHO** : Gibridologicheskiĭ analiz mozaichnykh pigmentnykh mutantov
Chlamydomonas reinhardi. [Hybridological analysis of the mosaic pigment mutants
of *Chlamydomonas reinhardi*.] - In : Geneticheskie Aspekty Fotosinteza. Tezisy
Dokladov. P. 75. Donish, Dushanbe 1972. [In R.]

16095 - **FANICA-GAIGNIER, M., CLEMENT-METRAL, J.** : 5-aminolevulinic-acid synthetase of
Rhodopseudomonas spheroides Y. Purification and some properties. - Europe. J.
Biochem. *40* : 13 - 18, 1973.

16096 - **FANICA-GAIGNIER, M., CLEMENT-METRAL, J.** : 5-aminolevulinic-acid synthetase of
Rhodopseudomonas spheroides Y. Kinetic mechanism and inhibition by ATP. - Europe. J. Biochem. *40* : 19 - 24, 1973.

16097 - **FANICA-GAIGNIER, M., CLÉMENT-MÉTRAL, J.** : Cellular compartmentation of two species of δ-aminolevulinic acid synthetase in a facultative photohetero-trophic
bacterium (*Rps. spheroides* Y.). - Biochem. biophys. Res. Commun. *55* : 610 - 615,
1973.

16098 - **FANICA-GAIGNIER, M., CLÉMENT-MÉTRAL, J.** : Are both fractions of *Rhodopseudomonas spheroides* Y ALA synthetase isoenzymes ? - Enzyme *16* : 94 - 100, 1973.

16099 - **FAST, A.W., MOSS, B., WETZEL, R.G.** : Effects of artificial aeration on the
chemistry and algae of two Michigan lakes. - Water Resources Res. *9* : 624 -
- 647, 1973. [Primary production.]

*16100 - **FATALIEVA, S.M., GOSTIMSKIĬ, S.A.** : Issledovanie khlorofil'nogo mutanta gorokha, utrativshego sposobnost' k pogloshcheniyu zheleza. [Chlorophyll mutant of
pea with impaired iron uptake.] - In : Geneticheskie Aspekty Fotosinteza. Tezisy Dokladov. P. 91. Donish, Dushanbe 1972. [In R.]

16101 - **FAUST, M.A., GANTT, E.** : Effect of light intensity and glycerol on the growth,
pigment composition, and ultrastructure of *Chroomonas* sp. - J. Phycol. *9* :
489 - 495, 1973.

16102 - **FAY, P., KULASOORIYA, S.A.** : A simple apparatus for the continuous culture of
photosynthetiç micro-organisms. - Brit. phycol. J. *8* : 51 - 57, 1973.

16103 - **FEDENKO, E.P.** : Vliyanie monokhromaticheskogo sveta na obrazovanie bakterial'-
nogo protokhlorofillida zelenym mutantom *Rhodopseudomonas palustris*. [The influence of monochromatic light on the synthesis of bacterial protochlorophyllide by the green mutant of *Rhodopseudomonas palustris*.] - Izv. Akad. Nauk SSSR,
Ser. biol. *1973* : 126 - 132, 1973. [In R, ab : E.]

16104 - **FEDINA, I.S., VAKLINOVA, S.G.** : Light-induced oxygen uptake by etioplasts of
barley and maize in the process of greening. - Dokl. bolg. Akad. Nauk *26* :
415 - 418, 1973.

16105 - **FEDINA, I.S., VAKLINOVA, S.G.** : Influence of oxygen on the assimilation of
$^{14}CO_2$ in isolated chloroplasts. - Dokl. bolg. Akad. Nauk *26* : 1681 - 1684,
1973.

16106 - **FEDOROV, V.D., DAUDA, T.A.** : Sezonnye izmeneniya pishchevoĭ konkurentsii i 'fi-
toplanktonnykh organizmov. [Seasonal changes of feeding competition in phyto-
planktonic organisms.] - Zh. obshch. Biol. *34* : 646 - 653, 1973. [Ps; in R,
ab : E.]

16107 - **FEDOSEEVA, G.P.** : K voprosu o biosinteze mannita u vysshikh rasteniĭ (na primere sel'dereya i sireni). [Mannitol biosynthesis in higher plants (on the example of celery and lilac).] - In : Voprosy Regulyatsii Fotosinteza. Vol. 3.
Pp. 101 - 108, 163. Ural'sk. gos. Univ., Sverdlovsk 1973. [In R.]

16108 - **FEDOSEEVA, G.P.** : Svetovye i temperaturnye krivye fotosinteza ogurtsa. [Light
and temperature curves of cucumber photosynthesis.] - Zap. sverdlovsk. Otd.
vsesoyuz. bot. Obshchestva *6* : 36 - 42, 1973. [In R.]

16109 - **FEDTKE, C.** : Effects of the herbicide methabenzthiazuron on the physiology of
wheat plants. - Pestic. Sci. *4* : 653 - 664, 1973. [Ps, Chl.]

16110 - **FEE, E.J.** : A numerical model for determining integral primary production and
its application to Lake Michigan. - J. Fish. Res. Board Can. *30* : 1447 - 1468,
1973.

16111 - **FEE, E.J.** : Modelling primary production in water bodies : a numerical approach
that allows vertical inhomogeneities. - J. Fish. Res. Board Can. *30* : 1469 -
- 1473, 1973.

*16112 - **FEHER, G., OKAMURA, M.Y., RAYMOND, J.A., STEINER, L.A.** : Subunit structure of
reaction centers from *Rhodopseudomonas spheroides*. - Biophys. Soc. annu. Meet.
Abstracts *15* : 38a, 1971.

16113 - **FEIGE, G.B.** : Beiträge zur Physiologie einheimischer Algen. 2. Untersuchungen
zur Kinetik der $^{14}CO_2$-Assimilation bei der Süßwasserrotalge *Lemanea fluviati-
lis* C.AG. - Z. Pflanzenphysiol. *69* : 290 - 292, 1973.

16114 - **FEKETE, G.** : A CO_2-koncentráció napi menetei tölgyesek légterében. [Daily
changes of CO_2 concentration in the air of oak forests.] - Bot. Közlem. *60* :
43 - 48, 1973. [In Hung., ab : E.]

16115 - **FEKETE, G., SZUJKÓ-LACZA, J.** : Leaf anatomical and photosynthetical reactions
of *Quercus pubescens* WILLD. to environmental factors in various ecosystems. I.
Leaf anatomical reactions. - Acta bot. Acad. Sci. hung. *18* : 59 - 89, 1973.

16116 - **FEKETE, G., SZUJKÓ-LACZA, J., HORVÁTH, G.** : Leaf anatomical and photosynthetical reactions of *Quercus pubescens* WILLD. to environmental factors in various
ecosystems. II. Photosynthetic activity. - Acta bot. Acad. Sci. hung. *18* :
281 - 293, 1973.

16117 - **FEKETE, M.A.R. DE, VIEWEG, G.H.** : Zur Synthese der Saccharose in Blättern von
Zea mays. - Ber. deut. bot. Ges. *86* : 227 - 231, 1973.

16118 - **FEL'DMAN, N.L., AGEEVA, O.A., LYUTOVA, M.I.** : Teploustoĭchivost' ferredoksina
iz zakalennykh progrevom list'ev gorokha *Pisum sativum* L. [Heat resistance of
ferredoxin from leaves of *Pisum sativum* L. hardened by warming up.] - Dokl.
Akad. Nauk SSSR *208* : 479 - 482, 1973. [In R.]

16119 - **FELIPPE, G.M., DALE, J.E.** : Effects of shading the first leaf of barley plants
on growth and carbon nutrition of tne stem apex. - Ann. Bot. *36* : 45 - 56,
1973.

16120 - **FELKER, P., IZAWA, S., GOOD, N.E., HAUG, A.** : Effects of electron transport inhibitors on millisecond delayed light emission from chloroplasts. - Biochim.
biophys. Acta *325* : 193 - 196, 1973.

16121 - **FELLER, U., ERISMANN, K.H.** : Wechselwirkungen zwischen Stickstoffquelle und
Ionenhaushalt bei *Lemna minor* L. unter Photosynthesebedingungen. - Verhandl.
schweiz. naturforsch. Ges. *1973* : 75 - 79, 1973.

16122 - **FEOFILOVA, E.P., LOZHNIKOVA, V.N., BEKHTEREVA, M.N., SAMOKHVALOV, G.I., CHAĬ-
LAKHYAN, M.Kh.** : Vliyanie trisporovykh kislot na rost, obrazovanie pigmentov
i dykhanie prorostkov gorokha. [Influence of trisporic acids on the growth,
pigment formation, and respiration of pea sprouts.] - Dokl. Akad. Nauk SSSR
208 : 483 - 486, 1973. [In R.]

16123 - FERGUSON, C.H.R., SIMON, E.W. : Membrane lipids in senescing green tissues. - J. exp. Bot. *24* : 307 - 316, 1973. [Chl.]

16124 - FERNANDEZ, J., SANZ, M. : Influencia de la radiación *gamma* sobre la actividad fotosintética de plántulas de cebada. I. Assimilacion de $^{14}CO_2$ por las hojas. [Influence of *gamma*-radiation on the photosynthetic activity of barley seedlings. I. Assimilation of $^{14}CO_2$ by leaves.] - Energia nuclear *17* : 437 - 445, 1973. [In Span.]

16125 - FERRARI, I., CAMURRI, L. : Densita'e biomassa dei popolamenti fitoplanctonici in un lago appenninico (Lago Santo Parmense) durante la copertura ghiacciata. [Density and biomass of phytoplankton populations in an Apennine lake (Lake Santo Parmense) during the winter freeze.] - Ist. Lombardo Accad. Sci. Lett. Rend. B *107* : 33 - 42, 1973. [Chl; in Ital., ab : E.]

16126 - FERRO, K., KRANZ, A.R. : Gene action of two ch-loci on the primary carotenoids in leaves of *Arabidopsis thaliana* (L.) HEYNH. - *Arabidopsis* Inform. Serv. *10* : 16 - 17, 1973.

16127 - FIALA, K. : Growth and production of underground organs of *Typha angustifolia* L., *Typha latifolia* L. and *Phragmites communis* TRIN. - Pol. Arch. Hydrobiol. *20* : 59 - 66, 1973.

16128 - FIALA, K. : Seasonal changes in the growth and total carbohydrate content in the underground organs of *Phragmites communis* TRIN. - In : HEJNÝ, S. (ed.) : Ecosystem Study on Wetland Biome in Czechoslovakia. Czechosl. IBP/PT-PP Report No.3. Pp. 107 - 110. Třeboň 1973.

16129 - FICK, G.W., WILLIAMS, W.A., LOOMIS, R.S. : Computer simulation of dry matter distribution during sugar beet growth. - Crop Sci. *13* : 413 - 417, 1973. [Leaf area determination.]

16130 - FIGUEIREDO, I.B. : Conceito geral sobre carotenóides. [General aspects on carotenoids.] - Bol. Inst. tecn. Alimentos *35* : 47 - 69, 1973. [In Port.]

16131 - FILATOV, G.V. : O prichinakh povysheniya fotosinteticheskoǐ deyatel'nosti u geterozisnykh gibridov. [Factors stimulating photosynthetic activity in heterosis hybrids.] - Sel'skokhoz. Biol. *8* : 658 - 661, 1973. [In R, ab : E.]

16132 - FILIPPIS, L.F. De, PALLAGHY, C.K. : Effect of light on the volume and ion relations of chloroplasts in detached leaves of *Elodea densa*. - Aust. J. biol. Sci. *26* : 1251 - 1265, 1973.

16133 - FILIPPOVA, L.A., MAMUSHINA, N.S. : Izuchenie fotosinteza i ottoka assimilyatov u khlorelly na svetu. [Studies of photosynthesis and photosynthates transfer in light in·*Chlorella*.] - In : Upravlenie Biosintezom Mikroorganizmov. P. 167. Krasnoyarsk 1973. [In R.]

16134 - FILIPPOVA, L.A., MAMUSHINA, N.S. : O dinamike peredvizheniya assimilyatov iz khloroplastov. [Dynamics of the movement of photosynthates from chloroplasts.] - Tr. biol.-pochv. Inst., dal'nevost. nauch. Tsentr Akad. Nauk SSSR *20* (123) : 26 - 32, 1973. [In R, ab : E.]

16135 - FILIPPOVA, L.A., MAMUSHINA, N.S., ZALENSKIǏ, O.V. : Vliyanie intensivnosti sveta na ottok assimilyatov iz khloroplastov u *Chlorella pyrenoidosa* CHICK. [Effect of illuminance on the transport of assimilates from chloroplasts of *Chlorella pyrenoidosa* CHICK.] - Bot. Zh. *58* : 1528 - 1534, 1973. [In R, ab : E.]

*16136 - FILIPPOVA, R.I., STREL'NIKOVA, T.R. : Osobennosti pigmentnoǐ sistemy rasteniǐ sakharnoǐ kukuruzy v svyazi s yavleniem geterozisa. [Peculiarities of pigment system in sweet maize in connection with heterosis.] - In : Geneticheskie Aspekty Fotosinteza. Tezisy Dokladov. Pp. 116 - 117. Donish, Dushanbe 1972. [In R.]

*16137 - FILIPPOVICH, I.I., ALINA, B.A., BEZSMERTNAYA, I.N., TONGUR, A.M., OPARIN, A.I.: Svyaz' beloksinteziruyushcheǐ sistemy so strukturoǐ khloroplastov. [Relationship between the protein-synthesizing system and the chloroplast structure.] - In : Geneticheskie Aspekty Fotosinteza. Tezisy Dokladov. Pp. 29 - 31. Donish, Dushanbe 1972. [In R.]

16138 - FINKEL'SHTEǏN, E.I., ALEKSEEV, É.V., KOZLOV, É.I. : Kinetika avtookisleniya tverdykh plenok β-karotina. [Kinetics of autooxidation of β-carotene solid films.] - Dokl. Akad. Nauk SSSR *208* : 1408 - 1411, 1973. [In R.]

16139 - **FIOLET, J.W.T., VAN DAM, K.** : The effects of tetraphenylboron on energy-linked reactions in spinach chloroplasts. - Biochim. biophys. Acta *325* : 230 - 239, 1973.

16140 - **FIRSOV, N.N., IVANOVSKIĬ, R.N.** : Znacheniya pH dlya rosta i okisleniya sul'-fida *Ectothiorhodospira shaposhnikovii* v avtotrofnykh i geterotrofnykh usloviyakh. [The dependence on pH value of *Ectothiorhodospira shaposhnikovii* growth and sulfide oxidation in autotrophic and heterotrophic conditions.] - Vestn. mosk. Univ., Ser. VI - Biol., Pochvoved. *28* (6) : 53 - 55, 1973. [In R, ab : E.]

16141 - **FISCHER, K., KRAMER, D., ZIEGLER, H.** : Elektronenmikroskopische Untersuchungen SO_2-begaster Blätter von *Vicia faba*. I. Beobachtungen an Chloroplasten mit akuter Schädigung. - Protoplasma *76* : 83 - 96, 1973.

16142 - **FISCHER, R.A.** : The effect of water stress at various stages of development on yield processes in wheat. - In : **SLATYER, R.O.** (ed.) : Plant Response to Climatic Factors. Pp. 233 - 241. Unesco, Paris 1973.

16143 - **FISHER, S.G., LIKENS, G.E.** : Energy flow in Bear Brook, New Hampshire : An integrative approach to stream ecosystem metabolism. - Ecol. Monogr. *43* : 421 - 439, 1973. [Ps.]

16144 - **FITZPATRICK, E.A., STERN, W.R.** : Net radiation estimated from global solar radiation. - In : **SLATYER, R.O.** (ed.) : Plant Response to Climatic Factors. Pp. 403 - 410. Unesco, Paris 1973.

16145 - **FLEISCHMAN, D.E., MAYNE, B.C.** : Chemically and physically induced luminescence as a probe of photosynthetic mechanisms. - In : **SANADI, D.R., PACKER, L.** (ed.): Current Topics in Bioenergetics. Vol. 5. Pp. 77 - 105. Academic Press, New York - London 1973.

16146 - **FLETCHER, J.S., NESIUS, K.K.** : Dark fixation of endogenously generated $^{14}CO_2$ *versus* exogenously provided $H^{14}CO_3^-$. - Plant Physiol. *51* (Suppl.) : 42, 1973.

16147 - **FLETCHER, R.A., TEO, C., ALI, A.** : Stimulation of chlorophyll synthesis in cucumber cotyledons by benzyladenine. - Can. J. Bot. *51* : 937 - 939, 1973.

*16148 - **FLOHÉ, L., MENZEL, H.** : The influence of glutathione upon light-induced high-amplitude swelling and lipid peroxide formation of spinach chloroplasts. - Plant Cell Physiol. *12* : 325 - 333, 1971.

16149 - **FLOROV, R.J.** : Approche thermodynamique pour évaluer l'utilisation de l'énergie solaire par les plantes. - In : The Sun in the Service of Mankind. Pp. V.14-1 - V.14-10. Unesco, Paris 1973.

16150 - **FLOROV, R.J.** : Un index bioclimatique complexe pour l'accumulation de la biomasse végétale. - In : **SLATYER, R.O.** (ed.) : Plant Response to Climatic Factors. Pp. 423 - 426. Unesco, Paris 1973.

16151 - **FOOTE, K.C., SCHAEDLE, M.** : Seasonal and diurnal field rates of photosynthesis and respiration in stems of *Populus tremuloides* MICHX. - Amer. J. Bot. *60* (4, Suppl.) : 24, 1973.

16152 - **FORGER, J.M. III, BOGORAD, L.** : Steps in the acquisition of photosynthetic competence by plastids of maize. - Plant Physiol. *52* : 491 - 497, 1973.

16153 - **FORK, D.C.** : Light-induced shifts in the absorption spectrum of carotenoids and chlorophyll *b* in the green alga *Ulva*. - Carnegie Inst. Year Book *72* : 374 - 376, 1973.

16154 - **FORK, D.C., HIYAMA, T.** : The photochemical reactions of photosynthesis in an alga exposed to extreme conditions. - Carnegie Inst. Year Book *72* : 384 - 388, 1973.

15155 - **FORTI, G.** : Photosynthetic electron transport. - In : **CHECCUCCI, A., WEALE, R.A.** (ed.) : Primary Molecular Events in Photobiology. Pp. 177 - 187. Elsevier, Amsterdam - London - New York 1973.

15156 - **FORTINI, S., GIORGI, B., GIACOMELLI, M., MANNINO, P., CORDISCHI, M.** : Modificazioni di enzimi carbossilativi, del contenuto proteico e del peso fogliare in plantule ottenute da linee ditelocentriche di frumento tenero, varietá "Chinese spring". [Modification of carboxylase activities, protein content and leaf weight in ditelocentric lines of bread wheat cv. Chinese Spring.] - Ann. Ist. sper. Cerealicolt. *4* : 133 - 144, 1973. [In Ital., ab : E.]

*16157 - Fotosintez Sel'skokhozyaÏstvennykh RasteniÏ Moldavii v Svyazi s Usloviyami Proizrastaniya. [Photosynthesis of Agricultural Plants of Moldaviya in Connection with Conditions of Growing.] - Redaktsionno-Izdat. Otdel Akad. Nauk moldav. SSR, Kishinev 1970. [In R.]

*16158 - FOTT, J. : Observations on primary production of phytoplankton in two fish ponds. - In : KAJAK, Z., HILLBRICHT-ILKOWSKA, A. (ed.) : Productivity Problems of Freshwaters. Pp. 673 - 683. Pol. Sci. Publ., Warszawa 1972.

*16159 - FOUZOVA, S. : Geneticheskaya izmenchivost' fotosinteticheskoÏ aktivnosti list'ev. [Genetic variation in photosynthetic activity of leaves.] - In : Geneticheskie Aspekty Fotosinteza. Tezisy Dokladov. P. 76. Donish, Dushanbe 1972. [In R.]

16160 - FOWLER, C.F., GRAY, B.H., NUGENT, N.A., FULLER, R.C. : Absorbance and fluorescence properties of the bacteriochlorophyll *a* reaction center complex and bacteriochlorophyll *a* protein in green bacteria. - Biochim. biophys. Acta *292* : 692 - 699, 1973.

16161 - FRĄCKOWIAK, B., DMOCHOWSKA, G., KANIUGA, Z. : Effect of the inhibitors on the nicotinamide nucleotides level in isolated chloroplasts. - Bull. Acad. pol. Sci., Sér. Sci. biol. *21* : 479 - 484, 1973.

16162 - FRĄCKOWIAK, B., KANIUGA, Z. : The relationship between non-cyclic electron transport and phosphorylation in the presence of ascorbate. - Photosynthetica *7* : 28 - 33, 1973.

16163 - FRĄCKOWIAK, D. : Luminescencja w badaniach fotosyntezy. [Luminescence in photosynthesis research.] - Postępy Fiz. *24* : 331 - 344, 1973. [In Pol., ab : E.]

16164 - FRĄCKOWIAK, D., GRABOWSKI, J. : Effect of pH on migration of excitation energy between biliproteins. - Photosynthetica *7* : 305 - 310, 1973.

16165 - FRĄCKOWIAK, D., GRABOWSKI, J. : Yield of energy transfer between chlorophyllin and biliproteins. - Photosynthetica *7* : 402 - 404, 1973.

16166 - FRĄCKOWIAK, D., MANIKOWSKI, H. : Effect of dark storage on absorption and fluorescence spectra of *Rhodospirillum rubrum* cells and fragments. - Photosynthetica *7* : 275 - 281, 1973.

16167 - FRĄCKOWIAK, D., WRÓBEL, D. : Energy transfer between chlorophyll *c* and chlorophyll *a*. - Biophys. Chem. *1* : 125 - 129, 1973.

16168 - FRADKIN, L.I., KOLYAGO, V.M., MORDACHEVA, G.S., ZEN'KO, A.A. : Razdelenie, spektral'nye i metabolicheskie svoÏstva pigment-lipoproteidnykh kompleksov khloroplastov. [Distribution, spectral and metabolic properties of pigment-lipoproteid complexes of chloroplasts.] - In : Formirovanie Pigmentnogo Apparata Fotosinteza. Pp. 50 - 87. Nauka i Tekhnika, Minsk 1973. [In R.]

16169 - FRADKIN, L.I., SHAÚCHUK, S.M. : Ab dakladnastsi dvukhkhvalevaga spektrafotametrychnaga vyznachênnya kantsêntratsyi khlarafilaǔ u sumesi. [Accuracy of two-wavelength spectrophotometric determination of chlorophyll concentration in mixture.] -Vestsi Akad. Navuk belarus.SSR, Ser. biyal. Navuk *1973* (1) : 79 - 86, 141, 1973. [In Belorus., ab : R.]

16170 - FRAGATA, M. : Cybernetic control of electron transfer and oxygen evolution in photosynthesis. - In : Ninth International Congress of Biochemistry. Stockholm, 1 - 7 July 1973. Abstract Book. P. 217. IUB, Stockholm 1973.

16171 - FRALICK, R.A., MATHIESON, A.C. : Ecological studies of *Codium fragile* in New England, USA. - Mar. Biol. *19* : 127 - 132, 1973. [Ps.]

16172 - FRANCIS, G.W., KNUTSEN, G., LIEN, T. : Loroxanthin from *Chlamydomonas reinhardti*. - Acta chem. scand. *27* : 3599 - 3601, 1973.

16173 - FRANK, A.B., POWER, J.F., WILLIS, W.O. : Effect of temperature and plant water stress on photosynthesis, diffusion resistance, and leaf water potential in spring wheat. - Agron. J. *65* : 777 - 780, 1973.

16174 - FREEMAN, T.P. : Developmental anatomy of epidermal and mesophyll chloroplasts in *Opuntia basilaris* leaves. - Amer. J. Bot. *60* : 86 - 91, 1973.

16175 - **FREEMAN, T.P., DUYSEN, M.E.** : Development of chloroplasts in water stressed
Triticum leaves. - Amer. J. Bot. *60* (4, Suppl.) : 6, 1973.

16176 - **FRENCH, C.S., TAKAMIYA, A., MURATA, T.** : Resolution of the low temperature absorption spectra of chlorophyll-protein complexes into their component bands.
- Carnegie Inst. Year Book *72* : 336 - 351, 1973.

16177 - **FRENYŐ, V., NINH, T.D.** : Examination of the toxic effect of copper salts in
maize. - Acta bot. Acad. Sci. hung. *18* : 91 - 94, 1973. [Chl.]

16178 - **FRESCO, L.F.M.** : A model for plant growth. Estimation of the parameters of
the logistic function. - Acta bot. neerl. *22* : 486 - 489, 1973.

16179 - **FREYMAN, S., CHARNETSKI, W.A., CROOKSTON, R.K.** : Role of leaves in the formation of seed in rape. - Can. J. Plant Sci. *53* : 693 - 694, 1973. [Ps.]

16180 - **FRIEDRICH, G., KATZFUSS, M., SALZER, J., BÜTTNER, R., SCHMIDT, S.** : Assimilation, distribution of substance, and development of yields in apple-trees. -
Proc. Res. Inst. Pomol., Skierniewice, Ser. E *1973* (3) : 41 - 50, 1973.

16181 - **FUHR, H., STAUFF, J.** : Chemiluminescence arising from the action of $^1\Delta g$-molecular oxygen on chlorophyll-a. - Z. Naturforsch. *28c* : 302 - 309, 1973.

16182 - **FUJITA, Y., SUZUKI, R.** : Studies on the Hill reaction of membrane fragments
of blue-green algae. III. Fluorescence characteristics of membrane fragments
of *Anabaena variabilis* and *Anabaena cylindrica*. - Plant Cell Physiol. *14* :
249 - 260, 1973.

16183 - **FUJITA, Y., SUZUKI, R.** : Studies on the Hill reaction of membrane fragments
of blue-green algae. IV. Carotenoid photobleaching induced by photosystem II
action. - Plant Cell Physiol. *14* : 261 - 273, 1973.

16184 - **FUNKHOUSER, E.A.** : Iron and the control of chlorophyll synthesis in *Euglena
gracilis*. - Diss. Abstr. Int. B *33* : 4690-B, 1973.

16185 - **FURUKAWA, A.** : Carbon dioxide compensation points in poplar plant. - J. Jap.
Forestry Soc. *55* (3) : 95 - 99, 1973.

16186 - **FURUKAWA, A.** : Photosynthesis and respiration in poplar plant in relation to
leaf development. - J. Jap. Forestry Soc. *55* (4) : 119 - 123, 1973.

16187 - **GÄCHTER, R., LUM-SHUE-CHAN, KEN, CHAU, Y.K.** : Complexing capacity of the nutrient medium and its relation to inhibition of algal photosynthesis by copper.
- Schweiz. Z. Hydrol. *35* : 252 - 261, 1973.

16188 - **GAEVSKIĬ, N.A., GOL'D, V.M., GRIGOR'EV, Yu.S.** : Fluorestsentsiya khlorofilla
i fotokhimicheskie reaktsii pri psevdotsiklicheskom (tsiklicheskom) transporte
èlektronov. [Chlorophyll fluorescence and photochemical reactions during the
pseudocyclic (cyclic) electron transport.] - Izv. sibir. Otd. Akad. Nauk SSSR,
Ser. biol. Nauk *3* (15) : 80 - 85, 1973. [in R, ab : E.]

16189 - **GAEVSKIĬ, N.A., GOL'D, V.M., GRIGOR'EV, Yu.S., PUZYR', A.P.** : Vykhod fluorestsentsii khlorofilla v izolirovannykh khloroplastakh i ego svyaz' s pervichnymi protsessami prevrashcheniya ènergii v fotosinteze. [Fluorescence yield of
chlorophyll in isolated chloroplasts and its relation to primary processes of
energy transformation in photosynthesis.] - In : Biologicheskaya Spektrofotometriya i Fitoaktinometriya. Pp. 34 - 35. Krasnoyarsk 1973. [in R.]

16190 - **GAJDOS, A., GAJDOS-TÖRÖK, M.** : Mechanism of the effect of exogenous glucose
on the biosynthesis of porphyrins by *Rhodopseudomonas spheroides*. - Enzyme
16 : 101 - 107, 1973.

16191 - **GALE, J.** : Experimental evidence for the effect of barometric pressure on photosynthesis and transpiration. - In : **SLATYER, R.O.** (ed.) : Plant Response to
Climatic Factors. Pp. 289 - 294. Unesco, Paris 1973.

16192 - **GALKIN, V.I.** : Sposob rascheta neobkhodimogo chisla povtornykh izmereniĭ v issledovaniyakh fotosinteza. [Method of evaluation of necessary number of repetitions in photosynthesis measurements.] - In : Metody Kompleksnogo Izucheniya
Fotosinteza. Vol. 2. Pp. 102 - 108. VASKhNIL, Leningrad 1973. [in R.]

16193 - **GALLAGHER, J.L.** : The significance of the surface film in plankton primary
production in a salt marsh. - J. Phycol. *9* (Suppl.) : 8, 1973.

16194 - **GALLAGHER, J.L., DAIBER, F.C.** : Diel rhythms in edaphic community metabolism in a Delaware salt marsh. - Ecology *54* : 1160 - 1163, 1973. [Ps.]

16195 - **GALLIARD, T.** : Phospholipid metabolism in photosynthetic plants. - In : ANSELL, G.B., HAWTHORNE, J.N., DAWSON, R.M.C. (ed.) : Form and Function of Phospholipids. 2nd Ed. Elsevier, Amsterdam - Oxford - New York 1973.

16196 - **GALLING, G., SALZMANN, C., SPIEβ, E.** : Synthese von Chlorophyll und Strukturelementen des Plastiden in *Chlorella* ohne Beteiligung der Chloroplasten-Ribosomen. - Planta *114* : 269 - 284, 1973.

16197 - **GALLOPIN, I.G., JOLLIFFE, P.A.** : Effects of low non-freezing temperatures on chlorophyll accumulation in corn and other grasses. - Crop Sci. *13* : 766 - 768, 1973.

16198 - **GALMICHE, J.M.** : Studies on the mechanism of glycerate 3-phosphate synthesis in tomato and maize leaves. - Plant Physiol. *51* : 512 - 519, 1973. [Ps.]

16199 - **GAMAYUNOVA, M.S., GRIGORA, M.Yu., OSTROVSKAYA, L.K.** : Vydelenie i kharakteristika soderzhashchikh fotosistemu I fragmentov iz tilakoidov stromy i gran khloroplastov gorokha. [Isolation and characteristic of Photosystem I containing fragments from stroma and grana thylakoids of pea chloroplasts.] - Fiziol. Biokhim. kul't. Rast. *5* : 456 - 461, 1973. [In R, ab : E.]

16200 - **GANF, G.G., VINER, A.B.** : Ecological stability in a shallow equatorial lake (Lake George, Uganda). - Proc. roy. Soc. London B *184* : 321 - 346, 1973. [Ps, Chl.]

16201 - **GANICHEVA, O.P.** : Aktivnost' karboangidrazy pri predposevnom obogashchenii semyan kormovoĭ svekly kobal'tom i tsinkom. [Carbonic anhydrase activity after pre-sowing treatment of beet seeds by cobalt and zinc.] - Uch. Zapiski gor'kov. gos. Univ., Ser. biol. *159* - Peredvizhenie Veshchestv i Metabolizm Rasteniĭ) : 92 - 96, 1973. [In R.]

16202 - **GANTT, E., LIPSCHULTZ, C.A.** : Energy transfer in phycobilisomes from phycoerythrin to allophycocyanin. - J. Phycol. *9* (Suppl.) : 19 - 20, 1973.

16203 - **GANTT, E., LIPSCHULTZ, C.A.** : Energy transfer in phycobilisomes from phycoerythrin to allophycocyanin. - Biochim. biophys. Acta *292* : 858 - 861, 1973.

16204 - **GAPONENKA, V.I., BALEVA, E.F., SHAŬCHUK, S.M., SHLYK, A.A.** : Prayaŭlenne fizika-khimichnaĭ getĕragennastsi khlarafilaŭ *a* i *b* u prysutnastsi trytonu X-100. [Expression of physico-chemical heterogeneity of chlorophyll *a* and *b* in the presence of *Triton X-100*.] - Vestsi Akad. Navuk belarus.SSR, Ser. biyal. Navuk *1973* (3) : 29 - 38, 1973. [In Belorus., ab : R.]

16205 - **GAPONENKA, V.I., NIKALAEVA, G.M., SHAŬCHUK, S.M., LASITSKAYA, T.U.** : Abnaŭlenne khlarafilu ŭ listsyakh pshanitsy, zakonchyŭshykh rost. [Renewal of chlorophyll in wheat leaves that have completed growth.] - Vestsi Akad. Navuk belarus. SSR, Ser. biyal. Navuk *1973* (1) : 40 - 46, 139, 1973. [In Belorus., ab : R.]

16206 - **GAPONENKA, V.I., NIKALAEVA, G.M., STANISHEŬSKAYA, A.M., SHAŬCHUK, S.M., BALEVA, E.F.** : Kol'kasts' labil'nykh padfondaŭ khlarafilaŭ *a* i *b* u zyalĕnaĭ rasline pry roznaĭ asvetlenastsi. [Contents of labile forms of chlorophyll *a* and *b* in a green plant under various illuminance.] - Vestsi Akad. Navuk belarus. SSR, Ser. biyal. Navuk *1973* (5) : 62 - 67, 1973. [In Belorus., ab : R.]

16207 - **GAPONENKA, V.I., SHAŬCHUK, S.M., LASITSKAYA, T.U.** : Dasledavanne abnaŭlennya khlarafilu ŭ shmatgadovykh listsyakh fikusa i aspidystry radyevuglyarodnym metadam. [Resynthesis of chlorophyll in sempervirent leaves of *Ficus* and *Aspidistra* studied with the ^{14}C method.] - Vestsi Akad. Navuk belarus. SSR, Ser. biyal. Navuk *1973* (2) : 43 - 51, 139, 1973. [In Belorus., ab : R.]

16208 - **GAPONENKO, A.K., BELETSKIĬ, Yu.D., ZHDANOV, Yu.A.** : Vliyanie razlichnykh srokov khraneniya semyan yachmenya, obrabotannykh N-nitrozo-N-metilmochevinoĭ pri razlichnykh usloviyakh obrabotki, na chastotu i spektr khlorofil'nykh mutatsiĭ. [Effect of different storage times of barley seeds treated with N-nitroso-N-methylurea under different conditions on the frequency and spectrum of chlorophyll mutations.] - Dokl. Akad. Nauk SSSR *210* : 1200 - 1202, 1973. [In R.]

16209 - **GARAY, A.S., CZÉGÉ, J., TOLVAJ, L., TÓTH, M., SZABÓ, M.** : Biological significance of molecular chirality in energy balance and metabolism. - Acta biotheor. *22* : 34 - 43, 1973. [Chl.]

16210 - **GARCÍA-PEREGRÍN, E., COLOMA, A., MAYOR, F.** : Intrachloroplastidic mevalonate-activating enzymes in plant cells. - Plant Sci. Lett. *1* : 367 - 373, 1973.

16211 - **GARMAN, W.H.** : Agriculture's place in the environment : Considerations for decision making. - J. environ. Qual. *2* : 327 - 333, 1973. [Ps.]

16212 - **GARRATT, J.R., PEARMAN, G.I.** : CO_2 concentration in the atmospheric boundary-layer over South-East Australia. - Atmos. Environ. *7* : 1257 - 1266, 1973. [Prediction model.]

16213 - **GARYAEV, P.P., VLADYCHENSKIĬ, A.S., DEYANOVA, S.A. KALOSHIN, P.M., POGLAZOV, B.F.** : Vydelenie i ochistka nekotorykh organicheskikh soedineniĭ iz dernovo-podzolistoĭ pochvy i vulkanicheskogo grunta. [Extraction and purification of some organic compounds from sod-podzolic soil and volcanic earth.] - Pochvovedenie *1973* (4) : 134 - 141, 1973. [Chl; in R, ab : E.]

16214 - **GASANOV, R.A., FRENCH, C.S.** : Chlorophyll composition and photochemical activity of photosystems detached from chloroplast grana and stroma lamellae. - Proc. nat. Acad. Sci. USA *70* : 2082 - 2085, 1973.

16215 - **GASSMAN, M.L.** : Absorbance and fluorescence properties of protochlorophyllide in etiolated bean leaves. - Biochem. biophys. Res. Commun. *53* : 693 - 702, 1973.

16216 - **GASSMAN, M.L.** : A reversible conversion of phototransformable protochlorophyll (ide)$_{650}$ to photoinactive protochlorophyll(ide)$_{633}$ by hydrogen sulfide in etiolated bean leaves. - Plant Physiol. *51* : 139 - 145, 1973.

16217 - **GASSMAN, M.L.** : The conversion of photoinactive protochlorophyllide$_{633}$ to phototransformable protochlorophyllide$_{650}$ in etiolated bean leaves treated with δ-aminolevulinic acid. - Plant Physiol. *52* : 590 - 594, 1973.

16218 - **GATES, D.M.** : Plant temperatures and energy budget. - In : PRECHT, H., CHRISTOPHERSEN, J., HENSEL, H., LARCHER, W. (ed.) : Temperature and Life. Pp. 87 - - 101. Springer-Verlag, Berlin - Heidelberg - New York 1973. [Ps.]

16219 - **GAUDILLÈRE, J.-P.** : [Une méthode de mesure de la photosynthèse et de la photorespiration in régime permanent. - Compt. rend. Acad. Sci. Paris, Sér. D *276* : 2081 - 2083, 1973.

16220 - **GAUSMAN, H.W.** : Light reflectance of *Peperomia* chloroplasts. - J. Rio Grande Valley hort. Soc. *27* : 86 - 89, 1973.

16221 - **GAUSMAN, H.W.** : Photomicrographic record of light reflected at 850 nanometers by cellular constituents of *Zebrina* leaf epidermis. - Agron. J. *65* : 504 - 505, 1973.

16222 - **GAUSMAN, H.W.** : Reflectance, transmittance, and absorptance of light by subcellular particles of spinach (*Spinacia oleracea* L.) leaves. - Agron. J. *65* : 551 - 553, 1973.

16223 - **GAUSMAN, H.W., ALLEN, W.A.** : Optical parameters of leaves of 30 plant species. - Plant Physiol. *52* : 57 - 62, 1973.

16224 - **GAUSMAN, H.W., ALLEN, W.A., CARDENAS, R., RICHARDSON, A.J.** : Reflectance discrimination of cotton and corn at four growth stages. - Agron. J. *65* : 194 - - 198, 1973. [Chl.]

16225 - **GAVRILĂ, L., CHIOȘILĂ, I.** : Cercetări privind influența unor factori biotici asupra producției primare planctonice în condiții experimentale. [Effect of some biotic factors on primary production of plankton under experimental conditions.] - Stud. Cercet. Biol., Ser. bot. *25* : 331 - 339, 1973. [In Roum., ab : E.]

16226 - **GAVRILENKO, V.F., GRABOVSKAYA, M.I., MEL'NIK, T.K.** : Vlivanie formy azotnogo pitaniya na sintez khlorofilla i protogema v rasteniyakh gorokha i kukuruzy. [Effect of the form of nitrogen nutrition on the synthesis of chlorophyll and protoheme in plants of pea and maize.] - Tr. VSKhIZO *68* (Voprosy Fiziologii Rasteniĭy · 2⁰ - 3⁰ ·1973, Fin R.]

16227 - **GAVRILOVA, V.A.** : Izmenenie kislotno-osnovnogo ravnovesiya v srede pri foto-vosstanovlenii i fotookislenii khlorofilla i ego analogov. [Changes in acido--basic balance of medium at photoreduction and photooxidation of chlorophyll and its analogues.] - In : Problemy Vozniknoveniya i Sushchnosti Zhizni. Pp. 134 - 140. Nauka, Moskva 1973. [In R.]

16228 - **GENCHEV, S.** : Vliyanie na giberelinovata, β-indolilotsetnata i 2,4-dichlor-fenoksiotsetnata kiselina, v"rkhu biosintezata na plastidnite pigmenti pri etiolirani ponitsi luk (*Allium cepa* L.). [Effect of gibberellic, β-indolyl-acetic and 2,4-dichlorphenoxyacetic acids on the biosynthesis of plastid pigments in etiolated shoots of *Allium cepa* L.] - In : 2 Nats. Konf. po Bot., 1969. Pp. 499 - 505. Sofiya 1973. [In Bulg.]

16229 - **GERASIMENKO, T.V.** : Zavisimost' fotosinteza ot temperatury u rastenii tundr ostrova Vrangelya. [Dependence of photosynthesis on temperature in tundra plants of Wrangel Island.] - Bot. Zh. *58* : 493 - 504, 1973. [In R, ab : E.]

16230 - **GERASIMENKO, T.V., ZALENSKII, O.V.** : Sutochnaya i sezonnaya dinamika fotosinteza u rastenii ostrova Vrangelya. [Diurnal and seasonal dynamics of photosynthesis in plants of Wrangel Island.] - Bot. Zh. *58* : 1655 - 1666, 1973. [In R.]

16231 - **GERHARDT, B.** : Untersuchungen zur Funktionsänderung der Microbodies in den Keimblättern von *Helianthus annuus* L. - Planta *110* : 15 - 28, 1973. [Chl.]

*16232 - **GERSTER, R.** : Analyse par spectrométrie de masse de composés enrichis en ^{17}O et ^{18}O. - In : Comptes Rendus d'un Colloque International sur les Isotopes de l'Oxygène. Pp. 40 - 50. Inst. nat. Sci. Tech. nucl., Gif-sur-Yvette 1972.

*16233 - **GERSTER, R.** : Echange isotopique entre CO_2 et H_2O dans des systèmes hétérogènes. - In : Comptes Rendus d'un Colloque International sur les Isotopes de l' Oxygène. Pp. 130 - 137. Inst. nat. Sci. Tech. nucl., Gif-sur-Yvette 1972.

*16234 - **GERSTER, R.** : Mesure au moyen de CO_2 enrichi en carbone 13 et oxygène 18 des effets isotopiques lors de la fixation photosynthétique de CO_2. - In : Comptes Rendus d'un Colloque International sur les Isotopes de l'Oxygène. Pp. 231 - - 237. Inst. nat. Sci. Tech. nucl., Gif-sur-Yvette 1972.

*16235 - **GERSTER, R., DUPUY, J., GUERIN de MONTGAREUIL, P.** : Origine de l'oxygène photosynthétique. - In : Comptes Rendus d'un Colloque International sur les Isotopes de l'Oxygène. Pp. 175 - 188. Inst. nat. Sci. Tech. nucl., Gif-sur-Yvette 1972.

16236 - **GEUS-KRUYT, M. De, SEGAL, S.** : Notes on the productivity of *Stratiotes aloides* in two lakes in the Netherlands. - Pol. Arch. Hydrobiol. *20* : 195 - 205, 1973.

16237 - **GHILDIYAL, M.C., SINHA, S.K.** : Variation in photorespiration in wheat genotypes using glycine-1-^{14}C decarboxylation technique. - Indian J. exp. Biol. *11* : 207 - 209, 1973.

16238 - **GIAQUINTA, R.T., DILLEY, R.A., ANDERSON, B.** : Light-induced potentiation of water oxidation inhibition by chemical modification of spinach chloroplast membrane proteins. - Plant Physiol. *51* (Suppl.) : 67, 1973.

16239 - **GIAQUINTA, R.T., DILLEY, R.A., ANDERSON, B.J.** : Light potentiation of photosynthetic oxygen evolution inhibition by water soluble chemical modifiers. - Biochem. biophys. Res. Commun. *52* : 1410 - 1417, 1973.

*16240 - **GIBBS, M.** : Assimilyatsiya CO_2 izolirovannymi khloroplastami. [Assimilation of CO_2 by isolated chloroplasts.] - In : Geneticheskie Aspekty Fotosinteza. Tezisy Dokladov. P. 46. Donish, Dushanbe 1972. [In R.]

16241 - **GIBBS, M., AVRON, M.** : Chloroplastic phosphoribulokinase and CO_2 fixation. - Plant Physiol. *51* (Suppl.) : 41, 1973.

B16242 - **GIESE, A.C.** (ed.) : Photophysiology. Current Topics in Photobiology and Photochemistry. Vol. 8. - Academic Press, New York - London 1973.

16243 - **GIESKES, W.W.C., van BENNEKOM, A.J.** : Unreliability of the ^{14}C method for estimating primary productivity in eutrophic Dutch coastal waters. - Limnol. Oceanogr. *18* : 494 - 495, 1973.

16244 - **GIFFORD, R.M., BREMNER, P.M., JONES, D.B.** : Assessing photosynthetic limitation to grain yield in a field crop. - Aust. J. agr. Res. *24* : 297 - 307, 1973.

16245 - **GIFFORD, R.M., MARSHALL, C.** : Photosynthesis and assimilate distribution in *Lolium multiflorum* LAM. following differential tiller defoliation. - Aust. J. biol. Sci. *26* : 517 - 526, 1973.

16246 - **GIFFORD, R.M., MUSGRAVE, R.B.** : Stomatal role in the variability of net CO_2 exchange rates by two maize inbreds. - Aust. J. biol. Sci. *26* : 35 - 44, 1973.

16247 - **GILBERT, J.J., ALLEN, H.L.** : Chlorophyll and primary productivity of some green, freshwater sponges. - Int. Rev. ges. Hydrobiol. *58* : 633 - 658, 1973.

*16248 - **GILES, K.L.** : The control of chloroplast division in *Funaria hygrometrica* II. The effects of kinetin and indole-acetic acid on nucleic acids. - Plant Cell Physiol. *12* : 447 - 450, 1971.

*16249 - **GILES, K.L., TAYLOR, A.O.** : The control of chloroplast division in *Funaria hygrometrica* I. Patterns of nucleic acid, protein and lipid synthesis. - Plant Cell Physiol. *12* : 437 - 445, 1971.

*16250 - **GILLER, Yu.E., ABDULLAEVA, S.K., KRASICHKOVA, G.V., YUSUPOVA, G.A., YUKHANANO-VA, L.N.** : Iskusstvennye pigment-belkovye kompleksy - model' molekulyarnoĭ organizatsii i nekotorykh funktsional'nykh svoĭstv pigmentnoĭ sistemy fotosinteticheskogo apparata. [Synthetic pigment-protein complexes - a model of the molecular organization and some functional properties of pigment system in the photosynthetic apparatus.] - In : Geneticheskie Aspekty Fotosinteza. Tezisy Dokladov. Pp. 86 - 87. Donish, Dushanbe 1972. [In R.]

16251 - **GILLER, Yu.E., ASOEVA, L.M.** : Osobennosti énergeticheskogo vzaimodeĭstviya fotosinteticheskikh pigmentov v plastidakh mutantnykh form vodorosli *Chlorella*. [Peculiarities of photosynthetic pigments energetic interaction in chloroplasts of *Chlorella* mutants.] - Biofizika *18* : 299 - 306, 1973. [In R, ab : E.]

16252 - **GIMMLER, H.** : Correlations between photophosphorylation and light-induced conformational changes of chloroplasts in whole cells of the halophilic green alga *Dunaliella parva*. - Z. Pflanzenphysiol. *68* : 289 - 307, 1973.

16253 - **GIMMLER, H.** : The effect of FCCP on the fluorescence induction of chlorophyll in DBMIB-treated intact cells of the unicellular green alga *Dunaliella parva*. - Z. Pflanzenphysiol. *68* : 385 - 390, 1973.

16254 - **GINSBURG, R., LAZAROFF, N.** : Ultrastructural development of *Nostoc muscorum* A. - J. gen. Microbiol. *75* : 1 - 9, 1973. [Thylakoids.]

16255 - **GINZO, H.D., LOVELL, P.H.** : Aspects of the comparative physiology of *Ranunculus bulbosus* L. and *Ranunculus repens* L. II. Carbon dioxide assimilation and distribution of photosynthates. - Ann. Bot. *37* : 765 - 776, 1973.

16256 - **GIRAULT, G., GALMICHE, J.-M., MICHEL-VILLAZ, M., THIERY, J.** : Comparative study of photophosphorylation coupling factor·ligand complex by circular dichroism and chemical isolation. - Europe. J. Biochem. *38* : 473 - 478, 1973.

16257 - **GIUDICI DE NICOLA, M., PIATTELLI, M., AMICO, V.** : Effect of continuous far red on betaxanthin and betacyanin synthesis. - Phytochemistry *12* : 2163 - 2166, 1973. [Chl.]

16258 - **GLADYSHEV, A.I.** : Biologicheskaya produktivnost' travyanistykh fitotsenozov poĭmy Amudar'i. [Biological productivity of herbaceous phytocenoses of the Amu Dar'ya flood lands.] - Izv. Akad. Nauk turkm. SSR, Ser. biol. Nauk *1973* (2) : 38 - 44, 1973. [In R, ab : E.]

16259 - **GLAGOLEVA, T.A., FILIPPOVA, L.A., MAMUSHINA, N.S., ZALENSKIĬ, O.V.** : Zavisimost' peredvizheniya assimilyatov iz khloroplastov ot funktsionirovaniya I i II fotosistem. [Dependence of translocation of photosynthates from chloroplasts on the functioning of photosystems I and II.] - Tr. biol.-pochv. Inst., dal'nevost. nauch. Tsentr Akad. Nauk SSSR *20* (123) : 33 - 38, 1973. [In R, ab : E.]

16260 - **GLAGOLEVA, T.A., ZALENSKIĬ, O.V.** : K voprosu o svyazi metabolizma ugleroda C^{14} u khlorelly s rabotoĭ I i II fotosistem. [Relationship between the ^{14}C-metabolism and activity of photosystems I and II in *Chlorella*.] - In : Upravlenie Biosintezom Mikroorganizmov. Pp. 111 - 113. Krasnoyarsk 1973. [In R.]

16261 - **GLAZER, A.N., FANG, S.** : Chromophore content of blue-green algal phycobiliproteins. - J. biol. Chem. *248* : 659 - 662, 1973.

16262 - **GLAZER, A.N., FANG, S.** : Formation of hybrid proteins from the α and β subunits of phycocyanins of unicellular and filamentous blue-green algae. - J. biol. Chem. *248* : 663 - 671,. 1973.

16263 - **GLAZER, A.N., FANG, S., BROWN, D.M.** : Spectroscopic properties of *C*-phycocyanin and of its α and β subunits. - J. biol. Chem. *248* : 5679 - 5685, 1973.

,16264 - **GLIKMAN, T.S., DUBINA, M.L., YANOVSKAYA, E.V.** : Vliyanie ingibitorov reaktsii Khila na fotovosstanovlenie Mn (III)-feoforbida gidroksil'nymi ionami. [Effect of Hill reaction inhibitors on photoreduction of manganese (III) - pheophorbide with hydroxyl ions.] - Fiziol. Biokhim. kul't. Rast. *5* : 263 - 266, 1973. [In R, ab : E.]

16265 - **GLIKMAN, T.S., ZAVGORODNYAYA, L.N.** : O vozmozhnosti fotookisleniya vody marganetssoderzhashchimi proizvodnymi khlorofilla. Okislenie vosstanovlennykh produktov khinonami. [Possibility of water photooxidation by manganese-containing chlorophyll derivatives. Oxidation of reduced products by quinones.] - Biokhimiya *38* : 101 - 105, 1973. [In R, ab : E.]

16266 - **GLINSKIĬ, V.P., SAMUILOV, V.D., SKULACHEV, V.P.** : Fotoindutsirovannye reaktsii pigmentov nesernoĭ purpurnoĭ bakterii *Rhodospirillum rubrum* v oblasti 400 - - 600 nm. [Photo-induced reactions of pigments of the non-sulphur purple bacterium *Rhodospirillum rubrum* in the region of 400 - 600 nm.] - Izv. Akad. Nauk SSSR, Ser. biol. *1973* : 93 - 98, 1973. [In R, ab : E.]

16267 - **GLOAGUEN, J.-C., TOUFFET, J.** : La biomasse végétale aérienne et la répartition des éléments biogènes essentiels dans quelques types de landes littorales armoricaines. - Compt. rend. Acad. Sci. Paris, Sér. D *277* : 505 - 508, 1973.

16268 - **GLOOSCHENKO, W.A., MOORE, J.E., VOLLENWEIDER, R.A.** : Chlorophyll *a* distribution in Lake Huron and its relationship to primary productivity. - In : Proceedings of the 16th Conference of Great Lakes Research. Pp. 40 - 49. Int. Ass. Great Lakes Res., Ann Arbor 1973.

16269 - **GLOVER, J.** : The dark respiration of sugar-cane and the loss of photosynthate during the growth of a crop. - Ann. Bot. *37* : 845 - 852, 1973.

16270 - **GMELIG MEYLING, H.D.** : Effect of light intensity, temperature and daylength on the rate of leaf appearance of maize. - Neth. J. agr. Sci. *21* : 68 - 76, 1973. [Growth analysis.]

16271 - **GODAVARI, H.R., BADOUR, S.S., WAYGOOD, E.R.** : Metabolism of glyoxylate in leaf extracts. - Plant Physiol. *51* (Suppl.) : 7, 1973.

16272 - **GODZIEMBA-CZYŻ, J.** : Certain aspects of the chemotaxic reaction of chloroplasts in *Funaria hygrometrica*. - Acta Soc. Bot. Pol. *42* : 453 - 459, 1973.

16273 - **GOEDHEER, J.C.** : Chlorophyll *a* forms in *Phaeodactylum tricornutum* : Comparison with other diatoms and brown algae. - Biochim. biophys. Acta *314* : 191 - 201, 1973.

16274 - **GOEDHEER, J.C.** : Fluorescence polarization and pigment orientation in photosynthetic bacteria. - Biochim. biophys. Acta *292* : 665 - 676, 1973.

16275 - **GOFFER, J., NEUMANN, J.** : Evidence for system I mediated non-cyclic photophosphorylation in chloroplasts. - FEBS Lett. *36* : 61 - 64, 1973.

16276 - **GOINS, D.J., REYNOLDS, R.J., SCHIFF, J.A., BARNETT, W.E.** : A cytoplasmic regulatory mutant of *Euglena* : constitutivity for the light-inducible chloroplast transfer RNAs. - Proc. nat. Acad. Sci. USA *70* : 1749 - 1752, 1973.

16277 - **GOL'D, V.M.** : Pervichnye protsessy fotosinteza i deĭstvie kachestvennogo sostava sveta. [Primary processes of photosynthesis and action of light quality.] - In : Upravlenie Skorost'yu i Napravlennost'yu Biosinteza u Rasteniĭ. Pp. 13- -14. Krasnoyarsk 1973. [In R.]

16278 - **GOL'D, V.M.** : Voprosy ékologicheskoĭ fiziologii tenelyubivykh i tenevynoslivykh rasteniĭ. [Problems of the ecological physiology of shade-requiring and shade-enduring plants.] - Probl. Ékol. *3* : 105 - 112, 1973. [Ps, Chl; in R, ab : E.]

16279 - **GOL'D, V.M., GAEVSKIĬ, N.A., BOTKINA, T.I., GRIGOR'EV, Yu.S.** : Aktivnost' fotokhimicheskikh reaktsiĭ u aspidistry na sinem i krasnom svetu. [Activity of

photochemical reactions in *Aspidistra* on blue and red light.] - Fiziol. Rast.
20 : 539 - 544, 1973. [In R, ab : E.]

16280 - **GOL'D, V.M., GAEVSKIĬ, N.A., GRIGOR'EV, Yu.S.** : Izuchenie svoĭstv vikasola kak
kofaktora ėlektrontransportnoĭ tsepi fotosinteza. [Characteristics of vicasol
as a cofactor of photosynthetic electron-transfer chain.] - Biokhimiya *38* :
906 - 908, 1973. [In R, ab : E.]

16281 - **GOL'D, V.M., GAEVSKIĬ, N.A., GRIGOR'EV, Yu.S., GOLEVA, N.G., YANSBERG, N.V.** :
Svetovye krivye vosstanovleniya NADF na sinem i krasnom svetu v khloroplastakh gorokha. [Light curves of NADP reduction in pea chloroplasts under blue
and red light.] - In : Informatsionnye Materialy Akad. Nauk SSSR, Sibir.
Otd., Sibir. Inst. Fiziol. Biokhim. Rast., Irkutsk *11* : 38 - 40, 1973. [In R.]

16282 - **GOLOMAZOVA, G.M.** : Vliyanie uglekisloty na fotosinteticheskuyu aktivnost' seyantsev khvoĭnykh drevesnykh porod. [Effect of CO_2 on the photosynthetic activity of seedlings of coniferous woody plants.] - In : Upravlenie Skorost'yu
i Napravlennost'yu Biosinteza u Rasteniĭ. Pp. 11 - 12. Krasnoyarsk 1973. [In
R.]

16283   **GOLOVANOVA, T.I., GOL'D, V.M.** : Voprosy teorii upravleniya vtorichnykh reaktsiĭ fotosinteza kofaktorami psevdotsiklicheskogo fotofosforilirovaniya i deĭ-
stviem kachestvennogo sostava sveta. [Problems of the theory of regulation of
secondary reactions of photosynthesis by co-factors of pseudocyclic photophosphorylation and by light quality.] - In : Upravlenie Skorost'yu i Napravlennost'yu Biosinteza u Rasteniĭ. Pp. 9 - 10. Krasnoyarsk 1973. [In R.]

16284 - **GOLUBTSEV, V.V., MOGILEVSKIĬ, G.L.** : Fotometricheskaya sistema dlya avtomati-
cheskoĭ zapisi spektrov opticheskogo pogloshcheniya. [Photometric system for
automatic recording of spectra of optical absorption.] - Zh. prikl. Spektroskop. *18* : 946 - 949, 1973. [In R.]

16285 - **GONCHARIK, M.N., IVANCHENKO, V.M., LEGENCHENKO, B.I.** : O spetsifichnosti vliyaniya ionov Cl na fotosintez rasteniĭ. [Specificity of the effect of Cl ions
on plant photosynthesis.] - In : Voprosy Soleustoĭchivosti Rasteniĭ. Pp. 149 -
- 155. Fan, Tashkent 1973. [In R.]

16286 - **GOODING, L.R., ROY, H., JAGENDORF, A.T.** : Immunological identification of nas-
cent subunits of wheat ribulose diphosphate carboxylase on ribosomes of both
chloroplast and cytoplasmic origin. - Arch. Biochem. Biophys. *159* : 324 - 335,
1973.

16287 - **GOODMAN, P.J.** : Physiological and ecotypic adaptations of plants to salt desert conditions in Utah. - J. Ecol. *61* : 473 - 494, 1973. [Growth analysis.]

16288 - **GOODMAN, R.A., OLDFIELD, E., ALLERHAND, A.** : Assignments in the natural-abun-
dance carbon-13 nuclear magnetic resonance spectrum of chlorophyll *a* and a
study of segmental motion in neat phytol. - J. amer. chem. Soc. *95* : 7553 -
- 7558, 1973.

16289 - **GOODWIN, T.W.** : Carotenoids. - In : MILLER, L.P. (ed.) : Phytochemistry. Vol.
I. The Process and Products of Photosynthesis. Pp. 112 - 142. Van Nostrand
Reinhold Company, New York - Cincinnati - Toronto - London - Melbourne 1973.

16290 - **GOODWIN, T.W.** : Carotenoids and biliproteins. - In : STEWART, W.D.P. (ed.) :
Algal Physiology and Biochemistry. Bot. Monogr. Vol. 10. Pp. 176 - 205. Blackwell sci. Publ., Oxford - London - Edinburgh - Melbourne 1973.

16291 - **GOPAL, B.** : A survey of the Indian studies on ecology and production of wetland and shallow water communities. - Pol. Arch. Hydrobiol. *20* : 21 - 29, 1973.

16292 - **GOPAL, N.H.** : Distribution of iron, chlorophyll, and heme enzymes in the life
span of groundnut. - Bot. Gaz. *134* : 100 - 103, 1973.

16293 - **GÓRAL, I.** : Distribution of radioactive products of photosynthesis in Scots
pine (*Pinus silvestris* L.) seedlings during the first vegetation season. -
Acta Soc. Bot. Pol. *42* : 541 - 553, 1973.

16294 - **GÓRAL, I.** : Incorporation of radioactive products of photosynthesis into lig-
nin and cellulose of Scots pine (*Pinus silvestris* L.) seedlings. - Acta Soc.
Bot. Pol. *42* : 555 - 565, 1973.

16295 - **GORCHEIN, A.** : Control of magnesium-protoporphyrin chelatase activity in *Rho-dopseudomonas spheroides*. Role of light, oxygen, and electron and energy transfer. - Biochem. J. *134* : 833 - 845, 1973.

16296 - **GORDEEV, V.I., GORSHKOV, V.K., EVSTIGNEEV, V.B., D'YACHENKO, A.P.** : Fotookislenie khlorofilla *a* v étanole pri impul'snom osveshchenii v zavisimosti ot pH sredy. [pH dependence of chlorophyll *a* photooxidation under flash illumination in ethanol.] - Biofizika *18* : 631 - 636, 1973. [In R, ab : E.]

16297 - **GORDON, D.C. Jr., PROUSE, N.J.** : The effects of three oils on marine phytoplankton photosynthesis. - Mar. Biol. *22* : 329 - 333, 1973.

*16298 - **GORDON, W.** : A dielectric dispersion technique for measuring the ionic permeability of internal membranes of isolated chloroplasts. - J. Membrane Biol. *8* : 97 - 107, 1972.

*16299 - **GORDON, W.** : The effect of illuminating spinach chloroplasts on their membrane permeability, measured by a dielectric dispersion technique. - J. Membrane Biol. *10* : 193 - 205, 1972.

16300 - **GORDON, W.** : The magnitude and mechanism of the passive permeability of cane chloroplast internal membranes to various ions, measured by dielectric dispersion. - J. Membrane Biol. *12* : 385 - 397, 1973.

16301 - **GORDON, W.** : Use of dielectric phenomena in measuring the capacitance and permeability of biological membranes, with special reference to chloroplast internal membranes. - T.-I.-T. J. Life Sci. *3* : 127 - 137, 1973.

16302 - **GORHAM, E., SOMERS, M.G.** : Seasonal changes in the standing crop of two montane sedges. - Can. J. Bot. *51* : 1097 - 1108, 1973.

16303 - **GORKOM, H.J. van, DONZE, M.** :·Charge accumulation in the reaction center of photosystem 2. - Photochem. Photobiol. *17* : 333 - 342, 1973.

16304 - **GORLANOV, N.A.** : Izmenenie khlorofilla i ego svoĭstv v list'yakh ukorenyayushchikhsya cherenkov fasoli, vyrashchennykh iz γ-obluchennykh semyan. [Changes in chlorophyll and its properties in leaves of rooted bean shoots grown from γ-irradiated seeds.] - Radiobiologiya *13* : 634 - 636, 1973. [In R, ab : E.]

*16305 - **GOSTIMSKIĬ, S.A.** : Vliyanie mutatsiĭ yadra na organizatsiyu fotosinteticheskogo apparata. [Effect of nuclear mutations on the organization of photosynthetic apparatus.] - In : Geneticheskie Aspekty Fotosinteza. Tezisy Dokladov. Pp. 57 - 58. Donish, Dushanbe 1972. [In R.]

16306 - **GOSTIMSKIĬ, S.A.** : Novyĭ mutant gorokha s narushennymi reaktsiyami fotosinteza. [A new mutant of pea defected in photosynthetic reactions.] - Genetika *9* (5) : 166 - 167, 1973. [In R, ab : E.]

16307 - **GOUDRIAAN, J.** : A calculation model and descriptive formulas for the extinction and reflection of radiation in leaf canopies. - In : The Sun in the Service of Mankind. Pp. V.6-1 - V.6-10. Unesco, Paris 1973.

16308 - **GOULD, J.M., IZAWA, S.** : Studies on the energy coupling sites of photophosphorylation. I. Separation of Site I and Site II by partial reactions of the chloroplast electron transport chain. - Biochim. biophys. Acta *314* : 211 - 223, 1973.

16309 - ⊦**GOULD, J.M., IZAWA, S.** : Photosystem-II electron transport and phosphorylation with dibromothymoquinone as the electron acceptor. - Europe. J. Biochem. *37* : 185 - 192, 1973.

16310 - **GOULD, J.M., IZAWA, S., GOOD, N.E.** : A site of photophosphorylation close to photosystem II. - Fed. Proc. *32* : 632Abs, 1973.

16311 - **GOULD, J.M., ORT, D.R.** : Studies on the energy coupling sites of photophosphorylation. III. The different effects of methylamine and ADP plus phosphate on electron transport through coupling sites I and II in isolated chloroplasts. - Biochim. biophys. Acta *325* : 157 - 166, 1973.

16312 - **GOULD, J.M., WINGET, G.D.** : A membrane-bound alkaline inorganic pyrophosphatase in isolated spinach chloroplasts. - Arch. Biochem. Biophys. *154* : 606 - 613, 1973.

16313 - **GOVINDJEE, PAPAGEORGIOU, G., RABINOWITCH, E.** : Chlorophyll fluorescence and photosynthesis. - In : GUILBAULT, G.G. (ed.) : Practical Fluorescence Theory, Methods, and Techniques. Pp. 543 - 575. M. Dekker Inc., New York 1973.

16314 - **GRABOWSKI, A.** : The biomass, organic matter contents and calorific values of macrophytes in the lakes of the Szeszupa drainage area. - Pol. Arch. Hydrobiol. *20* : 269 - 282, 1973.

16315 - **GRACE, J.** : An apparatus for the study of leaf canopy optics. - J. appl. Ecol. *10* : 57 - 61, 1973.

16316 - **GRACE, J., THOMPSON, J.R.** : The after-effect of wind on the photosynthesis and transpiration of *Festuca arundinacea*. - Physiol. Plant. *28* : 541 - 547, 1973.

16317 - **GRACE, J., WOOLHOUSE, H.W.** : A physiological and mathematical study of the growth and productivity of a *Calluna-Sphagnum* community. II. Light interception and photosynthesis in *Calluna*. - J. appl. Ecol. *10* : 63 - 76, 1973.

16318 - **GRACE, J., WOOLHOUSE, H.W.** : A physiological and mathematical study of the growth and productivity of a *Calluna-Sphagnum* community. III. Distribution of photosynthate in *Calluna vulgaris* L. HULL. - J. appl. Ecol. *10* : 77 - 91, 1973.

16319 - **GRAHL, H., WILD, A.** : Lichtinduzierte Veränderungen im Photosynthese-Apparat von *Sinapis alba*. - Ber. deut. bot. Ges. *86* : 341 - 349, 1973.

16320 - **GRAY, B., GANTT, E.** : Phycobilisome isolation from a blue-green alga *Nostoc* sp. - J. Phycol. *9* (Suppl.) : 19, 1973.

16321 - **GRAY, B.H., LIPSCHULTZ, C.A., GANTT, E.** : Phycobilisomes from a blue-green alga *Nostoc* species. - J. Bacteriol. *116* : 471 - 478, 1973.

16322 - **GRAY, J.C., KEKWICK, R.G.O.** : Mevalonate kinase in green leaves and etiolated cotyledons of the French bean *Phaseolus vulgaris*. - Biochem. J. *133* : 335 - - 347, 1973. [Chl.]

16323 - **GRAY, J.C., KEKWICK, R.G.O.** : Synthesis of the small subunit of ribulose 1,5--diphosphate carboxylase on cytoplasmic ribosomes from greening bean leaves. - FEBS Lett. *38* : 67 - 69, 1973.

16324 - **GREEF, J.A. De, CAUBERGS, R.** : Studies on greening of etiolated seedlings. II. Leaf greening by phytochrome action in the embryonic axis. - Physiol. Plant. *28* : 71 - 76, 1973.

16325 - **GREEF, J.A. De, VERBELEN, J.P.** : Physiological stress and crystallites in leaf plastids of *Phaseolus vulgaris* L. - Ann. Bot. *37* : 593 - 596, 1973.

16326 - **GREENE, D.M., WALTON, D.W.H., CALLAGHAN, T.V.** : Standing crop in a *Festuca* grassland on South Georgia. - In : BLISS, L.C., WIELGOLASKI, F.E. (ed.) : Primary Production and Production Processes, Tundra Biome. Pp. 191 - 194. Tundra Biome Steering Comm., Edmonton 1973.

16327 - **GREGORY, P., BRADBEER, J.W.** : Plastid development in primary leaves of *Phaseolus vulgaris* : The light-induced development of the chloroplast cytochromes. - Planta *109* : 317 - 326, 1973.

16328 - **GREGORY, P., RACKER, E.** : The relationship between the conformational change and the subunit structure of CF_1. - In : Ninth International Congress of Biochemistry. Stockholm, 1 - 7 July 1973. Abstract Book. P. 238. IUB, Stockholm 1973.

16329 - **GREPPIN, H., HORWITZ, B.A., HORWITZ, L.P.** : Light-stimulated bioelectric response of spinach leaves and photoperiodic induction. - Z. Pflanzenphysiol. *68* : 336 - 345, 1973. [Ps.]

16330 - **GRIBANOVSKI-SASSU, O.** : Biosynthesis of keto-karotenoids in *Dictyococcus cinnabarinus*. - Ann. Ist. super. Sanità *9* : 225 - 232, 1973.

16331 - **GRIFFITHS, D.J.** : Factors affecting the photosynthetic capacity of laboratory cultures of the diatom *Phaeodactylum tricornutum*. - Mar. Biol. *21* : 91 - 97, 1973.

16332 - **GRIGORA, M.Yu., GAMAYUNOVA, M.S., SILAEVA, A.M.** : Vydelenie razlichnykh fraktsii izolirovannykh khloroplastov mezofilla kukuruzy. [Isolation of different fractions of maize mesophyll chloroplasts.] - Fiziol. Biokhim. kul't. Rast. *5* : 144 - 148, 1973. [In R, ab : E.]

16333 - **GRIGOR'EV, Yu.S., GOL'D, V.M., GAEVSKIĬ, N.A., BELONOG, N.P.** : Izuchenie induk-tsionnykh perekhodov fluorestsentsii u razlichnykh grupp rasteniĭ. [Induction transitions of fluorescence studied in various plants.] - Fiziol. Rast. *20* : 747 - 752, 1973. [In R, ab : E.]

16334 - **GRIMME, L.H., BOARDMAN, N.K.** : Separation of photosynthetic systems I and II from a chloroplast preparation from *Chlorella*. - Hoppe-Seyler's Z. physiol. Chem. *354* : 1499 - 1502, 1973.

16335 - **GRINTAL', A.R.** : Vliyanie temperatury na intensivnost' fotosinteza *Laminaria saccharina* (L.) LAM. [Effect of temperature on photosynthetic rate in *Laminaria saccharina* (L.) LAM.] - Bot. Zh. *58* : 1361 - 1367, 1973. [In R.]

16336 - **GRINYUS, L.L., IL'INA, M.D., MILEĬKOVSKAYA, E.I., SKULACHEV, V.P., TIKHONOVA, G.V.** : Obrazovanie électricheskogo potentsiala, sopryazhennoe s perenosom é-lektronov v membrannykh chastitsakh bakteriĭ *Micrococcus lysodeikticus* i khlo-roplastov gorokha. [Formation of electric potential coupled with electron transport in membrane particles of *Micrococcus lysodeikticus* and pea chloro-plasts.] - Biokhimiya *38* : 1153 - 1162, 1973. [In R, ab : E.]

16337 - **GROB, E.C., EICHENBERGER, W.** : Beiträge zum Stickstoffmetabolismus in grünen Pflanzen. I. Die reversible Plastidenumwandlung in glucosestimulierten Kultu-ren von *Spirodela oligorrhiza*, eine Folge von Stickstoffmangel. - Experientia *29* : 398 - 400, 1973.

16338 - **GRODZINSKI, B., COLMAN, B.** : Loss of photosynthetic activity in two blue-green algae as a result of osmotic stress. - J. Bacteriol. *115* : 456 - 458, 1973.

16339 - **GROEN, J.** : Photosynthesis of *Calendula officinalis* L. and *Impatiens parviflo-ra* DC., as influenced by light intensity during growth and age of leaves and plants. - Med. Landbouwhogeschool Wageningen *73-8* : 1 - 128, 1973.

16340 - **GROMET-ELHANAN, Z., LEISER, M.** : Interchangeability of the membrane potential with the pH gradient of *Rhodospirillum rubrum* chromatophores. A comparison of the effect of sulfate, thiocyanate, and perchlorate. - Arch. Biochem. Biophys. *159* : 583 - 589, 1973.

16341 - **GROSS, E.L., HESS, S.C.** : Monovalent cation-induced inhibition of chlorophyll *a* fluorescence : Antagonism by divalent cations. - Arch. Biochem. Biophys. *159* : 832 - 836, 1973.

16342 - **GROSS, J., GABAI, M., LIFSHITZ, A., SKLARZ, B.** : Carotenoids in pulp, peel and leaves of *Persea americana*. - Phytochemistry *12* : 2259 - 2263, 1973.

16343 - **GROSSMAN, D., CRESSWELL, C.F.** : Influence of nitrogen supply in nutrient media on carbon dioxide compensation point (Γ) of C-4 photosynthetic plants. - South Afr. J. Sci. *69* : 244 - 246, 1973.

16344 - **GROUZIS, J.-P.** : Relations entre photophosphorylation cyclique et absorption du calcium dans les chloroplastes isolés de feuilles de Lupin (*Lupinus luteus* L.). - Physiol. vég. *11* : 643 - 654, 1973.

*16345 - **GROZOV, D.N.** : Opticheskie parametry list'ev yabloni pri razlichnykh urovnyakh azotnogo pitaniya. [Optical parameters of apple tree leaves under various mi-neral nutrition.] - In : Fotosintez Odnoletnikh i Mnogoletnikh Rasteniĭ. Pp. 45 - 59, 138 - 139. Shtiintsa, Kishinev 1972. [In R.]

*16346 - **GROZOV, D.N.** : Fotosintez i opticheskie svoĭstva list'ev yabloni pri pal'met-tnoĭ formirovke krony. [Photosynthesis and optical properties of apple tree leaves in palmette canopy.] - In : Fotosintez Odnoletnikh i Mnogoletnikh Raste-niĭ. Pp. 59 - 72, 139. Shtiintsa, Kishinev 1972. [In R.]

16347 - **GRUBER, P.J., BECKER, W.M., NEWCOMB, E.H.** : The development of microbodies and peroxisomal enzymes in greening bean leaves. - J. Cell Biol. *56* : 500 - 518, 1973.

16348 - **GRUMBACH, K.H., LICHTENTHALER, H.K.** : Der Verlauf der Lipochinon- und Pigment-synthese bei einer experimentell induzierten Chloroplastendegeneration in grü-nen Keimlingen von *Hordeum vulgare* L. - Z. Naturforsch. *28c* : 439 - 445, 1973.

16349 - **GRUNEWALD, W.** : Accuracy and errors of the pO_2 measurement by means of the platinum electrode and its calibration *in vivo*. - In : GROSS, J.F., KAUFMANN,

R., WETTERER, E. (ed.) : Modern Techniques in Physiological Science. Pp. 309 - - 320. Academic Press, London - New York 1973.

16350 - GRÜNSFELDER, M., SIMONIS, W. : Aktive und inaktive Phosphataufnahme in Blatt- zellen von *Elodea densa* bei hohen Phosphat-Außenkonzentrationen. - Planta *115*: 173 - 186, 1973. [Ps inhibitors.]

16351 - GUDKOV, N.D., STOLOVITSKIĬ, Yu.M., EVSTIGNEEV, V.B. : Impul'snaya fotoprovo- dimost' rastvorov khlorofilla i ego analogov. Fotoprovodimost' éfirnykh rast- vorov khlorofilla *a*. [Flash-conductivity of solutions of chlorophyll and its analogues. Photoconductivity of diethylether solutions of chlorophyll *a*.] - Biofizika *18* : 807 - 812, 1973. [In R, ab : E.]

16352 - GUÉRIN-DUMARTRAIT, E., HOARAU, J., LECLERC, J.-C., SARDA, C. : Effets de quel- ques conditions d'éclairement, notamment de la lumière rouge, sur la compositi- on pigmentaire et la structure de *Porphyridium* sp. (LEWIN). - Phycologia *12* : 119 - 130, 1973.

16353 - GUÉRIN-DUMARTRAIT, E., LECLERC, J.-C., HOARAU, J. : Effects de la lumière rou- ge et de la carence azote sur la composition pigmentaire (phycoérythrine, holo- chromes chlorophylliens) et l'émission d'O_2 photosynthétique de *Porphyridium* sp. - Arch. Hydrobiol. *39* (Suppl.) : 317 - 332, 1973.

16354 - GULANYAN, S.A., KURELLA, G.A., AVER'YANOV, A.A., TITOVA, Z.V. : Tekhnika iz- gotovleniya élektrodnoĭ sistemy dlya opredeleniya partsial'nogo davleniya mole- kulyarnogo kisloroda v zhidkoĭ ili gazoobraznoĭ srede. [Technique of manufac- turing the electrode system for the determination of partial pressure of mole- cular oxygen in liquid or gaseous media.] - Nauch. Dokl. vyssh. Shkoly, biol. Nauki *16* (1) : 131 - 136, 1973. [In R.]

16355 - GULYAEV, B.A., VEDENEEV, E.P., KARNEEVA, N.V., LITVIN, F.F. : Razlozhenie slozhnykh spektral'nykh krivykh biologicheskikh ob"ektov na komponenty s ispol zovaniem proizvodnykh spektrov. II. Analiz spektrov pogloshcheniya fotosinte- ticheskikh struktur. [Resolution of complex spectral curves of biological ob- jects into components using the derivative spectra. II. Analysis of absorption spectra of photosynthetic structures.] - Nauch. Dokl. vyssh. Shkoly, biol. Na- uki *16* (10) : 48 - 56, 1973. [In R.]

15356 - GULYAEV, B.I. : Osobennosti gazoobmena CO_2 u razlichnykh vidov rasteniĭ. [Spe- cific features of CO_2 gaseous exchange in different plant species.] - Dokl. Akad. Nauk SSSR *210* : 240 - 243, 1973. [In R.]

16357 - GUNDERSEN, K. : *In situ* determination of primary production by means of a new incubator ISIS. - Helgoländer wiss. Meeresuntersuchungen *24* : 465 - 475, 1973.

16358 - GUPTA, R.K. : Comparison between optical properties of maize leaves and soy- bean leaves. - Indian J. agr. Sci. *43* : 41 - 44, 1973.

*16359 - GUSEV, M.V. : Sravnitel'naya fiziologiya sinezelenykh vodorosleĭ. [Comparative physiology of blue-green algae.] - Uspekhi Mikrobiol. *3* : 74 - 103, 1966. [Ps; in R.]

16360 - GUTEL'MAKHER, B.L. : Radioavtograficheskiĭ metod opredeleniya otnositel'nogo znacheniya otdel'nykh vidov vodorosleĭ v pervichnoĭ produktsii planktona. [Ra- dioautographic method for determining the relative significance of some plant species in the primary production of plankton.] - Gidrobiol. Zh. *9* (1) : 103 - - 107, 1973. [In R.]

16361 - GUZEMAN, O.L.J., SYKES, A. : On the relationship between singlet and triplet absorption spectra of conjugated polyenes. - Photochem. Photobiol. *18* : 339 - - 341, 1973. [Car.]

16362 - GVARDIYAN, V.N., SHABEL'SKAYA, É.F., GODNEV, T.N. : Soderzhanie i sootnoshenie zelenykh pigmentov u rasteniĭ razlichnykh vidov pri prodolzhitel'nom polnom zatemnenii. [Green pigments content and ratio in plants of different species at long-term darkening.] - Vestsi Akad. Navuk belarus.SSR, Ser. biyal. Navuk *1973* (1) : 113 - 117, 1973. [In R.]

16363 - GYURJÁN, I. : A kloroplasztisz genetikája. II. A plastisz fehérjeszintetizáló rendszere. [Genetics of chloroplasts. II. The course of synthesis of plastid proteins.] - Biol. Közl. *21* : 127 - 141, 1973. [In Hung.]

16364 - **HAAN, G.A. DEN, WARDEN, J.T., DUYSENS, L.N.M.** : Kinetics of the fluorescence yield of chlorophyll a_2 in spinach chloroplasts at liquid nitrogen temperature during and following a 16 µs flash. - Biochim. biophys. Acta *325* : 120 - 125, 1973.

16365 - **HABERMANN, H.M.** : Evidence for two photoreactions and possible involvement of phytochrome in light-dependent stomatal opening. - Plant Physiol. *51* : 543 - - 548, 1973. [Chl.]

16366 - **HABESHAW, D.** : Translocation and the control of photosynthesis in sugar beet. - Planta *110* : 213 - 226, 1973.

16367 - **HACKETT, C.** : An exploration of the carbon economy of the tobacco plant. I. Inferences from a simulation. - Aust. J. biol. Sci. *26* : 1057 - 1071, 1973.

16368 - **HAEDER, D.P., NULTSCH, W.** : Negative photo-phobotactic reactions in *Phormidium uncinatum*. - Photochem. Photobiol. *18* : 311 - 317, 1973. [Ps.]

16369 - **HAEDER, H.-E., MENGEL, K., FORSTER, H.** : The effect of potassium on transloca-tion of photosynthates and yield pattern of potato plants. - J. Sci. Food Agr. *24* : 1479 - 1487, 1973.

16370 - **HAEHNEL, W.** : Electron transport between plastoquinone and chlorophyll a_1 in chloroplasts. - Biochim. biophys. Acta *305* : 618 - 631, 1973.

16371 - **HAEHNEL, W.** : The reaction kinetics of plastocyanin *in situ* and its function in chloroplasts. - In : Ninth International Congress of Biochemistry. Stock-holm, 1 - 7 July 1973. Abstract Book. P. 217. IUB, Stockholm 1973.

16372 - **HAGAR, W.G., FORD, G.A., FRENCH, C.S.** : Rate measuring circuit for improved action spectra. - Carnegie Inst. Year Book *72* : 361 - 364, 1973. [Ps.]

16373 - **HAHN, P.B., WECHTER, M.A., JOHNSON, D.C., VOIGT, A.F.** : Sodium tungsten bronze as a potentiometric indicating electrode for dissolved oxygen in aqueous solu-tion. - Anal. Chem. *45* : 1016 - 1021, 1973.

16374 - **HALE, C.N., WHITBREAD, R.** : The translocation of ^{14}C-labelled assimilates by dwarf bean plants infected with *Pseudomonas phaseolicola* (BURK). DOWS. - Ann. Bot. *37* : 473 - 480, 1973.

16375 - **HALL, D.O.** : Origin and development of life in relation to solar energy. - Anais Acad. bras. Ciênc. *45* (Suppl. - Novas Tendencias em Fotobiologia) : 71 - - 83, 1973. [Ps.]

16376 - **HALL, D.O., CAMMACK, R., RAO, K.K.** : Ferredoxins in the evolution of photosyn-thetic systems from anaerobic bacteria to higher plants. - Space Life Sci. *4* : 455 - 468, 1973.

16377 - **HALL, D.O., CAMMACK, R., RAO, K.K.** : The plant ferredoxins and their relation-ship to the evolution of ferredoxins from primitive life. - Pure appl. Chem. *34* : 553 - 577, 1973.

16378 - **HALL, R.H.** : Cytokinins as a probe of developmental processes. - Annu. Rev. Plant Physiol. *24* : 415 - 444, 1973. [Ps.]

16379 - **HALL, R.L., KUNG, M.C., FU, M., HALES, B.J., LOACH, P.A.** : Comparison of pho-totrap complexes from chromatophores of *Rhodospirillum rubrum, Rhodopseudomo-nas spheroides,* and the R-26 mutant of *Rhodopseudomonas spheroides.* - Photo-chem. Photobiol. *18* : 505 - 520, 1973.

16380 - **HALLDAL, P.** : Accessory pigments and enhancement effects. - In : CHECCUCCI, A., WEALE, R.A. (ed.) : Primary Molecular Events in Photobiology. Pp. 147 - 175. Elsevier, Amsterdam - London - New York 1973.

16381 - **HALLDAL, P., HALLDAL, K.** : Phytoplankton, chlorophyll, and submarine light con-ditions in Kings Bay, Spitsbergen, July 1971. - Norw. J. Bot. *20* : 99 - 108, 1973.

16382 - **HAMID, M.A.K., BADOUR, S.S.** : The effects of microwaves on green algae. - J. Microwave Power *8* : 267 - 273, 1973. [Ps, Chl.]

16383 - **HAMILL, A.S., PENNER, D.** : Interaction of alachlor and carbofuran. - Weed Sci. *21* : 330 - 335, 1973. [Ps.]

16384 - HAMILL, A.S., PENNER, D. : Chlorbromuron-carbofuran interaction in corn and barley. - Weed Sci. *21* : 335 - 338, 1973. [Ps.]

16385 - HAMILL, A.S., PENNER, D. : Butylate and carbofuran interaction in barley and corn. - Weed Sci. *21* : 339 - 342, 1973. [Ps.]

16386 - HAMMER, U.T. : Eutrophication and its alleviation in the Upper Qu'Appelle River system, Saskatchewan. - In : Proceedings of the Symposium on the Lakes of Western Canada. Pp. 352 - 369. Univ. Alberta, Water Resources Centre, Edmonton 1973. [Primary production.]

16387 - HAMMER, U.T., WALKER, K.F., WILLIAMS, W.D. : Derivation of daily phytoplankton production estimates from short-term experiments in some shallow, eutrophic Australian saline lakes. - Aust. J. mar. Freshwater Res. *24* : 259 - 266, 1973.

16388 - HAMPP, R., ZIEGLER, H., ZIEGLER, I. : Die Wirkung von Bleiionen auf die $^{14}CO_2$- -Fixierung und die ATP-Bildung von Spinatchloroplasten. - Biochem. Physiol. Pflanzen *164* : 126 - 134, 1973.

16389   HAMPP, R., ZIEGLER, R., ZIEGLER, I. : Der Einfluß von Bleiionen auf Enzyme des reduktiven Pentosephosphatcyclus. - Biochem. Physiol. Pflanzen *164* : 588 - 595, 1973.

16390 - HAMPTON, R.E. : Photosynthetic pigments in *Peltigera canina* (L.) WILLD. from sun and shade habitats. - Bryologist *76* : 543 - 545, 1973.

16391 - HAN, T.-W., ELEY, J.H. : Glycolate excretion by *Anacystis nidulans* : effect of HCO_3^- concentration, oxygen concentration and light intensity. - Plant Cell Physiol. *14* : 285 - 291, 1973.

16392 - HAND, D.W. : A null balance method for measuring crop photosynthesis in an airtight daylit controlled-environment cabinet. - Agr. Meteorol. *12* : 259 - 270, 1973.

16393 - HANEBUTH, W.F., RASCHKE, K. : Stomatal aperture in *Zea mays* controlled by light or CO_2 ? - Plant Physiol. *51* (Suppl.) : 9, 1973.

16394 - HANNAH, R.P., SIMMONS, A.T., MOSHIRI, G.A. : Nutrient-productivity relationships in a bayou estuary. - J. Water Pollut. Control Fed. *45* : 2508 - 2520, 1973. [Ps.]

16395 - HANNAN, P.J., THOMPSON, N.P., PATOUILLET, C., WILKNISS, P.E. : ^{203}Hg as tracer in pigments of *Phaeodactylum tricornutum*. - Int. J. appl. Rad. Isotopes *24* : 665 - 670, 1973.

16396 - HANOWER, P., BRZOZOWSKA, J. : Influence d'un choc hydrique sur l'activité de la phosphatase acide chez le Cotonnier (*Gossypium*). - Physiol. vég. *11* : 385 - - 394, 1973. [Chloroplast.]

16397 - HAN SAN KU, McCURRY, S.D., TOLBERT, N.E. : Phosphogluconate phosphatase from bean leaves. - Plant Physiol. *51* (Suppl.) : 41, 1973.

16398 - HANSEN, P., GRAUSLUND, J. : ^{14}C-studies on apple trees. VIII. The seasonal variation and nature of reserves. - Physiol. Plant. *28* : 24 - 32, 1973. [Ps.]

16399 - HANSEN, T.A., VELDKAMP, H. : *Rhodopseudomonas sulfidophila*, nov. spec., a new species of the purple nonsulfur bacteria. - Arch. Mikrobiol. *92* : 45 - 58, 1973. [Chl, Car.]

16400 - HANSEN, W.R. : Net photosynthesis and evapotranspiration of field-grown soybean canopies. - Diss. Abstr. Int. B *33* : 4619-B - 4620-B, 1973.

16401 - HANSMANN, E. : Pigment analysis. - In : STEIN, J.R. (ed.) : Handbook of Phycological Methods. Culture Methods and Growth Measurements. Pp. 359 - 368. Cambridge Univ. Press, London 1973.

*16402 - HANSMANN, E.W., LANE, C.B., HALL, J.D. : A direct method of measuring benthic primary production in streams. - Limnol. Oceanogr. *16* : 822 - 826, 1971.

16403 - HANSON, W.D. : Changes in efficiencies and numbers of chloroplasts associated with divergent selections for juvenile productivity in *Zea mays* (L.). - Crop Sci. *13* : 386 - 387, 1973. [Ps, Chl.]

16404 - HANSON, W.D., GRIER, R.E. : Rates of electron transfer and of non-cyclic photophosphorylation for chloroplasts isolated from maize populations selected for differences in juvenile productivity and in leaf widths. - Genetics *75* : 247 - 257, 1973.

16405 - **HARDT, H., MALKIN, S.** : Oscillations of the triggered luminescences of isolated chloroplasts preilluminated by short flashes. - Photochem. Photobiol. *17*: 433 - 440, 1973.

16406 - **HARDWICK, K., BAKER, N.R.** : *In vivo* measurement of chlorophyll content of leaves. - New Phytol. *72* : 51 - 54, 1973.

16407 - **HARDY, J.T.** : Phytoneuston ecology of a temperate marine lagoon. - Limnol. Oceanogr. *18* : 525 - 533, 1973. [Ps, Chl.]

16408 - **HARDY, R.W.F., HAVELKA, U.D.** : Symbiotic N_2 fixation : multifold enhancement by CO_2-enrichment of field-grown soybeans. - Plant Physiol. *51* (Suppl.) : 35, 1973. [Ps.]

16409 - **HARGRAVE, B.T.** : Coupling carbon flow through some pelagic and benthic communities. - J. Fish. Res. Board Can. *30* : 1317 - 1326, 1973. [Primary production.]

16410 - **HARI, P., LUUKKANEN, O.** : Effect of water stress, temperature, and light on photosynthesis in alder seedlings. - Physiol. Plant. *29* : 45 - 53, 1973.

16411 - **HARNISCHFEGER, G.** : Chloroplast degradation in ageing cotyledons of pumpkin. - J. exp. Bot. *24* : 1236 - 1246, 1973.

16412 - **HARPER, L.A., BAKER, D.N., BOX, J.E. Jr., HESKETH, J.D.** : Carbon dioxide and the photosynthesis of field crops : a metered carbon dioxide release in cotton under field conditions. - Agron. J. *65* : 7 - 11, 1973.

16413 - **HARPER, L.A., BOX, J.E. Jr., BAKER, D.N., HESKETH, J.D.** : Carbon dioxide and the photosynthesis of field crops. A tracer examination of turbulent transfer theory. - Agron. J. *65* : 574 - 578, 1973.

16414 - **HARRIS, D.G.** : Photosynthesis, diffusion resistance and relative plant water content of cotton as influenced by induced water stress. - Crop Sci. *13* : 570 - 572, 1973.

16415 - **HARRIS, E.H., PRESTON, J.F., EISENSTADT, J.M.** : Amino acid incorporation and products of protein synthesis in isolated chloroplasts of *Euglena gracilis*. - Biochemistry *12* : 1227 - 1234, 1973.

16416 - **HARRIS, G.P.** : Diel and annual cycles of net plankton photosynthesis in Lake Ontario. - J. Fish. Res.Board Can. *30* : 1779 - 1787, 1973.

16417 - **HARRIS, G.P., LOTT, J.N.A.** : Light intensity and photosynthetic rates in phytoplankton. - J. Fish. Res. Board Can. *30* : 1771 - 1778, 1973.

16418 - **HARRIS, J.B., ARNOTT, H.J.** : Effects of senescence on chloroplasts of the tobacco leaf. - Tissue Cell *5* : 527 - 544, 1973.

*16419 - **HARRIS, N., DODGE, A.D.** : The effect of paraquat on flax cotyledon leaves : changes in fine structure. - Planta *104* : 201 - 209, 1972. [Chloroplast.]

16420 - **HARTMAN, F.C., WELCH, M.H., NORTON, I.L.** : 3-bromo-2-butanone 1,4-bisphosphate as an affinity label for ribulosebisphosphate carboxylase. - Proc. nat. Acad. Sci. USA *70* : 3721 - 3724, 1973.

16421 - **HARTMANN, E., GEISSLER, G.** : Über den Ureidstoffwechsel beim Laubmoosprotonema von *Funaria hygrometrica* L. (SIBTH.). I. Der Einfluß von verschiedenfarbigem Licht. - Biochem. Physiol. Pflanzen *164* : 614 - 623, 1973. [Chl.]

16422 - **HARTMANN, K.M., COHNEN UNSER, I.** : Carotenoids and flavins versus phytochrome as the controlling pigment for blue-UV-mediated photoresponses. - Z. Pflanzenphysiol. *69* : 109 - 124, 1973.

16423 - **HARTT, C.E.** : Mechanism of Translocation in Sugarcane. - Harold L. Lyon Arboretum Lecture No. 4. 40 pp. Univ. Hawaii, Honolulu 1973. [Ps.]

16424 - **HARVEY, D.M.** : The translocation of ^{14}C-photosynthate in *Pisum sativum* L. - Ann. Bot. *37* : 787 - 794, 1973.

16425 - **HASLETT, B.G., CAMMACK, R., WHATLEY, F.R.** : Quantitative studies on ferredoxin in greening bean leaves. - Biochem. J. *136* : 697 - 703, 1973.

16426 - **HASPEL-HORVATOVIČ, E., FRIČ, F., HORIČKOVÁ, B.** : Verkürzte Methode zur Isolierung radiochemisch reinen Chlorophylls. - Experientia *29* : 507 - 508, 1973.

16427 - **HASPELOVÁ-HORVATOVIČOVÁ, A.** : Pigmentveränderungen von Weizen nach Mehltauinfektion (*Triticum sativum* L. - *Erysiphe graminis* f. sp. *tritici* MARCHAL). - Biológia (Bratislava) *28* : 247 - 252, 1973.

16428 - **HASSELT, P.R. VAN** : Photo-oxidative damage to triphenyltetrazoliumchloride (TTC) reducing capacity of *Cucumis* leaf discs during chilling. - Acta bot. neer. *22* : 546 - 552, 1973. [Chl, Car.]

16429 - **HASTINGS, J.W.** : Rhythms in dinoflagellates. - In : PÉREZ-MIRAVETE, A. (ed.) : Behavior of Microorganisms. Pp. 267 - 281. Plenum Press, London 1973. [Ps.]

16430 - **HATCH, M.D.** : An assay for PEP carbokinase in crude tissue extracts. - Anal. Biochem. *52* : 280 - 285, 1973.

16431 - **HATCH, M.D.** : Separation and properties of leaf aspartate aminotransferase and alanine aminotransferase isoenzymes operative in the C_4 pathway of photosynthesis. - Arch. Biochem. Biophys. *156* : 207 - 214, 1973.

16432 - **HATCH, M.D., BOARDMAN, N.K.** : Biochemistry of photosynthesis. - In : BUTLER, G.W., BAILEY, R.W. (ed.) : Chemistry and Biochemistry of Herbage. Vol. 2. Pp. 25 - 55. Academic Press, London - New York 1973.

16433 - **HATCH, M.D., KAGAWA, T.** : Enzymes and functional capacities of mesophyll chloroplasts from plants with C_4-pathway photosynthesis. - Arch. Biochem. Biophys. *159* : 842 - 853, 1973.

16434 - **HATCH, M.D., MAU, S.-L.** : Activity, location, and role of aspartate aminotransferase and alanine aminotransferase isoenzymes in leaves with C_4 pathway photosynthesis. - Arch. Biochem. Biophys. *156* : 195 - 206, 1973.

*16435 - **HAUPT, W.** : Perception of light direction in oriented displacements of cell organelles. - Acta protozool. *11* : 179 - 188, 1972. [Chloroplast.]

16436 - **HAUPT, W.** : Regulation der Chloroplastenverteilung in der Zelle durch Lichtintensität und Lichtrichtung. - Ber. deut. bot. Ges. *86* : 403 - 406, 1973.

16437 - **HAUPT, W.** : Role of light in chloroplast movement. - BioScience *23* : 289 - 296, 1973.

*16438 - **HAUSKA, G.** : Die Entkopplung der Photophosphorylierung durch Ammoniumchlorid. - Forschungsber. Ruhr-Univ. Bochum *1972* : 343, 1972.

16439 - **HAUSKA, G., TREBST, A., DRABER, W.** : Lipophilicity and catalysis of photophosphorylation. II. Quinoid compounds as artificial carriers in cyclic photophosphorylation and photoreductions by Photosystem I. - Biochim. biophys. Acta *305* : 632 - 641, 1973.

16440 - **HAUSKA, G.A.** : Die Orientierung des Reaktionszentrums vom Photosystem I in der Chloroplastenmembran. - Hoppe-Seyler's Z. physiol. Chem. *354* : 1200, 1973.

16441 - **HAWCROFT, D.M., SHORT, K.C.** : Some experiments on the light reaction of photosynthesis. - J. biol. Educ. *7* (5) : 23 - 28, 1973.

16442 - **HEAL, O.W.** : Primary production studies at Moor House (U.K.). - In : BLISS, L.C., WIELGOLASKI, F.E. (ed.) : Primary Production and Production Processes, Tundra Biome. Pp. 137 - 139. Tundra Biome Steering Comm., Edmonton 1973.

16443 - **HEATH, R.L.** : Ethyl red as a probe into the mechanism of light-driven proton translocation by isolated chloroplasts. I. The spectral shift of ethyl red and membrane conformational changes. - Biochim. biophys. Acta *292* : 444 - 458, 1973.

16444 - **HEATH, R.L.** : The energy state and structure of isolated chloroplasts : the oxidative reactions involving the water-splitting step of photosynthesis. - Int. Rev. Cytol. *34* : 49 - 101, 1973.

16445 - **HEATH, R.L., COULSON, C.L., CHIMIKLIS, P.** : The coupling of the Coulter Counter to a pulse height analyzer for a rapid monitor of size distributions of cells and organelles. - Anal. Biochem. *53* : 555 - 563, 1973. [Chloroplasts.]

16446 - **HEBER, U.** : Elektronentransport zum Sauerstoff- und ATP-Verbrauch in der Photosynthese. - Ber. deut. bot. Ges. *86* : 187 - 195, 1973.

16447 - HEBER, U. : Stoichiometry of reduction and phosphorylation during illumination of intact chloroplasts. - Biochim. biophys. Acta *305* : 140 - 152, 1973.

16448 - HEBER, U., TYANKOVA, L., SANTARIUS, K.A. : Effects of freezing on biological membranes *in vivo* and *in vitro*. - Biochim. biophys. Acta *291* : 23 - 37, 1973. [Ps.]

16449 - HEICHEL, G.H. : Screening for slow photorespiration in *Nicotiana tabacum* L. - Plant Physiol. *51* (Suppl.) : 42, 1973.

16450 - HEICHEL, G.H., TURNER, N.C. : Physiological responses of dogwood (*Cornus florida*) to infestation by the dogwood borer (*Thamnosphecia scitula*). - Ann. appl. Biol. *75* : 401 - 408, 1973. [Ps.]

16451 - HEISE, K.-P., JACOBI, G. : The correlation of lipid release and photochemical activities in isolated spinach chloroplasts. - Z. Naturforsch. *28c* : 120 - 127, 1973.

16452 - HEISE, K.-P., JACOBI, G. : Vergleichende Untersuchungen über die Lipidzusammensetzung von Etioplasten und Chloroplasten aus *Pisum* sowie von Chloroplasten aus einer Mutante von *Nicotiana*. - Planta *111* : 137 - 148, 1973.

B16453 - HEJNÝ, S. (ed.) : Ecosystem Study on Wetland Biome in Czechoslovakia. - Czechosl. IBP/PT-PP Rep. No. 3. Czechosl. Acad. Sci., Třeboň 1973. [Ps.]

16454 - HELDT, H.W., WERDAN, K. : pH Änderungen im Chloroplasten-Stroma, erzeugt durch energieabhängigen Protonentransport über die Thylakoidmembran. - Ber. deut. bot. Ges. *86* : 203 - 208, 1973.

16455 - HELDT, H.W., WERDAN, K. : The pH gradient between the thylakoid space and the stroma of intact spinach chloroplasts in correlation to photophosphorylation. - In : Ninth International Congress of Biochemistry. Stockholm, 1 - 7 July 1973. Abstract Book. P. 232. IUB, Stockholm 1973.

16456 - HELDT, H.W., WERDAN, K., MILOVANCEV, M., GELLER, G. : Alkalization of the chloroplast stroma caused by light-dependent proton flux into the thylakoid space. - Biochim. biophys. Acta *314* : 224 - 241, 1973.

*16457 - HENDRICH, W. : Reakcje fotoredukcji chlorofilu *a* i związków pokrewnych sterowane aminami jako modele reakcji biologicznych. [Reactions of photoreduction of chlorophyll *a* and related compounds controlled by amines as models of biological reactions.] - Monogr. biochem. *23* : 1 - 72, 1971. [In Pol.]

16458 - HENNINGSEN, K.W., BOARDMAN, N.K. : Development of photochemical activity and the appearance of the high potential form of cytochrome *b*-559 in greening barley seedlings. - Plant Physiol. *51* : 1117 - 1126, 1973.

16459 - HENNINGSEN, K.W., BOYNTON, J.E., von WETTSTEIN, D., BOARDMAN, N.K. : Nuclear genes controlling chloroplast development in barley. - In : POLLAK, J.K., LEE, J.W. (ed.) : The Biochemistry of Gene Expression in Higher Organisms. Pp. 457- - 478. Aust. & New Zeal. Book Co., Sydney 1973.

16460 - HENNINGSEN, K.W., KAHN, A., HOUSSIER, C. : Circular dichroism of protochlorophyllide and chlorophyllide holochrome subunits. - FEBS Lett. *37* : 103 - 108, 1973.

16461 - HEPBURN, A.G. : The effect of shade on the photosynthetic pigments of needles on trees of *Pinus sylvestris* ssp. *scotica* grown under field conditions. - Trans. bot. Soc. Edinburgh *41* : 461 - 467, 1973.

16462 - HERATH, H.M.W., ORMROD, D.P. : Temperature and sulphur nutrition effects on the CO_2 assimilation rates of barley, peas and rape. - Plant Soil *38* : 525 - 530, 1973.

16463 - HERMSEN, J.G.T., RAMANNA, M.S., VOGEL, J. : The location of a recessive gene for chlorophyll deficiency in diploid *Solanum tuberosum* by means of trisomic analysis. - Can. J. Genet. Cytol. *15* : 807 - 813, 1973.

*16464 - HERMSEN, J.G.T., WANINGE, J. : Attempts to localize the gene Ch_1 for hybrid chlorosis in wheat. - Euphytica *21* : 204 - 208, 1972.

16465 - HERNÁNDEZ-GIL, R., SCHAEDLE, M. : Functional and structural changes in senescing *Populus deltoides* (BARTR.) chloroplasts. - Plant Physiol. *51* : 245 - 249, 1973.

16466 - **HERNÁNDEZ MÉNDEZ, M.T.** : Estudio quimico de la *Avena sativa* L: Evolucion de rendimiento (sustancia seca), clorofilas *a* y *b*, y sustancias minerales. [Chemical study on oats : Development of yield (dry matter), chlorophylls *a* and *b*, and mineral substances.] - In : Tesis de Ciencias 1971-72 (Acta Salmanticensia, Ciencias 45). Pp. 1 - 28. Salamanca 1973. [In Span.]

16467 - **HERODEK, S., OLÁH, J.** : Primary production in the frozen Lake Balaton. - Ann. Inst. Biol. (Tihany) *40* : 197 - 206, 1973. [Chl.]

16468 - **HERODEK, S., TAMÁS, G.** : The primary production of phytoplankton in Lake Balaton April - September 1972. - Ann. Inst. Biol. (Tihany) *40* : 207 - 218, 1973.

*16469 - **HERRMANN, R.G.** : Chloroplastengröße und inkorporierte ^{3}H-Thymidinmenge. Autoradiographische Studien zur Frage : Gibt es genetisch mehrwertige Chloroplasten ? - Ber. deut. bot. Ges. *81* : 332, 1968.

16470 - **HEWSON, R.T., ROBERTS, H.A.** : Some effects of weed competition on the growth of onions. - J. hort. Sci. *48* : 51 - 57, 1973. [Ps, Chl.]

16471 - **HEWSON, R.T., ROBERTS, H.A.** : Effects of weed competition for different periods on the growth and yield of red beet. - J. hort. Sci. *48* : 281 - 292, 1973. [Chl.]

16472 - **HICKMAN, M.** : The standing crop and primary productivity of the phytoplankton of Abbot's Pond, north Somerset. - J. Ecol. *61* : 269 - 287, 1973.

16473 - **HIGGINBOTHAM, K.O., STRAIN, B.R.** : The influence of canopy position on net photosynthesis in loblolly pine. - Amer. J. Bot. *60* (4, Suppl.) : 24, 1973.

16474 - **HIGUCHI, M.** : Inhibition by adenine compounds of the induced formation of photosynthetic pigments in *Rhodopseudomonas spheroides* cells under dark-semiaerobic conditions. - Plant Cell Physiol. *14* : 51 - 60, 1973.

16475 - **HILLER, R.G., GENGE, S., PILGER, D.** : Relationship between chlorophyll *b*, pigment-protein complex II and Hill activity in chloroplasts. - Proc. aust. biochem. Soc. *6* : 50, 1973.

16476 - **HILLER, R.G., PILGER, D., GENGE, S.** : Photosystem II activity and pigment-protein complexes in flashed bean leaves. - Plant Sci. Lett. *1* : 81 - 88, 1973.

16477 - **HINCHMAN, R.R.** : A permanent iodine stain-mountant combination for starch in plant tissues. - Stain Technol. *48* : 344 - 346, 1973.

16478 - **HINCKLEY, T.M.** : Responses of black locust and tomato plants after water stress. - HortScience *8* : 405 - 407, 1973. [Ps.]

16479 - **HIND, G., McCARTY, R.E.** : The role of cation fluxes in chloroplast activity. - In : GIESE, A.C. (ed.) : Photophysiology. Current Topics in Photobiology and Photochemistry. Vol. 8. Pp. 113 - 156. Academic Press, New York - London 1973.

*16480 - **HIPKE, H.** : Der Pigmentgehalt induzierter Mutanten von *Pisum sativum* als Grundlage einer neuen Farbcharakterisierung und seine Beziehung zur Plastidengröße. - Theor. appl. Genet. *40* : 341 - 344, 1970.

16481 - **HIRANPRADIT, H., FOY, C.L.** : Retardation of leaf senescence in maize by subtoxic levels of bromacil, fluometuron, and atrazine. - Bot. Gaz. *134* : 26 - - 31, 1973. [Ps, Chl.]

*16482 - **HIROSE, S., YAMASHITA, K., SHIBATA, K.** : Formation of thiol groups in spinach chloroplasts by illumination. - Plant Cell Physiol. *12* : 775 - 778, 1971.

16483 - **HIRTZ, R.-D., MENKE, W.** : Not freezing water in the lamellar system of chloroplasts. - Z. Naturforsch. *28c* : 230, 1973.

16484 - **HISATAKE, M., NAKANISHI, M.** : On the changes of chlorophylls and qualities of pickled cucumbers during storage. - J. Food Sci. Technol. (Tokyo) *20* : 429 - - 431, 1973.

16485 - **HIYAMA, T.** : A computer resolution of complex kinetics. - Carnegie Inst. Year Book *72* : 371 - 373, 1973. [Ps.]

16486 - **HIYAMA, T.** : Electron transport in green plant photosynthesis. - What's new Plant Physiol. *5* (7) : 1 - 5, 1973.

16487 - **HIYAMA, T., FORK, D.C.** : Some recent improvements of a high-speed difference spectrophotometer. - Carnegie Inst. Year Book *72* : 368 - 371, 1973. [Ps.]

16488 - **HOARAU, J., LECLERC, J.C.** : Low temperature absorption spectrum studies : light-induced pigment changes in *Porphyridium* cultures. - Photochem. Photobiol. *17* : 403 - 412, 1973.

16489 - **HOBSON, L.A.** : The effects of interactions of light intensity, daylength, and temperature on division rates of three species of marine unicellular algae. - J. Phycol. *9* (Suppl.) : 4 - 5, 1973. [Ps.]

16490 - **HOBSON, L.A., MENZEL, D.W., BARBER, R.T.** : Primary productivity and sizes of pools of organic carbon in the mixed layer of the ocean. - Mar. Biol. *19* : 298 - 306, 1973.

16491 - **HOCHAPFEL, A., LECOIN, D., VIOVY, R.** : Orientation de colorants dans un cristal liquide nématique. - Compt. rend. Acad. Sci. Paris, Sér. C *276* : 221 - 224, 1973. [Chl.]

16492 - **HOCHMAN, A., CARMELI, C.** : ATPase and ATP-P$_i$ exchange activities in *Chromatium* strain D chromatophores. - Photosynthetica *7* : 238 - 245, 1973.

16493 - **HODÁŇOVÁ, D.** : Structure and development of sugar beet canopy. III. Chlorophyll characteristics. - Photosynthetica *7* : 338 - 344, 1973.

16494 - **HODKINSON, I.D.** : The population dynamics and host plant interactions of *Strophingia ericae* (CURT.) (*Homoptera : Psylloidea*). - J. anim. Ecol. *42* : 565 - - 583, 1973. [Ps.]

16495 - **HOFFMANN, G.** : Periodik des Substanzmassenzuwachses von Wurzeln und Sproßorganen junger Forstgehölze. - Flora *162* : 126 - 133, 1973.

16496 - **HOFFMANN, G., LYR, H.** : Charakterisierung des Wachstumsverhaltens von Pflanzen durch Wachstumsschemata. - Flora *162* : 81 - 98, 1973. [Ps.]

16497 - **HOFFMANN, P.** : Vergleichende pigmentphysiologische Untersuchungen an CCC-behandelten Weizenkeimpflanzen. - Photosynthetica *7* : 213 - 225, 1973.

16498 - **HOLDEN, M.** : Chloroplast pigments in plants with the C_4-dicarboxylic acid pathway of photosynthesis. - Photosynthetica *7* : 41 - 49, 1973.

*16499 - **HOLE, C.C., DODGE, A.D.** : The interaction of a gibberellin and a kinin in the control of chlorophyll synthesis. - Experientia *28* : 1113, 1972.

16500 - **HOLLISTER, T.A., WALSH, G.E.** : Differential responses of marine phytoplankton to herbicides : oxygen evolution. - Bull. environ. Contamination Toxicol. *9* : 291 - 295, 1973. [Ps.]

16501 - **HOLMES, D.P.** : Effects of defoliation on chlorophyll loss in senescing wheat inflorescence and on grain maturation. - Can. J. Plant Sci. *53* : 499 - 500, 1973.

16502 - **HOLUB, Z.** : Veränderungen der Photosyntheseintensität von Weizenblättern unter dem Einfluss von Natriumfluorid. - Biológia (Bratislava) *28* : 253 - 262, 1973.

16503 - **HOMANN, P.H.** : Inhibition on the reducing side of photosystem II by carbonylcyanide *m*-chlorophenylhydrazone and lithium 3,5-diiodosalicylate. - Europe. J. Biochem. *33* : 247 - 252, 1973.

16504 - **HOMÈS, J.** : Modifications ultrastructurales des chloroplastes de protocormes d'Orchidée cultivés *in vitro* en présence de saccharose. - J. Microscop. (Paris) *17* : 66a, 1973.

16505 - **HOMÈS, J., VANSÉVEREN-VAN ESPEN, N.** : : Quelques formes de plastes induites par le milieu de culture dans des protocormes d'Orchidées cultivés *in vitro*. - Bull. Soc. roy. Bot. Belg. *106* : 117 - 121, 1973.

16506 - **HONEYCUTT, R.C., MARGULIES, M.M.** : Protein synthesis in *Chlamydomonas reinhardi*. Evidence for synthesis of proteins of chloroplastic ribosomes on cytoplasmic ribosomes. - J. biol. Chem. *248* : 6145 - 6153, 1973.

16507 - **HOOBER, J.K., STEGEMAN, W.J.** : Control of the synthesis of a major polypeptide of chloroplast membranes in *Chlamydomonas reinhardi*. - J. Cell Biol. *56* : 1 - 12, 1973. [Chl.]

16508 - **HOPE, A.B.** : Physics and photosynthesis. - Australas. Bull. Med. Phys. Biophys. *57* : 3 - 22, 1973.

16509 - HORAK, A., ZALIK, S. : Development of photosystem I and photosystem II activities in a virescent mutant of barley. - Plant Physiol. *51* (Suppl.) : 66, 1973.

16510 - HORI, T. : Comparative studies of pyrenoid ultrastructure in algae of the *Monostroma* complex. - J. Phycol. *9* : 190 - 199, 1973.

16511 - HORIE, T. : Photosynthesis and growth of plant canopy in relation to solar radiation climate. - In : The Sun in the Service of Mankind. Pp. V.9-1 - V.9-10. Unesco, Paris 1973.

16512 - HORIO, T., YAMASHITA, J., NISHIKAWA, K., KAKUNO, T., HOSOI, K., SUZUKI, J., YOSHIMURA, S. : Systems for hydrolysis of ATP and pyrophosphate in chromatophores from *Rhodospirillum rubrum*. - In : MAKOTO, N., PACKER, L. (ed.) : Organization of Energy-Transducing Membranes. Pp. 239 - 249. Univ. Tokyo Press, Tokyo 1973.

16513 - HORVÁTH, I., SZALAY, L., SZÁSZ, K., RAAFAT, A. : The effect of the spectral composition of light on the metabolism. Utilization of light and the chlorophylls of *Sinapis alba*. - Acta biochim. biophys. Acad. Sci. hung. *8* : 161 - - 169, 1973.

16514 - HORVÁTH, M., NAGY, Gy., ROJIK, I. : Investigation into the 2,4-D effect on some metabolism indices in *Vicia faba* seedlings. - Acta bot. Acad. Sci. hung. *18* : 131 - 133, 1973. [Chl, Car.]

16515 - HOSOI, K., YOSHIMURA, S., SOE, G., KAKUNO, T., HORIO, T. : Competition between P_i and pH indicators in photosynthetic ATP formation in chromatophores of *Rhodospirillum rubrum*. - J. Biochem.(Tokyo) *74* : 1275 - 1278, 1973.

16516 - HOSTETTER, H.P. : Toxic effects of copper on algae determined by polarography. - J. Phycol. *9* (Suppl.) : 13, 1973.

16517 - HOUBA, V.J.G. : Effect of Nitrogen Dressings on Growth and Development of Sugar-Beet. - Agr. Res. Rep. (Wageningen) *791* : 1 - 65, 1973. [Growth analysis.]

16518 - HOVIOUS, J.C., CONWAY, R.A., GANZE, C.W. : Anaerobic lagoon pretreatment of petrochemical wastes. - J. Water Pollut. Control Fed. *45* : 71 - 84, 1973. [Ps.]

16519 - HOWARD, A. : Dependence of radiation sensitivity on oxygen tension in *Oedogonium* - I. Extra- and intracellular oxygen. - Rad. Bot. *13* : 293 - 296, 1973. [Ps.]

16520 - HOWARD, T.M. : Studies in the ecology of *Nothofagus cunninghamii* OERST. III. Two limiting factors : Light intensity and water stress. - Aust. J. Bot. *21* : 93 - 102, 1973. [Ps.]

16521 - HOWELL, R.K., KREMER, D.F. : The chemistry and physiology of pigmentation in leaves injured by air pollution. - J. environ. Qual. *2* : 434 - 438, 1973. [Chl.]

16522 - HOWES, C.D., STERN, A.I. : Photophosphorylation during chloroplast development in Red Kidney bean. II. Photophosphorylation and photoreduction appear concomitantly but initially are uncoupled. - Plant Physiol. *51* : 386 - 390, 1973.

16523 - HOZYO, Y. : [The callus formation on tissue explant derived from tuberous roots of sweet potato plants, *Ipomoea batatas* POIRET.] - Bull. nat. Inst. agr. Sci., Ser. D *24* : 1 - 33, 1973. [Ps; in Jap., ab : E.]

16524 - HOZYO, Y., KATO, S. : [The plant production of wild type plants in *Ipomoea trifida* (H.N.B.) DON.] - Bull. nat. Inst. agr. Sci., Ser. D *24* : 35 - 60, 1973. [In Jap., ab : E.]

16525 - HRADECKÁ, D., KVĚT, J. : Morphological and production characteristics of three clones of *Phragmites communis* TRIN. from the Nesyt area. - In : KVĚT, J. (ed.): Littoral of the Nesyt Fishpond. Studies ČSAV 15. Pp. 97 - 101. Academia, Praha 1973. [Growth analysis.]

16526 - HRADILÍK, J., GAMBURG, K.Z., NAZAREVA, G.D. : Effects of CCC and B-995 on growth and pigmentation of tobacco tissue in a suspension culture. - Biol. Plant. *15* : 223 - 228, 1973. [Car.]

16527 - HSIAO, T.C. : Plant responses to water stress. - Annu. Rev. Plant Physiol. *24:* 519 - 570, 1973.

16528 - HSIAO, T.C., ALLAWAY, W.G., EVANS, L.T. : Action spectra for guard cell Rb$^+$ uptake and stomatal opening in *Vicia faba*. - Plant Physiol. *51* : 82 - 88, 1973. [Ps.]

16529 - HUANG, J.-S., HUANG, P.-Y., GOODMAN, R.N. : Reconstitution of a membrane-like structure with structural proteins and lipids isolated from tobacco thylakoid membranes. - Amer. J. Bot. *60* : 80 - 85, 1973.

16530 - HUANG, J.W., KAPLAN, S. : Membrane proteins of *Rhodopseudomonas spheroides*. IV. Characterization of chromatophore proteins. - Biochim. biophys. Acta *307* : 317 - 331, 1973.

16531 - HUANG, J.W., KAPLAN, S. : Membrane proteins of *Rhodopseudomonas spheroides*. V. Additional chemical characterization of a pigment-lipid-associated protein isolated from chromatophores. - Biochim. biophys. Acta *307* : 332 - 342, 1973.

*16532 - HÜBEL, H. : Die Bestimmung der Primärproduktion des Phytoplanktons der Nord- -Rügenschen Boddengewässer unter Verwendung der Radiokohlenstoffmethode. - Int. Rev. ges. Hydrobiol. *53* : 601 - 633, 1968.

*16533 - HÜBEL, H. : Die Primärproduktion des Phytoplanktons in Brackgewässern unter- schiedlichen Salz- und Nährstoffgehaltes. - Limnologica *7* : 185 - 190, 1969.

16534 - HÜBEL, H. : Die Primärproduktion des Phytoplanktons in den Boddengewässern südlich der Halbinseln Darss und Zingst im Jahre 1972 unter besonderer Berück- sichtigung der Ergebnisse einer synoptischen Aufnahme. - Wiss. Z. Univ. Ros- tock, math.-naturwiss. Reihe *22* : 1101 - 1104, 1973.

16535 - HUBER, S.C., KANAI, R., EDWARDS, G.E. : Characteristics of malate decarboxyla- tion by bundle sheath cells of certain C_4 plants. - Plant Physiol. *51* (Suppl.): 6, 1973.

16536 - HUBER, S.C., KANAI, R., EDWARDS, G.E. : Decarboxylation of malate by isolated bundle-sheath cells of certain plants having the C_4-dicarboxylic acid cycle of photosynthesis. - Planta *113* : 53 - 66, 1973.

16537 - HUBER, W., SANKHLA, N., ZIEGLER, H. : Eco-physiological studies on Indian arid zone plants. I. Photosynthetic characteristics of *Pennisetum typhoides* (Burm. f.) STAPF & HUBBARD and *Lasiurus sindicus* HENR. - Oecologia *13* : 65 - 71, 1973.

16538 - HUDÁK, J. : Influence of boron on the differentiation of proplastids. - Acta Fac. Rerum nat. Univ. Comen., Physiol. Plant. *7* : 59 - 66, 1973.

16539 - HUEBNER, J.S., TIEN, H.T. : Large amplitude photo-voltage transients of bilayer lipid membranes in the presence of chlorophyllin. - J. Bioenerg. *4* : 469 - 478, 1973.

16540 - HUGHES, A.P. : A comparison of the effects of light intensity and duration on *Chrysanthemum morifolium* cv. Bright Golden Anne in controlled environments. I. Growth analysis. - Ann. Bot. *37* : 267 - 274, 1973.

16541 - HUGHES, A.P. : A comparison of the effects of light intensity and duration on *Chrysanthemum morifolium* cv. Bright Golden Anne in controlled environments. II. Ontogenetic changes in respiration. - Ann. Bot. *37* : 275 - 286, 1973. [Ps.]

16542 - HUME, D.J., CRISWELL, J.G. : Distribution and utilization of ^{14}C-labelled as- similates in soybeans. - Crop Sci. *13* : 519 - 524, 1973.

16543 - HUMPHREY, G.F. : Photosynthetic and respiratory rates and phosphorous content of algae grown at different phosphate levels. - Spl. Publ., mar. biol. Ass. India *1973* : 74 - 79, 1973.

16544 - HUMPHREY, G.F. : Photosynthetic and respiratory rates of marine micro-algae. - CSIRO mar. biochem. Unit annu. Rep. *1972-73* : 8 - 11, 1973.

16545 - HUNDING, C. : Diel variation in oxygen production and uptake in a microbenthic littoral community of a nutrient-poor lake. - Oikos *24* : 352 - 360, 1973.

16546 - HUNECK, S., SCHREIBER, K., JÄNICKE, S. : Inhaltsstoffe einiger Laubmoose. - Phytochemistry *12* : 2533 - 2534, 1973. [Car.]

16547 - HUNSIGI, G. : Growth analysis in cotton genotypes grown under different levels of fertility. - Indian J. agr. Sci. *43* : 690 - 693, 1973.

16548 - HUNTER, C.S., PEAT, W.E. : The effect of tomato aspermy virus on photosynthesis in the young tomato plant. - Physiol. Plant Pathol. 3 : 517 - 524, 1973.

*16549 - HURD, L.E., MELLINGER, V.M., WOLF, L.L., McNAUGHTON, S.J. : Stability and diversity at three trophic levels in terrestrial successional ecosystems. - Science 173 : 1134 - 1136, 1971. [Production.]

16550 - HURD, R.G. : Long-day effects on growth and flower initiation of tomato plants in low light. - Ann. appl. Biol. 73 : 221 - 228, 1973. [Ps, Chl.]

16551 - HUSÁK, Š., HEJNÝ, S. : Marginal plant communities of the Nesyt fishpond (South Moravia). - Pol. Arch. Hydrobiol. 20 : 461 - 467, 1973. [Productivity.]

16552 - HOUSSAINY, S.U. : Some difficulties in the determination of photosynthetic pigments in inland waters. - Aust. Soc. Limnol. Bull. 5 : 26 - 28, 1973.

16553 - HUZISIGE, H., YAMAMOTO, Y. : Analysis of photosystem II using particle II preparation. II. Action spectra for Hill activities and fluorescence properties of variously-treated particle II preparation. - Plant Cell Physiol. 14 : 953 - - 963, 1973.

16554 - HYNNINEN, P.H. : Chlorophylls. III. Keto-enol tautomerism of chlorophylls a and b. The nature of chlorophylls a' and b'. - Acta chem. scand. 27 : 1487 - - 1495, 1973.

16555 - HYNNINEN, P.H. : Chlorophylls. IV. Preparation and purification of some derivatives of chlorophylls a and b. - Acta chem. scand. 27 : 1771 - 1780, 1973.

16556 - HYNNINEN, P.H., ASSANDRI, S. : Chlorophylls. II. Allomerization of chlorophylls a and b. - Acta chem. scand. 27 : 1478 - 1486, 1973.

16557 - HYNNINEN, P.H., ELLFOLK, N. : Chlorophylls. I. Separation and isolation of chlorophylls a and b by multiple liquid-liquid partition. - Acta chem. scand. 27 : 1463 - 1477, 1973.

16558 - IBRAGIMOV, V.I., DUBROVIN, V.N., RUBIN, L.B. : Polyarograficheskaya ustanovka dlya issledovaniya vliyaniya sveta na dykhanie fotosinteziruyushchikh bakterii. [Polarographic apparatus for investigation of light influence on the respiration of photosynthetic bacteria.] - Vestn. mosk. Univ., Ser. VI - Biol., Pochvoved. 28 (2) : 54 - 58, 1973. [in R, ab : E.]

16559 - ICHINOSE, N., SASA, T. : Studies on chlorophyllase of *Chlorella protothecoides* III. Purification and catalytic properties. - Plant Cell Physiol. 14 : 1157 - - 1166, 1973.

16560 - IGNATIADES, L. : Studies on the factors affecting the release of organic matter by *Skeletonema costatum* (GREVILLE) CLEVE in field conditions. - J. mar. biol. Ass. UK 53 : 923 - 935, 1973. [Ps, Chl, Car.]

16561 - IGNATIADES, L., FOGG, G.E. : Studies on the factors affecting the release of organic matter by *Skeletonema costatum* (GREVILLE) CLEVE in culture. - J. mar. biol. Ass. UK 53 : 937 - 956, 1973. [Ps, Chl, Car.]

16562 - IKEDA, H., EMOTO, T. : Effect of temperature on vegetative growth in four ecotypes of *Paspalum distichum* L. - Proc. Crop Sci. Soc. Jap. 42 : 131 - 134, 1973. [Growth analysis.]

16563 - IKEGAMI, I., KATOH, S. : Studies on chlorophyll fluorescence in chloroplasts II. Effect of ferricyanide on the induction of fluorescence in the presence of 3-(3,4-dichlorophenyl)-1,1-dimethylurea. - Plant Cell Physiol. 14 : 829 - 836, 1973.

16564 - IKEGAMI, I., KATOH, S. : Studies on chlorophyll fluorescence in chloroplasts III. Effect of artificial electron donors for photosystem 2 on reoxidation of the fluorescence quencher, Q, in spinach chloroplasts. - Plant Cell Physiol. 14 : 837 - 850, 1973.

16565 - IKEHARA, N., NISHIMURA, M. : Anion-dependent changes of energy transfer in chloroplasts I. Effects of Tris salts on electron flow and phosphorylation. - Plant Cell Physiol. 14 : 61 - 75, 1973.

16566 - IKEHARA, N., NISHIMURA, M. : Anion-dependent changes of energy transfer in chloroplasts II. Relationships of the coupling factor to light-induced proton transport and absorbance change at 515 nm. - Plant Cell Physiol. 14 : 77 - 90, 1973.

16567 - ILANI, A., BERNS, D.S. : A theoretical model for electron transport through chlorophyll-containing bileaflet membranes. - Biophysik 9 : 209 - 224, 1973.

16568 - IL'INA, L.P., SIMONOVA, E.I. : Vliyanie preryvistogo sveta raznogo spektral'nogo sostava na zaklyuchitel'nuyu stadiyu biosinteza khlorofilla v étiolirovannykh rasteniyakh yachmenya. [Effect of intermittent light of different spectral composition on the final phase of chlorophyll biosynthesis in etiolated plants.] - In : Voprosy Botaniki, Zoologii i Pochvovedeniya. Vyp. 1. Pp. 29 - - 35. Tomsk 1973. [In R.]

16569 - IL'INA, M.D., BORISOV, A.Yu. : The quantitative estimation of subchloroplast particles enrichment by chlorophyll a of photosystem 2. - In : Ninth International Congress of Biochemistry. Stockholm, 1 - 7 July 1973. Abstract Book. P. 232. Stockholm 1973.

16570 - IL'INYKH, Z.G. : Raspredelenie assimilyatov iz list'ev raznykh yarusov u rastenii ogurtsa. [Translocation of photosynthates from leaves of different insertion in cucumber plants.] - Tr. biol.-pochv. inst., dal'nevost. nauch. Tsentr Akad. Nauk SSSR 20 (123) : 221 - 227, 1973. [In R, ab : E.]

16571 - IMAI, H., FUKUYAMA, M., YAMADA, Y., HARADA, T. : Comparative studies on the photosynthesis of higher plants. III. Differences in response to various factors affecting the photosynthetic rate between C-4 and C-3 plants. - Soil Sci. Plant Nutr. 19 : 61 - 71, 1973.

16572 - IMAI, I., SIEGEL, S.M. : A specific response to toxic cadmium levels in red kidney bean embryos. - Physiol. Plant. 29 : 118 - 120, 1973. [Chl.]

16573 - IMAMALIEV, A.I., KHODZHAEV, A.S., AGZAMOV, A. : Vliyanie defoliantov na fotosinteticheskii apparat rastenii khlopchatnika. [Effect of defoliants on the photosynthetic apparatus of cotton plants.] - Nauch. Tr. tashkent. sel'.-khoz. Inst. 37 (Voprosy Fiziologii, Biokhimii Khlopchatnika i Drugikh Sel'skokhozyaistvennykh Kul'tur) : 72 - 79, 1973. [In R.]

16574 - IMHOFF, C. : Action d'un inhibiteur de la photosynthèse, le 3-(3,4-dichlorophényl)-1,1-diméthylurée (DCMU), sur la floraison de l'Anagallis arvensis. - Compt. rend. Acad. Sci. Paris, Sér. D 276 : 3303 - 3306, 1973.

16575 - IMHOFF, V. : Synthesis of galactosides by chloroplasts isolated from pea leaves. - Hoppe-Seyler's Z. physiol. Chem. 354 : 1550 - 1554, 1973.

16576 - IMPENS, I.I. : Daytime distribution of energy sinks and sources and transfer processes within a sunflower canopy. - In : SLATYER, R.O. (ed.) : Plant Response to Climatic Factors. Pp. 357 - 367. Unesco, Paris 1973.

16577 - INADA, K. : Spectral dependence of growth and development of rice plant. I. Effects of the selective removal of spectral components from white light on the growth of seedlings. - Proc. Crop Sci. Soc. Jap. 42 : 63 - 71, 1973. [Chl, Car.]

16578 - INAYAMA, M., MURAKAMI, T. : [Effect of temperature on the translocation of photosynthates in cucumber plants I. Effect of raising-temperature on the assimilation and distribution of ^{14}C in cucumber seedlings.] - J. Jap. Soc. hort. Sci. 42 : 27 - 34, 1973. [In Jap., ab : E.]

16579 - INGRAM, L.O., Van BAALEN, C., CALDER, J.A. : Role of reduced exogenous organic compounds in the physiology of the blue-green bacteria (algae) : Photoheterotrophic growth of an "autotrophic" blue-green bacterium. - J. Bacteriol. 114 : 701 - 705, 1973.

16580 - INGRAM, L.O., CALDER, J.A., Van BAALEN, C., PLUCKER, F.E., PARKER, P.L. : Role of reduced exogenous organic compounds in the physiology of the blue-green bacteria (algae) : Photoheterotrophic growth of a "heterotrophic" blue-green bacterium. - J. Bacteriol. 114 : 695 - 700, 1973.

16581 - INOUE, A., NAKAMURA, H., ASAKAWA, S. : [Changes of a few constituents in purple laver (red alga Porphyra) after some periods of storage in frozen state.] - J. Fac. Fish. anim. Husb., Hiroshima Univ. 12 : 73 - 82, 1973. [Chl.; in Jap., ab : E.]

16582 - INOUE, Y., OGAWA, T., KAWAI, T., SHIBATA, K. : Analysis of rice mutants by low temperature-derivative spectrophotometry in relation to pigment compositions and photochemical activities. - Physiol. Plant. *29* : 390 - 395, 1973.

16583 - **INOUE, Y., OGAWA, T., SHIBATA, K.** : Light-induced spectral changes of *P*700 in the 800-nm region in *Anacystis* and spinach lamellae. - Biochim. biophys. Acta *305* : 483 - 487, 1973.

16584 - **INOUE, Y., SHIBATA, K.** : Light-induced chloroplast rearrangements and their action spectra as measured by absorption spectrophotometry. - Planta *114* : 341 - 358, 1973.

16585 - **IONESCU, A., MACOVEI, F.** : Data concerning the physiology and productivity of plants from zones with impure atmosphere. - Rev. roum. Biol., Sér. Bot. *18* : 97 - 108, 1973. [Ps, Chl.]

16586 - **IONESCU, A., MOSCALU, T.** : Potenţialul de producţie al plantelor în zonele cu atmosfera impurificată. [Plant yielding potential in the areas with impure atmosphere.] - Probleme agricole *1973* (5) : 27 - 35, 1973. [Chl; in Roum., ab: E, F, R.]

16587 - **IONESCU, A., NEAMU, G.** : Efectele biologice ale poluării şi problema mediului în zona Bîrseşti (Tg.-Jiu). [Biological effects of pollution in Bîrsesti area (Tg.-Jiu).] - Stud. Cercet. Biol., Ser. bot. *25* : 261 - 267, 1973. [Chl; in Roum., ab : F.]

16588 - **IORDANOV, I.** : Sravnitelna kharakteristika na fotosintetichnata aktivnost i razpredelenieto na ^{14}C na nyakoi vidove rasteniya, otgledani v oranzheriĭni usloviya. [Comparative characteristics of photosynthetic activity and ^{14}C distribution in some plant species cultivated under greenhouse conditions.] - Izv. Inst. Fiziol. Rast. Metodiĭ Popov b"lg. Akad. Nauk *18* : 205 - 215, 1973. [In Bulg., ab : E, R.]

16589 - **IORDANOV, I., CHICHEV, P.** : Vliyanie na khloramfenikola v"rkhu s"d"rzhanieto na plastidnite pigmenti, intenzivnostta na fotosintezata i razpredelenieto na C^{14} v lista s razlichno fiziologichno s"stoyanie. [Effect of chloramphenicol on the content of plastid pigments, photosynthetic rate and distribution of ^{14}C in leaves of different physiological states.] - In : Vtora Natsionalna Konferentsiya po Botanika 1969. Pp. 475 - 482. B"lg. Akad. Nauk., Inst. Bot., Sofia 1973. [In Bulg.]

16590 - **IRSCHIK, H., OELZE, J.** : Membrane differentiation in phototrophically growing *Rhodospirillum rubrum* during transition from low to high light intensity. - Biochim. biophys. Acta *330* : 80 - 89, 1973.

16591 - **ISEBRANDS, J.G., LARSON, P.R.** : Anatomical changes during leaf ontogeny in *Populus deltoides*. - Amer. J. Bot. *60* : 199 - 208, 1973. [Photosynthates.]

16592 - **ISHAG, H.M.** : Physiology of seed yield in field beans (*Vicia faba* L.). II. Dry-matter production. - J. agr. Sci. *80* : 191 - 199, 1973. [Growth analysis.]

16593 - **ISHIKURA, N.** : The changes in anthocyanin and chlorophyll content during the autumnal reddening of leaves. - Kumamoto J. Sci., Biol. *11* (2) : 43 - 50, 1973.

16594 - **ISHIURA, M., IWASA, K.** : Gametogesis in *Chlamydomonas* I. Effect of light on the induction of sexuality. - Plant Cell Physiol. *14* : 911 - 921, 1973. [Ps.]

16595 - **ISOBE, S.** : Statistics of sunflecks and its application to light flickering problems in plant communities. - In : The Sun in the Service of Mankind. Pp. V.10-1 - V.10-9. Unesco, Paris 1973.

16596 - **ITAI, C., BEN-ZIONI, A., ORDIN, L.** : Correlative changes in endogenous hormone levels and shoot growth induced by short heat treatments to the root. - Physiol. Plant. *29* : 355 - 360, 1973. [Ps.]

16597 - **ITO, A., UDAGAWA, T., UCHIJIMA, Z.** : [Phytometrical studies of crop canopies. II. Canopy structure of rice crops in relation to varieties and growing stage.] - Proc. Crop Sci. Soc. Jap. *42* : 334 - 342, 1973. [In Jap., ab : E.]

16598 - **ITO, T.** : [Plant growth and physiology of vegetable plants as influenced by carbon dioxide environment.] - Trans. Fac. Hort., Chiba Univ. *1973* (7) : 1 - - 134, 1973. [Ps; in Jap., ab : E.]

16599 - **ITOH, S., MURATA, N.** : Correlation between delayed light emission and fluores-

cence of chlorophyll *a* in system II particles derived from spinach chloroplasts.
- Photochem. Photobiol. *18* : 209 - 218, 1973.

16600 - **IVANCHENKO, V.M., KRUCHININA, S.S., MARSHAKOVA, M.I., URBANOVICH, T.A., LEGEN-
CHENKO, B.I., MIKUL'SKAYA, S.A., DOROZHKINA, L.N.** : O mekhanizme vliyaniya vodnogo rezhima assimilyatsionnoǐ tkani na intensivnost' fotosinteza. [Mechanism
of influence of water relations of the assimilatory tissue on the photosynthetic rate.] - In : VECHER, A.S. (ed.) : Fotosintez i Ustoǐchivost' Rasteniǐ.
Pp. 12 - 28. Nauka i Tekhnika, Minsk 1973. [In R.]

16601 - **IVANCHENKO, V.M., LEGENCHENKO, B.I.** : Mnogomernyǐ regressionnyǐ analiz zavisimosti intensivnosti fotosinteza rasteniǐ ot vodnogo defitsita assimilyatsionnoǐ tkani. [Multiplex regressive analysis of the dependence of photosynthetic
rate on water deficit in assimilatory tissue.] - Sel'skokhoz. Biol. *8* : 662 -
- 665, 1973. [In R, ab : E.]

16602 - **IVANCHENKO, V.M., LEGENCHENKO, B.I., GONCHARIK, M.N.** : Vodnyǐ defitsit assimilyatsionnoǐ tkani i intensivnost' fotosinteza v svyazi s usloviyami vodoobespechennosti rasteniǐ na torfyanoǐ pochve. [Water deficit of assimilatory tissue
and photosynthetic rate in relation to water supply to plants in peat soil.]
- In : VECHER, A.S. (ed.) : Fotosintez i Ustoǐchivost' Rasteniǐ. Pp. 3 - 11.
Nauka i Tekhnika, Minsk 1973. [In R.]

16603 - **IVANOV, A.F., RAKHTEENKO, L.I., MOISEENKO, E.I.** : Intensivnost' fotosinteza
drevesnykh rasteniǐ, introdutsirovannykh v Belorussii.[Photosynthetic rate of
woody plants introduced in Belorussia.] - In : VECHER, A.S. (ed.) : Fotosintez i Ustoǐchivost' Rasteniǐ. Pp. 29 - 32. Nauka i Tekhnika, Minsk 1973. [In
R.]

16604 - **IVANOV, A.V., GANAGO, A.O., RUBIN, L.B.** : O mekhanizme selektivnogo deǐstviya
izlucheniya rubinovogo lazera na fotosistemu 1 v khloroplastakh gorokha. [Me-
chanism of selective damage in photosystem 1 of pea chloroplasts after ruby
laser irradiation.] - Biofizika *18* : 1117 - 1119, 1973. [In R, ab : E.]

16605 - **IVANOV, O.V.** : Primenenie elektronnykh vesov v kompleksnykh issledovaniyakh
fotosinteza. [Use of electronic balance in complex research of photosynthesis.]
- In : Metody Kompleksnogo Izucheniya Fotosinteza. Vol. 2. Pp. 129 - 151.
VASKHNIL, Leningrad 1973. [In R.]

16606 - **IVANOV, O.V.** : Rele vremeni, ispol'zuemye v issledovaniyakh fotosinteza. [Ti-
mers in photosynthetic research.] - In : Metody Kompleksnogo Izucheniya Fotosinteza. Vol. 2. Pp. 152 - 161. VASKHNIL, Leningrad 1973. [In R.]

16607 - **IVANOV, V.V., KOLOSOVSKAYA, E.A., SID'KO, F.Ya., BELYANIN, V.N.** : Adsorbtsiya
vody v ěkstrakte khlorofilla. [Water adsorption in chlorophyll extract.] -
Mol. Biol. (Moskva) *7* : 368 - 373, 1973. [In R, ab : E.]

16608 - **IVANOVA, N.A.** : Fotosintetícheskaya funktsiya izolirovannykh list'ev kartofelya i ikh fragmentov pri starenii. [Photosynthetic function of detached potato
leaves and their fragments in the course of ageing.] - In : Voprosy Regulyatsii Fotosinteza. Vol. 3. Pp. 117 - 124, 164. Ural'. gos. Univ., Sverdlovsk
1973. [In R.]

16609 - **IVERSON, R.L., CURL, H. Jr.** : Action spectrum of photosynthesis for *Skeletonema costatum* obtained with carbon-14. - Physiol. Plant. *28* : 498 - 502, 1973.

16610 - **IWAMOTO, K., ARUGA, Y.** : [Distribution of the UV-absorbing substance in algae
with reference to the peculiarity of *Prasiola japonica* YATABE.] - J. Tokyo
Univ. Fish. *60* (1) : 43 - 54, 1973. [In Jap., ab : E.]

16611 - **IZAWA, S., GOULD, J.M., ORT, D.R., FELKER, P., GOOD, N.E.** : Electron transport
and photophosphorylation in chloroplasts as a function of the electron acceptor. III. A dibromothymoquinone-insensitive phosphorylation reaction associated with photosystem II. - Biochim. biophys. Acta *305* : 119 - 128, 1973.

16612 - **IZAWA, S., KRAAYENHOF, R., RUUGE, E.K., DeVAULT, D.** : The site of KCN inhibition in·the photosynthetic electron transport pathway. - Biochim. biophys. Acta *314* : 328 - 339, 1973.

16613 - **JACKSON, J.B., COGDELL, R.J., CROFTS, A.R.** : Some effects of *o*-phenanthroline on electron transport in chromatophores from photosynthetic bacteria. - Biochim. biophys. Acta *292* : 218 - 225, 1973.

16614 - **JACKSON, J.B., DUTTON, P.L.** : The kinetic and redox potentiometric resolution of the carotenoid shifts in *Rhodopseudomonas spheroides* chromatophores : their relationship to electric field alterations in electron transport and energy coupling. - Biochim. biophys. Acta *325* : 102 - 113, 1973.

16615 - **JACOBSON, B.S., KANNANGARA, C.G., STUMPF, P.K.** : Biosynthesis of α-linolenic acid by disrupted spinach chloroplasts. - Biochem. biophys. Res. Commun. *51* : 487 - 493, 1973.

16616 - **JACQUES, G., CAHET, G., FIALA, M., PANOUSE, M.** : Enrichissement de communautés phytoplanctoniques néritiques de Méditerranée Nord Occidentale. - J. exp. mar. Biol. Ecol. *11* : 287 - 295, 1973. [Chl.]

16617 - **JACQUES, G., MINAS, H.J., MINAS, M., NIVAL, P.** : Influence des conditions hivernales sur les productions phyto- et zooplanctoniques en Méditerranée Nord--Occidentale. II. Biomasse et production phytoplanctonique. - Mar. Biol. *23* : 251 - 265, 1973. [Chl.]

16618 - **JACQUES, R.** : Interaction entre la photosynthèse et l'action du phytochrome : conséquences pour la morphogénèse des plantes. - In : The Sun in the Service of Mankind. Pp. V.4-1 - V.4-10. Unesco, Paris 1973.

16619 - **JAGENDORF, A.T.** : The role of phosphate in photosynthesis. - In : GRIFFITH, E.J., BEETON, A., SPENCER, J.M., MITCHELL, L.T. (ed.) : Environmental Phosphorus Handbook. Pp. 381 - 392. John Wiley & Sons, New York 1973.

16620 - **JAHN, O.L.** : Degreening citrus fruit with postharvest applications of (2-chloroethyl)phosphonic acid (Ethephon). - J. amer. Soc. hort. Sci. *98* : 230 - 233, 1973. [Chl.]

16621 - **JAHNKE, L.S.** : Manganese uptake and restoration of photosynthesis in manganese deficient algae. - Plant Physiol. *51* (Suppl.) : 68, 1973.

16622 - **JANES, B.E.** : Variations in transpiration, net CO_2 assimilation and leaf water potential of pepper plants produced by changes in the root and top environment. - In : SLATYER, R.O. (ed.) : Plant Response to Climatic Factors. Pp. 193 - 199. Unesco, Paris 1973.

16623 - **JANES, B.E., GEE, G.W.** : Changes in transpiration, net carbon dioxide assimilation and leaf water potential resulting from application of hydrostatic pressure to roots of intact pepper plants. - Physiol. Plant. *28* : 201 - 208, 1973.

16624 - **JANSZ, E.R., MACLEAN, F.I.** : The effect of cold shock on the blue-green alga *Anacystis nidulans*. - Can. J. Microbiol. *19* : 381 - 387, 1973. [Ps.]

16625 - **JANSZ, E.R., MACLEAN, F.I.** : CO_2 fixation by the blue-green alga *Anacystis nidulans*. - Can. J. Microbiol. *19* : 497 - 504, 1973.

16626 - **JARECKA, M.** : Influence of light, water deficit and age of plant on the intensity of photosynthesis and air passage capacity in leaves of sugar beet (*Beta vulg.* var. *saccharifera*). - Hodowla Rośl., Aklimat. Nasienn. *17* : 329 - 357, 1973.

16627 - **JAVORNICKÝ, P.** : A field method for measuring the photosynthesis of snow and aerophytic algae. - Arch. Hydrobiol. *41* (Suppl. 3 - Algol. Stud. *8*) : 363 - - 371, 1973.

16628 - **JAVORNICKÝ, P., KOMÁRKOVÁ, J.** : The changes in several parameters of plankton primary productivity in Slapy Reservoir 1960 - 1967, their mutual correlations and correlations with the main ecological factors. - In : HRBÁČEK, J., STRAŠKRABA, M. (ed.) : Hydrobiological Studies. Vol. 2. Pp. 155 - 211. Academia, Praha 1973.

16629 - **JENNINGS, M.E., SEIJO, M.A.** : Comments on "On determination of BOD and parameters in polluted stream models from DO measurements only" by A.J. Koivo and G.R. Phillips. - Water Resources Res. *9* : 496 - 497, 1973.

16630 - **JENNINGS, R.C., EYTAN, G.** : Biogenesis of chloroplast membranes XIV. Inhomogeneity of membrane protein distribution in photosystem particles obtained from *Chlamydomonas reinhardi* Y-1. - Arch. Biochem. Biophys. *159* : 813 - 820, 1973.

16631 - **JENNINGS, R.C., OHAD, I.** : Biogenesis of chloroplast membranes. XII. The influence of chloramphenicol on chlorophyll fluorescence yield and chlorophyll
organisation in greening cells of a mutant of *Chlamydomonas reinhardi* y-1. -
Plant Sci. Lett. *1* : 3 - 9, 1973.

16632 - **JENSEN, A., SAKSHAUG, E.** : Studies on the phytoplankton ecology of the Trondheimsfjord. II. Chloroplast pigments in relation to abundance and physiological state of the phytoplankton. - J. exp. mar. Biol. Ecol. *11* : 137 - 155,
1973.

16633 - **JENSEN, K.F., ROBERTS, B.R.** : Effects of *Fusarium* canker on translocation in
yellow-poplar seedlings. - Physiol. Plant Pathol. *3* : 359 - 361, 1973. [Photo-
synthates.]

16634 - **JHAMB, S., ZALIK, S.** : Soluble and lamellar proteins in seedlings of barley
and its virescens mutant in relation to chloroplast development. - Can. J. Bot.
51 : 2147 - 2154, 1973. [Chl.]

16635 - **JODO, S.** : [Stomatal movement and water relations in crops. 2. Stomatal behavior of tobacco leaves of different ages and the influence of soil water short-
age.] - Proc. Crop Sci. Soc. Jap. *42* : 123 - 130, 1973. [In Jap., ab : E.]

16636 - **JOHAM, H.E., GOSSETT, D.R.** : The translocation of photosynthetically fixed
$^{14}CO_2$ as influenced by Ca deficiency. - Plant Physiol. *51* (Suppl.) : 45, 1973.

16637 - **JOHANSSON, B.C., BALTSCHEFFSKY, M.** : A purified coupling factor from *Rhodospirillum rubrum*. - In : Ninth International Congress of Biochemistry. Stockholm,
1 - 7 July 1973. Abstract Book. P. 219. IUB, Stockholm 1973.

16638 - **JOHN, D.M.** : Accumulation and decay of litter and net production of forest in
tropical West Africa. - Oikos *24* : 430 - 435, 1973.

16639 - **JOHNS, G.G., LAZENBY, A.** : Defoliation, leaf area index, and the water use of
four temperate pasture species under irrigated and dryland conditions. - Aust.
J. agr. Res. *24* : 783 - 795, 1973.

16640 - **JOHNSON, (Sister), C., BROWN, W.V.** : Grass leaf ultrastructural variations. -
Amer. J. Bot. *60* : 727 - 735, 1973. [Chloroplast.]

16641 - **JOHNSON, D.A., CALDWELL, M.M., TIESZEN, L.L.** : Photosynthesis in relation to
leaf water potential in three alpine plant species. - In : BLISS, L.C., WIEL-
GOLASKI, F.E. (ed.) : Primary Production and Production Processes. Tundra Biome. Pp. 205 - 210. Tundra Biome Steering Comm., Edmonton 1973.

16642 - **JOHNSON, G.R.** : Diallel analysis of leaf area heterosis and relationships to
yield in maize. - Crop Sci. *13* : 178 - 180, 1973. [Growth analysis.]

16643 - **JOHNSON, P.L., TIESZEN, L.L.** : Vegetative research in Arctic Alaska. - In :
BRITTON, M.E. (ed.) : Alaska Arctic Tundra. Pp. 169 - 198. Arctic Inst. N. Amer.
Tech. Paper No. 25. 1973. [Ps, Chl.]

*16644 - **JOHNSTON, A., SMOLIAK, S., WILSON, D.B.** : Seedling growth of four *Agropyron*
species. - Can. J. Plant Sci. *52* : 763 - 768, 1972. [Growth analysis.]

16645 - **JOLIOT, P., BENNOUN, P., JOLIOT, A.** : New evidence supporting energy transfer
between photosynthetic units. - Biochim. biophys. Acta *305* : 317 - 328, 1973.

16646 - **JOLIOT, P., JOLIOT, A.** : Different types of quenching involved in photosystem
II centers. - Biochim. biophys. Acta *305* : 302 - 316, 1973.

16647 - **JOLLIFFE, P.A., TREGUNNA, E.B.** : Environmental regulation of the oxygen effect
on apparent photosynthesis. - Plant Physiol. *51* (Suppl.) : 68, 1973.

16648 - **JOLLIFFE, P.A., TREGUNNA, E.B.** : Environmental regulation of the oxygen effect
on apparent photosynthesis in wheat. - Can. J. Bot. *51* : 841 - 853, 1973.

16649 - **JONES, H.G.** : Limiting factors in photosynthesis. - New Phytol. *72* : 1089 -
- 1094, 1973.

16650 - **JONES, H.G.** : Photosynthesis by thin leaf slices in solution. II. Osmotic stress
and its effects on photosynthesis. - Aust. J. biol. Sci. *26* : 25 - 33, 1973.

16651 - **JONES, H.G.** : Gas exchange in plant leaves having different transfer resistances through their two surfaces. - Aust. J. biol. Sci. *26* : 1045 - 1055, 1973.

16652 - JONES, H.G. : Moderate-term water stresses and associated changes in some photosynthetic parameters in cotton. - New Phytol. 72 : 1095 - 1105, 1973.

16653 - JONES, H.G., OSMOND, C.B. : Photosynthesis by thin leaf slices in solution. I.
Properties of leaf slices and comparison with whole leaves. - Aust. J. biol.
Sci. 26 : 15 - 24, 1973.

16654 - JONES, M.B. : Some observations on a circadian rhythm in carbon dioxide compensation in Bryophyllum fedtschenkoi. - Ann. Bot. 37 : 1027 - 1034, 1973.

16655 - JONES, O.T.G. : Chlorophyll. - In : MILLER, L.P. (ed.) : Phytochemistry. Vol.
I. The Process and Products of Photosynthesis. Pp. 75 - 111. Van Nostrand Reinhold Comp., New York - Cincinnati - Toronto - London - Melbourne 1973.

16656 - JOSEPH, B., GAUR, B.K., CHADHA, M.S., PATANKAR, A.V. : Stimulation of growth
in Ocimum kilimandscharicum by low-dose x-irradiation. - Aust. J. biol. Sci.
26 :349 - 355, 1973. [Chl.]

16657 - JOSHI, G.V., KAREKAR, M.D. : Pathway of $^{14}CO_2$ fixation in marine algae. - Proc.
Indian nat. Sci. Acad. 39 B : 489 - 493, 1973.

*16658 - JOY, K.W., HAGEMAN, R.H. : The purification and properties of nitrite reducta-
se from higher plants, and its dependence on ferredoxin. - Biochem. J. 100 :
263 - 273, 1966. [Ps.]

16659 - JOYARD, J., BISCH, A.M., FOURCY, A. : Le manganèse dans le chloroplaste. Données du problème et perspectives. - B.I.S.T. Commiss. Energie at. 183 : 37 -
- 48, 1973.

16660 - JUNGE, W., ECKHOF, A. : On the orientation of chlorophyll-a_1 in the functional
membrane of photosynthesis. - FEBS Lett. 36 : 207 - 212, 1973.

16661 - JUPIN, H. : Modification de l'importance relative des deux systèmes photochimiques chez la Diatomée Detonula sp. cultivée en lumière rouge. - Physiol.
vég. 11 : 507 - 517, 1973.

16662 - JUPIN, H.J. : Modifications pigmentaires et ultrastructurales chez la diatomée
Detonula sp. cultivée en lumière rouge. - Arch. Mikrobiol. 91 : 19 - 27, 1973.

16663 - KAFALIEVA, D.N., RUSKOVA, M.Kh. : Izmenenie fotokhimicheskoĭ aktivnosti khloroplastov inkubirovannykh digitoninom. [Change of photochemical activity of chloroplasts incubated with digitonin.] - Dokl. bolg. Akad. Nauk 26 : 267 - 268,
1973. [In R.]

16664 - KAGAWA, T., McGREGOR, D.I., BEEVERS, H. : Development of enzymes in the cotyledons of watermelon seedlings. - Plant Physiol. 51 : 66 - 71, 1973. [Ps, Chl.]

16665 - KAHN, J.S. : Euglena gracilis : Phenotypic resistance to 2,4-dinitrophenol. -
Arch. Biochem. Biophys. 159 : 646 - 650, 1973. [Ps, Chl.]

16666 - KAISER, W., URBACH, W. : Endogene cyclische Phosphorylierung in isolierten
Chloroplasten. - Ber. deut. bot. Ges. 86 : 213 - 226, 1973.

*16667 - KAKUNO, T., HORIO, T. : [Energy conversion in photosynthesis.] - Tampakushitsu,
Kakusan, Koso [Protein, nucl. Acid, Enzyme] 16 : 787 - 794, 1971. [In Jap.]

16668 - KAKUNO, T., HOSOI, K., HIGUTI, T., HORIO, T. : Electron and proton transport
in Rhodospirillum rubrum chromatophores. - J. Biochem.(Tokyo) 74 : 1193 - 1203, 1973.

*16669 - KALER, V.L. : Metabolicheskoe i épigeneticheskoe upravlenie biosintezom khlo-
rofilla. [Metabolic and epigenetic regulation of chlorophyll biosynthesis.] -
In : Geneticheskie Aspekty Fotosinteza. Tezisy Dokladov. Pp. 22 - 23. Donish,
Dushanbe 1972. [In R.]

16670 - KALER, V.L. : Svetovoe upravlenie fermentnoĭ sistemoĭ biosinteza khlorofilla
v vysshikh rasteniyakh. [Light control of the enzyme system of chlorophyll biosynthesis in higher plants.] - Tr. mosk. Obshch. Ispyt. Prirody, Ser. biol.49
(Problemy Biofotokhimii) : 243 - 249, 1973. [In R, ab : E.]

16671 - KALISZ, L. : Role of algae in sewage purification. Part I. Oxygen production.
- Pol. Arch. Hydrobiol. 20 : 389 - 412, 1973. [Ps.]

16672 - KALLIO, P., HEINONEN, S. : Ecology of Rhacomitrium lanuginosum (HEDW.) BRID. -
Rep. Kevo subarctic Res. Sta. 10 : 43 - 54, 1973. [Ps.]

16673 - KAL'NOĬ, P.G. : Biologiya rosta odnoletnikh seyantsev klena ostrolistnogo i
klena serebristogo.[Biology of growth of year-old Norway and silver maple seedlings.] - Nauch. Dokl. Vyssh. Shkoly, biol. Nauki 16 (10) : 78 - 83, 1973.[In R.]

*16674 - KAMANINA, M.S., ANISIMOV, A.A. : K voprosu o vliyanii élementov mineral'nogo
pitaniya na sostav transportiruemykh iz list'ev assimilyatov. [Effect of mineral nutrition elements on the composition of photosynthates transported out of
leaves.] - Uch. Zap. gor'k. gos. Univ., Ser. biol. 139 : 85 - 88, 1971. [In R.]

16675 - KAMEN, M.D. : Toward a comparative biochemistry of the cytochromes. - Protein,
Nucleic Acid, Enzyme [Tampakushitsu, Kakusan, Koso] 18 : 753 - 773, 1973. [Ps.]

16676 - KAMEN, M.D., KAKUNO, T., BARTSCH, R.G., HANNON, S. : Spin-state correlations
in near infrared spectroscopy of cytochrome c'. - Proc. nat. Acad. Sci. USA
70 : 1851 - 1854, 1973. [Ps bacteria.]

16677 - KAMIŃSKA, M. : The effect of apple tree proliferation disease on chlorophyll,
free amino acid and amide contents in the leaves of infected trees. - Phytopathol. Z. 76 : 142 - 148, 1973.

16678 - KAMIYA, A., MIYACHI, S. : [Metabolic regulation by light.] - Kagaku 43 : 621 -
- 625, 1973. [Ps; in Jap.]

16679 - KANAI, R., EDWARDS, G.E. : Enzymatic separation of mesophyll protoplasts and
bundle sheath cells from C₄ plants. - Naturwissenschaften 60 : 157 - 158, 1973.

16680 - KANAI, R., EDWARDS, G.E. : Separation of mesophyll protoplasts and bundle
sheath cells from maize leaves for photosynthetic studies. - Plant Physiol.
51 : 1133 - 1137, 1973.

16681 - KANAI, R., EDWARDS, G.E. : Purification of enzymatically isolated mesophyll
protoplasts from C_3, C_4, and crassulacean acid metabolism plants using an a-
queous dextran-polyethylene glycol two-phase system. - Plant Physiol. 52 :
484 - 490, 1973.

16682 - KANAI, R., GUTIERREZ, M., KU, S.B., EDWARDS, G.E. : Isolation of mesophyll protoplasts and bundle sheath cells from plants having the C₄-dicarboxylic acid
pathway of photosynthesis. - Plant Physiol. 51 (Suppl.) : 6, 1973.

16683 - KANNANGARA, C.G., JACOBSON, B.S., STUMPF, P.K. : Fat metabolism in higher
plants. LVII. A comparison of fatty acid-synthesizing enzymes in chloroplasts
isolated from mature and immature leaves of spinach. - Plant Physiol. 52 :
156 - 161, 1973.

16684 - KAO, O.H.W., BERNS, D.S., TOWN, W.R. : The characterization of C-phycocyanin
from an extremely halo-tolerant blue-green alga, Coccochloris elabens. - Biochem. J. 131 : 39 - 50, 1973.

16685 - KARAPETYAN, N.V. : Evolyutsiya pervichnykh protsessov fotosinteza. [Evolution
of primary processes of photosynthesis.] - In : Problemy Vozniknoveniya i
Sushchnosti Zhizni. Pp. 233 - 238. Nauka, Moskva 1973. [In R.]

16686 - KARAPETYAN, N.V., KLIMOV, V.V. : Priroda obratimogo i neobratimogo umen'she-
niya fluorestsentsii pri osveshchenii khloroplastov v vosstanovitel'nykh usloviyakh. [Nature of reversible and irreversible decrease of fluorescence during
illumination of the chloroplasts under reductive conditions.] - Fiziol. Rast.
20 : 545 - 553, 1973. [In R, ab : E.]

16687 - KARAPETYAN, N.V., KLIMOV, V.V., KRASNOVSKIĬ, A.A. : Light-induced changes in
the fluorescence yield of particles obtained by digitonin fragmentation of
chloroplasts. - Photosynthetica 7 : 330 - 337, 1973.

16688 - KARAPETYAN, N.V., KLIMOV, V.V., KRASNOVSKIĬ, A.A. : Peremennaya fluorestsentsiya digitoninovykh fragmentov khloroplastov. [Variable fluorescence of digitonin fragments of chloroplasts.] - Dokl. Akad. Nauk SSSR 211 : 729 - 732, 1973.
[In R.]

*16689 - KARAPETYAN, N.V., KLIMOV, V.V., LANG, F., KRASNOVSKIĬ, A.A. : Induktsiya fluo-
restsentsii normal'nykh i mutantnykh rastenii kukuruzy. [Induction of fluorescence in normal and mutant maize plants.] - In : Geneticheskie Aspekty Fotosinteza. Tezisy Dokladov. Pp. 89 - 90. Donish, Dushanbe 1972. [In R.]

16690 - KARAPETYAN, N.V., KRAKHMALEVA, I.N. : Dve fotosistemy purpurnykh bakterii

Chromatium minutissimum. [Two photosystems of the purple bacterium *Chromatium minutissimum.*] - Tr. mosk. Obshch. Ispyt. Prirody *49* (Problemy Biofotokhimii): 210 - 216, 1973. [In R, ab : E.]

16691 - **KARAPETYAN, N.V., KRAKHMALEVA, I.N., KRASNOVSKIĬ, A.A.** : Fotoprevrashcheniya bakteriokhlorofillov i tsitokhromov v khromatoforakh i bakteriyakh *Chromatium* v vosstanovitel'nykh usloviyakh. [Light-induced transformation of bacteriochlorophylls and cytochromes in *Chromatium* cells and chromatophores and under reductive conditions.] - Mol. Biol. (Moskva) *7* : 868 - 875, 1973. [In R, ab : E.]

16692 - **KÄRENLAMPI, L.** : Biomass and estimated yearly net production of the ground vegetation at Kevo. - In : BLISS, L.C., WIELGOLASKI, F.E. (ed.) : Primary Production and Production Processes, Tundra Biome. Pp. 111 - 114. Tundra Biome Steering Comm., Edmonton 1973.

16693 - **KARIKARI, S.K.** : Estimation of leaf area in papaya (*Carica papaya*) from leaf measurements. - Trop. Agr. *50* : 346, 1973.

*16694 - **KARIMOV, Kh.Kh., CHERNER, R.I., RAKHMONOV, A.** : Fotosinteticheskaya deyatel'-nost' zimnevegetiruyushchikh rasteniĭ v svyazi s geneticheskoĭ obuslovlennost'-yu ustoĭchivosti fotosinteticheskogo apparata k deĭstviyu nizkikh temperatur. [Photosynthetic activity of winter-vegetating plants in relation to the genetic basis of resistance of the photosynthetic apparatus to the action of low temperatures.] - In : Geneticheskie Aspekty Fotosinteza. Tezisy Dokladov. Pp. 103 - 104. Donish, Dushanbe 1972. [In R.]

16695 - **KARIYA, K., TSUNODA, S.** : Chloroplast characters and the photosynthetic rate of cultivated *Brassica* species. - Tohoku J. agr. Res. *24* : 1 - 13, 1973.

16696 - **KARLMARK, B., SOHTELL, M.** : The determination of bicarbonate in nanoliter samples. - Anal. Biochem. *53* : 1 - 11, 1973.

16697 - **KARN, R.C., HUDOCK, G.A.** : A photorepressible isozyme of malic enzyme in *Euglena gracilis* strain Z. - J. Protozool. *20* : 316 - 320, 1973.

16698 - **KARNACHUK, R.A., POSTOVALOVA, V.M., FROLOVA, N.M.** : O biosinteze adeninnukleo-tidov v rasteniyakh v svyazi s protsessom fotoassimilyatsii glyukozy pri deĭ-stvii sveta razlichnogo sostava. [Biosynthesis of adenine nucleotides in plants in relation to glucose photoassimilation in light of various spectral composition.] - In : Upravlenie Skorost'yu i Napravlennost'yu Biosinteza u Rasteniĭ. Pp. 17 - 18. Krasnoyarsk 1973. [In R.]

16699 - **KARPILOV, Yu.S.** : Regulyatsiya metabolizma i intensivnosti fotosinteza azotom i fosforom. [Control of metabolism and photosynthetic rate by nitrogen and phosphorus.] - In : Upravlenie Skorost'yu i Napravlennost'yu Biosinteza u Rasteniĭ. P. 19. Krasnoyarsk 1973. [In R.]

16700 - **KARPOVA, G.Ya., SIVTSEV, M.V., KHILIK, L.A.** : Fiziologicheskie osobennosti *Nepeta transcaucasica* GROSSH. v usloviyakh orosheniya. [Physiological peculiarities of irrigated *Nepeta transcaucasica* GROSSH.] - Rast. Resursy *9* : 242 - 250, 1973. [Chl, Car; in R.]

16701 - **KARTASHOV, I.M., MAKAROV, A.D., MAL'YAN, A.N.** : Izuchenie pervichnogo vzaimo-deĭstviya komponentov protsessa fotofosforilirovaniya s khloroplastami meto-dom adsorbtsii. [Primary interaction between photophosphorylation components and chloroplasts studied by adsorption method.] - Biofizika *18* : 272 - 278, 1973. [In R, ab : E.]

16702 - **KARUNEN, P.** : Studies on moss spores II. Production of chlorophylls in germinating *Polytrichum commune* HEDW. spores. - J. exp. Bot. *24* : 1186 - 1188, 1973.

16703 - **KARYAGIN, Yu.G., TOLSTENKO, L.A.** : Vliyanie mineral'nykh udobreniĭ na funktsio-nal'noe sostoyanie kluben'kovykh bakteriĭ i urozhaĭnost' soi. [Effect of mineral fertilizers on functional state of nodule bacteria and soybean production.] - Mikrobiologiya *42* : 931 - 936, 1973. [Ps, Chl; in R, ab : E.]

16704 - **KASEMIR, H., OBERDORFER, U., MOHR, H.** : A twofold action of phytochrome in controlling chlorophyll *a* accumulation. - Photochem. Photobiol. *18* : 481 - 486, 1973.

16705 - **KASPERBAUER, M.J., PEASLEE, D.E.** : Morphology and photosynthetic efficiency of tobacco leaves that received end-of-day red or far red light during development. - Plant Physiol. *52* : 440 - 442, 1973.

16706 - **KASS, L.B., PAOLILLO, D.J. Jr.** : The light requirement for chloroplast replication in *Polytrichum* spores. - Amer. J. Bot. *60* (4 Suppl.) : 8, 1973.

16707 - **KASSNER, R.J., YANG, W.** : The redox potentials of the two-iron plant and algal
ferredoxins. An electrostatic model. - Biochem. J. *133* : 283 - 287, 1973.

*16708 - **KAS'YANENKO, A.G.** : Fenotipy i genotipy khlorofil'nykh mutantov. [Phenotypes
and genotypes of chlorophyll mutants.] - In : Geneticheskie Aspekty Fotosinteza. Tezisy Dokladov. Pp. 61 - 62. Donish, Dushanbe 1972. [In R.]

*16709 - **KATO, T., NAKAMURA, S.-I.** : [Studies on the physiological role of rutin and
its application to vegetable crops. I. Effect of foliage application of rutin
on the watermelon plants.] - Res. Rep. Kôchi Univ. *21* (agr. Sci. 11) : 91 -
- 96, 1972. [Ps; in Jap., ab : E.]

16710 - **KATO, T., YOSHIHIRO, M., NAKAYAMA, N.** : [Studies on the physiological role of
rutin and its application to vegetable crops. II. Effect of foliage application
of rutin on the tuberous root formation in sweet potato.] - Res. Rep. Kôchi
Univ. *22* (agr. Sci. 12) : 107 - 114, 1973. [Ps; in Jap., ab : E.]

16711 - **KATOH, S.** : [Energy conversion and electron transport in photosynthesis.] -
Kagaku to Seibutsu [Chemistry and Life] *11* : 482 - 489, 1973. [In Jap.]

16712 - **KATTERMAN, F.R.H., ENDRIZZI, J.E.** : Studies on the 70S ribosomal content of a
plastid mutant in *Gossypium hirsutum*. - Plant Physiol. *51* : 1138 - 1139, 1973.
[Chloroplast.]

16713 - **KATZ, J.J.** : Chlorophyll interactions and light conversion in photosynthesis.
- Naturwissenschaften *60* : 32 - 39, 1973.

16714 - **KATZ, J.J., JANSON, T.R.** : Chlorophyll-chlorophyll interactions from ^{1}H and
^{13}C nuclear magnetic resonance spectroscopy. - Ann. New York Acad. Sci. *206* :
579 - 603, 1973.

16715 - **KATZ, J.J., NORRIS, J.R. Jr.** : Chlorophyll and light energy transduction in
photosynthesis. - In : **SANADI, D.R., PACKER, L.** (ed.) : Current Topics in Bioenergetics. Vol. 5. Pp. 41 - 75. Academic Press, New York - London 1973.

16716 - **KAUFMANN, M.R.** : Design, calibration, and use of a porometer for conifers. -
Plant Physiol. *51* (Suppl.) : 8, 1973.

16717 - **KAVAL'CHUK, R.A., VECHAR, A.S.** : Lipidny sastaŭ khlaraplastaŭ prarostkaŭ zhy-
ta. [Lipid composition of chloroplasts of rye seedlings.] - Vestsi Akad. Navuk belorus.SSR,Ser. biyal. Navuk *1973* (4) : 20 - 28, 1973. [Chl, Car; in Belorus.]

16718 - **KAVEH, D., HAREL, E.** : Light-induced changes in the pattern of protein synthesis during the early stages of greening of etiolated maize leaves. - Plant Physiol. *51* : 671 - 676, 1973.

16719 - **KAWASHIMA, K., YAMANISHI, T.** : Thermal degradation of β-carotene. - J. agr.
chem. Soc. Jap. [Nippon Nogei Kagaku Kaishi] *47* : 79 - 81, 1973. [In Jap., ab :
E.]

16720 - **KAZARYAN, V.O., DAVTYAN, V.O., CHILINGARYAN, A.A.** : O roli korneĭ v aktivatsii
zhiznedeyatel'nosti list'ev v faze generativnogo razvitiya rasteniĭ. [The role
of roots in the activation of leaf viability during the generative growth phase of plants.] - Fiziol. Rast. *20* : 700 - 707, 1973. [Ps; in R, ab : E.]

16721 - **KE, B.** : The primary electron acceptor of photosystem I. - Biochim. biophys.
Acta *301* : 1 - 33, 1973.

16722 - **KE, B., BEINERT, H.** : Evidence for the identity of P430 of Photosystem I and
chloroplast-bound iron-sulfur protein. - Biochim. biophys. Acta *305* : 689 -
- 693, 1973.

16723 - **KE, B., GARCIA, A.F., VERNON, L.P.** : Light-induced absorption changes in *Chromatium* subchromatophore particles exhaustively extracted with non-polar solvents. - Biochim. biophys. Acta *292* : 226 - 236, 1973.

16724 - **KE, B., HANSEN, R.E., BEINERT, H.** : Oxidation-reduction potentials of bound
iron-sulfur proteins of photosystem I. - Proc. nat. Acad. Sci. USA *70* : 2941 -
- 2945, 1973.

16725 - **KEENAN, J.D.** : Effects of inorganic carbon, ortho-phosphate and pH on rates of photosynthesis and respiration in the blue-green alga, *Anabaena flos-aquae* (LYNGBYE) DE BRÉB. - Diss. Abstr. Int. B *33* : 4833-B, 1973.

16726 - **KEENAN, J.D.** : Response of *Anabaena* to pH, carbon, and phosphorus. - J. environ. eng. Div., ASCE *99* : 607 - 620, 1973. [Ps.]

16727 - **KEISTER, D.L., FLEISCHMAN, D.E.** : Nitrogen fixation in photosynthetic bacteria. - In : GIESE, A.C. (ed.) : Photophysiology. Current Topics in Photobiology and Photochemistry. Vol. 8. Pp. 157 - 183. Academic Press, New York - London 1973.

16728 - **KELLER, J.H., BACHOFEN, R.** :|Über die Bindung von Mangan an Chloroplastenfragmente. - Ber. schweiz. bot. Ges. *83* : 66 - 74, 1973.

16729 - **KELLER, T.** : CO_2 exchange of bark of deciduous species in winter. - Photosynthetica *7* : 320 - 324, 1973.

16730 - **KELLER, T.** : Über die schädigende Wirkung des Fluors. - Schweiz. Z. Forstwesen *124* : 700 - 706, 1973. [Ps.]

16731 - **KELLER, T.** : Zur Phytotoxizität staubförmiger Fluor-Verbindungen. - Staub-Reinhalt. Luft *33* : 395 - 397, 1973. [Ps.]

16732 - **KELLER, T., BEDA-PUTA, H.** : Zum winterlichen Gaswechsel unbelaubter Sprossachsen. - Schweiz. Z. Forstwesen *124* : 433 - 441, 1973. [Ps, Chl.]

16733 - **KELLY, D.P., POOLE, P.C., OWEN, G.H.** : Responses of *Chlorobium thiosulfatophilum* to exogenous organic nutrients. - J. gen. Microbiol. *75* : XX - XXI, 1973. [Ps.]

16734 - **KELLY, G.J., GIBBS, M.** : Nonreversible D-glyceraldehyde 3-phosphate dehydrogenase of plant tissues. - Plant Physiol. *52* : 111 - 118, 1973.

16735 - **KELLY, G.J., GIBBS, M.** : A mechanism for the indirect transfer of photosynthetically reduced nicotinamide adenine dinucleotide phosphate from chloroplasts to the cytoplasm. - Plant Physiol. *52* : 674 - 676, 1973.

16736 - **KENNEDY, R.A., LAETSCH, W.M.** : Relationship between leaf development and primary photosynthetic products in the C_4 plant *Portulaca oleracea* L. - Planta *115* : 113 - 124, 1973.

16737 - **KENNEDY, R.A., LAETSCH, W.M.** : Changes of primary products of photosynthesis in developing leaves of the C_4 plant *Portulaca oleracea*. - Plant Physiol. *51* (Suppl.) : 7, 1973.

16738 - **KENNER, G.W., McCOMBIE, S.W., SMITH, K.M.** : Pyrroles and related compounds. Part XXIV. Separation and oxidative degradation of chlorophyll derivatives. - J. chem. Soc. Perkin Trans. I *21* : 2517 - 2523, 1973.

16739 - **KENNEY, R.L., FISHER, G.S.** : Preparation of trans-pinocarveol and myrtenol. - Ind. Eng. Chem., Prod. Res. Develop. *12* : 317 - 319, 1973. [Chl.]

16740 - **KERESZTES, Á.** : Correlations of structure and function in normal and mutant *Tradescantia* chloroplasts. - Mikroskopie *29* : 48, 1973.

16741 - **KERESZTES, Á., FALUDI-DÁNIEL, Á.** : Ultrastructure, pigment content and photosynthetic activity of the normal and mutant chloroplasts in developing *Tradescantia* leaves. - Acta biol. Acad. Sci. hung. *24* : 175 - 189, 1973.

16742 - **KESSEL, M., MacCOLL, R., BERNS, D.S., EDWARDS, M.R.** : Electron microscope and physical chemical characterization of *C*-phycocyanin from fresh extracts of two blue-green algae. - Can. J. Microbiol. *19* : 831 - 836, 1973.

16743 - **KESSLER, E.** : Effect of anaerobiosis on photosynthetic reactions and nitrogen metabolism of algae with and without hydrogenase. - Arch. Mikrobiol. *93* : 91 - - 100, 1973.

16744 - **KESSLER, E., ZUMFT, W. G.** : Effect of nitrite and nitrate on chlorophyll fluorescence in green algae. - Planta *111* : 41 - 46, 1973.

16745 - **KESTEMONT, P.** : Production primaire de la strate arborée d'une hêtraie à fétuques. - Bull. Soc. roy. Bot. Belg. *106* : 305 - 316, 1973.

16746 - **KETIKU, A.O., OYENUGA, V.A.** : Changes in the carbohydrate constituents of yam tuber (*Dioscorea rotundata*, POIR) during growth. - J. Sci. Food Agr. *24* : 367 · - 373, 1973. [Ps.]

*16747 - KETSKHOVELI, É.N., DZHAPARIDZE, I.G. : [The pigment content of green and red
leaves of the agrestial myrobalan plum.] - Soobshch. Akad. Nauk gruz. SSR
62 : 409 - 412, 1971. [Chl; in Georg., ab : E, R.]

*16748 - KETSKHOVELI, É.N., DZHAPARIDZE, I.G. : Dykhanie list'ev antotsiansoderzhashchikt
i zelenykh form nekotorykh drevesnykh rasteniĭ. [Respiration of anthocyan-con-
taining and green forms of some woody plants.] - Soobshch. Akad. Nauk gruz.
SSR *62* : 661 - 664, 1971. [In R, ab : E, Georg.]

*16749 - KETSKHOVELI, É.N., DZHAPARIDZE, I.G. : Dykhanie zelenykh i krasnykh list'ev
nekotorykh vechnozelenykh rasteniĭ. [Respiration of green and red leaves of
some sempervirent plants.] - Soobshch. Akad. Nauk gruz. SSR *68* : 669 - 672,
1972. [Chl; in R, ab : E, Georg.]

 16750 - KHAGEMANN, R., GERRMANN, F., BËRNER, T. : Ispol'zovanie plastidnykh i gennykh
mutatsiĭ vysshikh rasteniĭ v issledovanii geneticheskogo kontrolya funktsii
plastid. [The use of plastid and gene mutants of higher plants in studying the
genetic control of plastid functions.] - In : Geneticheskie Aspekty Fotosinteza. Tezisy Dokladov. Pp. 31 - 33. Donish, Dushanbe 1972. [In R.]

 16751 - KHALIFA, M.A. : Effects of nitrogen on leaf area index, leaf area duration,
net assimilation rate, and yield of wheat. - Agron. J. *65* : 253 - 256, 1973.

 16752 - KHANIN, Ya.D. : Khod fiziologicheskikh protsessov v molodom vinogradnom rastenii pod vozdeĭstviem molibdena, vnesennogo pod matochniki podvoya. [The course
of physiological processes in young grapevine plant affected by molybdenum
applied on mother plantation of rootstocks.] - Tr. kishinev. sel'skokhoz. Inst.
M.V. Frunze *118* (Vinogradarstvo) : 34 - 42, 104, 1973. [Chl; in R.]

 16753 - KHANNA, R., SINHA, S.K. : Change in the predominance from C_4 to C_3 pathway
following anthesis in *Sorghum*. - Biochem. biophys. Res. Commun. *52* : 121 -
- 124, 1973.

 16754 - KHAZANOV, V.S., KUZNETSOVA, G.K. : Ob additivnosti deĭstviya raznospektral'-
nykh izlucheniĭ pri fotosinteze. [Additivity of action of radiation of various spectra in photosynthesis.] - Nauch. Dokl. vyssh. Shkoly, biol. Nauki *16*
(2) : 78 - 81, 1973. [In R.]

 16755 - KHAZANOV, V.S., KUZNETSOVA, G.K. : Ob spektral'noĭ éffektivnosti izlucheniya
v obespechenii fotosinteza. [Spectral effectivity of radiation in photosynthesis.] - Nauch. Dokl. vyssh. Shkoly, biol. Nauki *16* (7) : 84 - 87, 1973.
[In R.]

 16756 - KHMARA, L.A. : Znachenie margantsa dlya struktury khloroplastov u rasteniĭ gorokha. [Significance of manganese for chloroplast structure in pea plants.]
- Fiziol. Biokhim. kul't. Rast. *5* : 512 - 515, 1973. [In R, ab : E.]

 16757 - KHODASEVICH, É.V., ARNAUTOVA, A.I., MEL'NIKOVA, L.M. : Fotosinteticheskaya aktivnost' i biosintez pigmentov u khvoĭnykh v svyazi s obratimoi degradatsieĭ
fonda khlorofillov *a* i *b*. [Photosynthetic activity and biosynthesis
in conifers in relation to reversible degradation of chlorophyll *a* and *b*
pools.] - In : Formirovanie Pigmentnogo Apparata Fotosinteza. Pp. 130 - 142.
Nauka i Tekhnika, Minsk 1973. [In R.]

 16758 - KHODASEVICH, É.V., MEL'NIKOVA, L.M., ARNAUTOVA, A.I., GODNEV, T.N. : Formirovanie i sostoyanie fonda pigmentov u khvoĭnykh proyavlyayushchikh fenomen
"sezonnogo obestsvechivaniya". [Formation and state of pigment pool in conifers with the phenomenon of "seasonal bleaching".] - Biokhimiya - mezhved. Sb.
1 : 146 - 149, 1973. [In R.]

 16759 - KHODASEVICH, É.V., MEL'NIKOVA, L.M., ARNAUTOVA, A.I., GODNEV, T.N. : Sostoyanie pigment-belkovogo kompleksa u khvoĭnykh v svyazi s obratimoĭ degradatsieĭ
fonda khlorofillov *a* i *b*. [State of conifer pigment-protein complex with respect to reversible degradation of chlorophyll *a* and *b* fund.] - Dokl. Akad.
Nauk belorus. SSR *17* (1) : 80 - 83, 1973. [In R.]

 16760 - KHODOS, V.N. : Osobennosti obrazovaniya i pereraspredeleniya aminokislot v
list'yakh pshenitsy v zavisimosti ot temperatury. [Peculiarities of formation
and redistribution of amino acids of wheat leaves with temperature.] - Tr.
biol.-pochv. Inst., dal'nevost. nauch. Tsentr Akad. Nauk SSSR *20* (123) : 59 -
- 65, 1973. [In R, ab : E.]

16761 - **KHODOS, V.N., GRODZINSKIĬ, D.M.** : Vliyanie ékzogennykh veshchestv na raspredelenie éndogennykh metabolitov v list'yakh pshenitsy. [Effect of exogenous
substances on distribution of endogenous metabolites in wheat leaves.] - Tr.
biol.-pochv. Inst., dal'nevost. nauch. Tsentr Akad. Nauk SSSR *20* (123) : 66 -
- 70, 1973. [In R, ab : E.]

*16762 - **KHODZHAEV, A.-S., RAKHMANKULOVA, M.E., RODIMTSEVA, N.E.** : Fotosinteticheskaya
funktsiya i struktura khloroplastov razlichnykh sortov khlopchatnika. [Photo-
synthetic function and structure of chloroplasts in different cotton cultivars.] - In : Geneticheskie Aspekty Fotosinteza. Tezisy Dokladov. P. 118. Donish, Dushanbe 1972. [In R.]

16763 - **KHODZHAEV, A.S., RODIMTSEVA, N.E., YULDASHEVA, Z., PAĬZULLAEVA, D.** : Vozmozh-
nost' izucheniya na semyadolyakh potentsial'nykh sposobnosteĭ fotosinteticheskogo apparata khlopchatnika. [Possibility of studying potential ability of the
photosynthetic apparatus of cotton using cotyledons.] - Dokl. Akad. Nauk uz.
SSR *1973* (10) : 45 - 46, 1973. [In R.]

*16764 - **KHODZHIEV, A., VOSKRESENSKAYA, N.P.** : Regulyatsiya sinim svetom aktivnosti nekotorykh fermentov glikolatnogo puti fotosinteticheskogo metabolizma ugleroda. [Regulation by the blue light of the activity of some enzymes of the glycolate pathway of the photosynthetic metabolism of carbon.] - In : Geneticheskie Aspekty Fotosinteza. Tezisy Dokladov. Pp. 49 - 50. Donish, Dushanbe 1972.
[In R.]

16765 - **KHRYANIN, V.N.** : Vliyanie gibberellina na soderzhanie alkaloidov i khlorofilla v lekarstvennykh rasteniyakh. [Effect of gibberellin on the contents of alkaloids and chlorophyll in medical plants.] - Nauch. Dokl. vyssh. Shkoly, biol.
Nauki *16* (1) : 78 - 80, 1973. [In R.]

16766 - **KHVAN, A.V.** : Ustoĭchivost' soi k pereuvlazhneniyu pochvy v usloviyakh Amur-
skoĭ oblasti i rol' mikroélementov v étom protsesse. [Resistance of soybean
to surplus soil moisture in Amur region and the role of microelements in this
process.] - Uch. Zap. DVGU *61* (Ustoĭchivost' Rasteniĭ k Pereuvlazhneniyu Pochvy na Dal'nem Vostoke i Deĭstvie v Étikh Usloviyakh Mikroélementov). Pp. 112 -
- 123. Vladivostok 1973. [Ps, Chl; in R.]

16767 - **KIBALENKO, A.P., KHOMLYAK, M.M., VELYKA, S.L.** : Znachennya boru v azotnomu ob-
mini ta syntezi bilkiv u roslyn. [Significance of boron in nitrogen metabolism
and protein synthesis in plants.] - Dopov. Akad. Nauk ukr. RSR, Ser. B : Geol.,
Geofiz., Khim., Biol. *35* : 457 - 463, 1973. [In Ukr., ab : E, R.]

16768 - **KICHIGIN, A.A.** : Vliyanie nekotorykh tipov pochv na soderzhanie karotina v
zlakakh. [Effect of certain soil types on the carotene content in grasses.] -
Rast. Resursy *9* (1) : 88 - 93, 1973. [In R.]

16769 - **KIEFER, D.A.** : Cellular chlorophyll *a* fluorescence in phytoplankton. - Diss.
Abstr. Int. B *33* : 5420-B, 1973.

16770 - **KIEFER, D.A.** : Fluorescence properties of natural phytoplankton populations. -
Mar. Biol. *22* : 263 - 269, 1973.

16771 - **KIEFER, D.A.** : Chlorophyll *a* fluorescence in marine centric diatoms : responses
of chloroplasts to light and nutrient stress. - Mar. Biol. *23* : 39 - 46, 1973.

16772 - **KIHARA, T.** : The role of water in photosynthetic electron transfer. - In :
Ninth International Congress of Biochemistry. Stockholm, 1 - 7 July 1973. Abstract Book. P. 242. IUB, Stockholm 1973.

16773 - **KIHARA, T., McCRAY, J.A.** : Water and cytochrome oxidation-reduction reactions.
- Biochim. biophys. Acta *292* : 297 - 309, 1973.

16774 - **KIMBALL, S.L., SALISBURY, F.B.** : Ultrastructural changes of plants exposed to
low temperatures. - Amer. J. Bot. *60* : 1028 - 1033, 1973. [Chloroplasts.]

16775 - **KIMIMURA, M., KATOH, S.** : Studies on electron transport associated with photosystem I. III. The reduction sites of various Hill oxidants in the photosynthetic electron transport system. - Biochim. biophys. Acta *325* : 167 - 174,
1973.

16776 - **KIMOR, B.** : Plankton relations of the Red Sea, Persian Gulf and Arabian Sea. -
In : ZEITSCHEL, B. (ed.) : The Biology of the Indian Ocean. Pp. 221 - 232.
Springer-Verlag, Berlin - Heidelberg - New York 1973. [Chl.]

16777 - **KIMURA, T., MIZOKAMI, A., HASHIMOTO, T.** : [The red tide that caused severe damage to the fishery resources in Hiroshima Bay : Outline of its occurrence and the environmental conditions.] - Bull. Plankton Soc. Jap. *19* : 82 - 96, 1973. [Chl; in Jap., ab : E.]

16778 - **KINDL, H., MAJUNKE, G.** : Glyoxysomale Enzyme in Laubblättern von *Lens culinaris*. - Hoppe-Seyler's Z. physiol. Chem. *354* : 999 - 1005, 1973.

16779 - **KINERSON, R.S.** : Fluxes of visible and net radiation within a forest canopy. - J. appl. Ecol. *10* : 657 - 660, 1973.

16780 - **KINERSON, R.S. Jr., FRITSCHEN, L.J.** : Modeling air flow through vegetation. - Agr. Meteorol. *12* : 95 - 104, 1973.

16781 - **KING, R.W., ZEEVAART, J.A.D.** : Floral stimulus movement in *Perilla* and flower inhibition caused by noninduced leaves. - Plant Physiol. *51* : 727 - 738, 1973. [Ps.]

16782 - **KIPPS, A., BOULTER, D.** : Carbon transfer from the bloom node leaf to the fruit of *Vicia faba* L. - New Phytol. *72* : 1293 - 1297, 1973.

16783 - **KIRBY, E.J.M.** : The control of leaf and ear size in barley. - J. exp. Bot. *24* : 567 - 578, 1973. [Growth analysis.]

*B16784 - **KIRICHENKO, E.B.** (ed.) : Metody Vydeleniya Khloroplastov. [Methods of Chloroplast Isolation.] - Akad.Nauk SSSR, Pushchino-na-Oke 1970. [In R.]

*B16785 - **KIRICHENKO, E.B.** (ed.) : Metody Issledovaniya Struktury Fotosinteticheskogo Apparata. [Methods of Studying Structure of the Photosynthetic Apparatus.] - Akad. Nauk SSSR, Pushchino-na-Oke 1972. [In R.]

*16786 - **KIRICHENKO, E.B., VASIL'EVA, V.T.** : Nekotorye aspekty formirovaniya khloroplastov razlichnogo tipa. [Some aspects of formation of different-type chloroplasts.] - In : Geneticheskie Aspekty Fotosinteza. Tezisy Dokladov. Pp. 23 - - 24. Donish, Dushanbe 1972. [In R.]

16787 - **KIRK, J.T.O.** : Development of photosynthetic induction transients in greening leaves of wheat and French bean. - Aust. J. biol. Sci. *26* : 277 - 280, 1973.

16788 - **KIRYUKHIN, V.P.** : Izuchenie fotosinteticheskoĭ deyatel'nosti rasteniĭ i ikh dykhaniya. [Studying photosynthetic activity of plants and their respiration.] - Nauch. Tr. nauch.-issled. Inst. kartof. Khoz. *17* (Voprosy Fiziologo-Biokhimicheskikh Issledovaniĭ po Kul'ture Kartofelya) : 85 - 113, 1973. [In R.]

16789 - **KISAKI, T., HIRABAYASHI, S., YANO, N.** : Effect of the age of tobacco leaves on photosynthesis and photorespiration. - Plant Cell Physiol. *14* : 505 - 514, 1973.

16790 - **KITAGAWA, H.** : Coloring of Satsuma mandarin (*Citrus unshu* MARC.) with ethylene. - Jap. agr. Res. Quart. *7* : 43 - 46, 1973. [Chl.]

16791 - **KITAGAWA, H., TARUTANI, T.** : [Studies on the coloring of Satsuma mandarin (*Citrus unshiu* MARC.). III. The relation of fruit condition and coloring by the treatment of sealing with ethylene for 15 hours.] - J. jap. Soc. hort. Sci. [Engei Gakkai Zasshi] *42* : 65 - 69, 1973. [Chl; in Jap., ab : E.]

16792 - **KITAJIMA, M., BUTLER, W.L.** : *C*-550 in Photosystem II subchloroplast particles. - Biochim. biophys. Acta *325* : 558 - 564, 1973.

16793 - **KITAJIMA, M., OGAWA, T., INOUE, Y., SHIBATA, K.** : Sites of electron donation by alkylhydroquinones in the electron transport chain of spinach chloroplasts. - Plant Cell Physiol. *14* : 787 - 790, 1973.

16794 - **KJELVIK, S.** : Biomass and production in a willow thicket and a sub-alpine birch forest, Hardangervidda, Norway. - In : BLISS, L.C., WIELGOLASKI, F.E. (ed.) : Primary Production and Production Processes, Tundra Biome. Pp. 115 - 122. Tundra Biome Steering Comm., Edmonton 1973.

16795 - **KLEIN, R.M.** : Determining radiant energy in different wavelengths present in white light. - HortScience *8* : 210 - 211, 1973.

*16796 - **KLEINHOFS, A., WARNER, R.L.** : Effect of temperature on chlorophyll content of induced viable *chlorina* mutants of Himalaya barley. - Barley Genet. Newslett. *2* : 41 - 42, 1972.

16797 - **KLINGE, H.** : Root mass estimation in lowland tropical rain forests of central
        Amazonia, Brazil. I. Fine root masses of a pale yellow latosol and a giant hu-
        mus podzol. - Trop. Ecol. *14* : 29 - 38, 1973.

16798 - **KLINGE, H., RODRIGUES, W.A.** : Biomass estimation in a central Amazonian rain
        forest. - Acta cient. venezol. *24* : 225 - 237, 1973.

16799 - **KLOB, W., KANDLER, O., TANNER, W.** : The role of cyclic photophosphorylation
        in vivo. - Plant Physiol. *51* : 825 - 827, 1973.

16800 - **KLUETER, H.H., BAILEY, W.A., BOLTON, P.N., KRIZEK, D.T.** : Light and tempera-
        ture for maximum photosynthesis in a cucumber leaf. - Trans. ASAE *16* : 142 -
        - 144, 1973.

16801 - **KLUGE, M., HEININGER, B.** : Untersuchungen über den Efflux von Malat aus den
        Vacuolen der assimilierenden Zellen von *Bryophyllum* und mögliche Einflüsse
        dieses Vorganges auf den CAM. - Planta *113* : 333 - 343, 1973.

16802 - **KLUGE, M., LANGE, O.L., v. EICHMANN, M., SCHMID, R.** : Diurnaler Säurerhythmus
        bei *Tillandsia usneoides* : Untersuchungen über den Weg des Kohlenstoffs sowie
        die Abhängigkeit des CO_2-Gaswechsels von Lichtintensität, Temperatur und Was-
        sergehalt der Pflanze. - Planta *112* : 357 - 372, 1973.

16803 - **KLYACHKO-GURVICH, G.L., RUDOVA (ZHUKOVA), T.S., KOVANOVA, E.S., SEMENENKO, V.
        E.** : Vliyanie imidazola na obmen zhirnykh kislot pri vosstanovlenii kletok
        khlorelly posle azotnogo golodaniya. [Effect of imidazole on fatty acid meta-
        bolism after recovery of *Chlorella* cells from nitrogen starvation.] - Fiziol.
        Rast. *20* : 326 - 332, 1973. [Chl; in R, ab : E.]

16804 - **KLYACHKO-GURVICH, G.L., RUDOVA (ZHUKOVA), T.S., KUZNETSOVA, G.P.** : K izuche-
        niyu roli zhirnykh kislot lipidov khlorelly pri vosstanovlenii kletok posle
        azotnogo golodaniya. [Role of lipid fatty acids from *Chlorella* during cell
        recovery after nitrogen starvation.] - Fiziol. Rast. *20* : 114 - 122, 1973. [In
        R, ab : E.]

16805 - **KNAFF, D.B.** : Contrasting effects of plastocyanin on the photoreduction and
        photooxidation of cytochrome *f* in chloroplasts. - Biochim. biophys. Acta *292* :
        186 - 192, 1973.

16806 - **KNAFF, D.B.** : Light-induced oxidation-reduction reactions in a cell-free pre-
        paration from the blue-green alga *Nostoc muscorum* : the role of cytochrome *f*,
        cytochrome b_{558}, *C*550, and P700 in noncyclic electron transport. - Biochim.
        biophys. Acta *325* : 284 - 296, 1973.

16807 - **KNAFF, D.B., BUCHANAN, B.B., MALKIN, R.** : Effect of oxidation-reduction po-
        tential on light-induced cytochrome and bacteriochlorophyll reactions in
        chromatophores from the photosynthetic green bacterium *Chlorobium.* - Biochim.
        biophys. Acta *325* : 94 - 101, 1973.

16808 - **KNAFF, D.B., MALKIN, R.** : The oxidation-reduction potentials of electron car-
        riers in chloroplast photosystem 1 fragments. - Arch. Biochem. Biophys. *159* :
        555 - 562, 1973.

16809 - **KNIEVEL, D.P.** : Procedure for estimating ratio of live to dead root dry matter
        in root core samples. - Crop Sci. *13* : 124 - 126, 1973.

16810 - **KNIGHT, D.H.** : Leaf area dynamics of a shortgrass prairie in Colorado. - Ecology
        54 : 891 - 896, 1973.

16811 - **KNOBLOCH, K., MAYER, F.** : Energy coupling in a non-vesicular cell-free sys-
        tem. - In : Ninth International Congress of Biochemistry. Stockholm, July
        1 - 7, 1973. P. 237. [UB, Stockholm 1973. [Ps.]

16812 - **KNOOP, B.** : Untersuchungen zum Regenerationsmechanismus bei *Funaria hygromet-
        rica* SIBTH. I. Die Auslösung der Caulonemaregeneration. - Z. Pflanzenphysiol.
        70 : 22 - 33, 1973. [Ps.]

16813 - **KNOWLES, A.** : Energy transfer and photosensitised reactions. - Chem. Industry
        1973 : 1058 - 1062, 1973.

16814 - **KNOWLES, F.C.** : Discovery of an enzymic transformation of ribulose-5-phosphate
        (Ru5P) to a compound other than ribose-5-phosphate (R5P) or xylulose-5-phos-
        phate (Xu5P) by ribosephosphate isomerase (RPI) from spinach chloroplasts. -
        Plant Physiol. *51* (Suppl.) : 41, 1973.

16815 - **KNOX, R.S.** : Transfer of electronic excitation energy in condensed systems. - - In : CHECCUCCI, A., WEALE, R.A. (ed.) : Primary Molecular Events in Photobiology. Pp. 45 - 77. Elsevier, Amsterdam - London - New York 1973. [Ps.]

16816 - **KNYPL, J.S.** : Effects of growth retarding compounds on chlorophyll accumulation and nitrate reductase activity in nitrate induced cucumber cotyledons. - Acta Soc. Bot. Pol. *42* : 431 - 439, 1973.

16817 - **KOBAYASHI, M., TCHAN, Y.T.** : Treatment of industrial waste solutions and production of useful by-products using a photosynthetic bacterial method. - Water Res. *7* : 1219 - 1224, 1973.

16818 - **KOBAYASHI, Y., NISHIMURA, M.** : Studies on ion transport in cells of photosynthetic bacteria. I. The key role of anions in determining the direction and magnitude of hydrogen ion flux. - J. Biochem. *74* : 1217 - 1226, 1973.

16819 - **KOBAYASHI, Y., NISHIMURA, M.** : Studies on ion transport in cells of photosynthetic bacteria. II. Analysis of reversed hydrogen ion change. - J. Biochem. *74* : 1227 - 1232, 1973.

16820 - **KOBAYASHI, Y., NISHIMURA, M.** : Studies on ion transport in cells of photosynthetic bacteria. III. The influence of uncouplers on hydrogen ion change. - J. Biochem. *74* : 1233 - 1238, 1973.

16821 - **KOBZA, F.** : Vliv různého stupně redukce postranních os na fotosyntetickou produktivitu skleníkových okurek. [Photosynthetic productivity of forced cucumbers as affected by various levels of reduction of lateral axes.] - Rostlinná Výroba *19* : 983 - 994, 1973. [Growth analysis; in Czech, ab : E, R.]

16822 - **KOCHUBEĬ, S.M.** : O primenimosti razlichnykh fizicheskikh modeleĭ v raschetakh perenosa ênergii v fotosinteticheskoĭ edinitse. [Suitability of different physical models for calculating energy transfer in the photosynthetic unit.] - Tr. mosk. Obshch. Ispyt. Prirody *49* (Problemy Biofotokhimii) : 112 - 115, 1973. [In R, ab : E.]

16823 - **KOCHUBEĬ, S.M., OSTROVSKAYA, L.K.** : Ferricyanide-induced absorption changes associated with oxidation of photosystem 1 centres. - Photosynthetica *7* : 252 - 256, 1973.

16824 - **KOCMANOVÁ, J., MINÁŘ, J.** : The influence of different ratios of iron upon the sorption of mineral elements and the growth of maize. - Scr. Fac. Sci. nat. Univ. Purkynianae brun., Biol. *3* (2) : 53 - 70, 1973. [Growth analysis, Chl.]

16825 - **KOFRANEK, A.M., ROBINSON, M.** : Tables for calculating desired light flux densities for horticultural crops. - Scientia Horticulturae *1* : 263 - 269, 1973.

16826 - **KOH, S., KUMURA, A.** : [Studies on matter production in wheat plant. I. Diurnal changes in carbon dioxide exchange of wheat plant under ᶠield conditions.] - Proc. Crop Sci. Soc. Jap. *42* : 227 - 235, 1973. [In Jap., ab : E.]

16827 - **KÖHLE, U., WINKLER, S.** : Produktion und Konkurrenzverhältnisse der Flechten am Märchensee bei Tübingen (SW-Deutschland). - Beitr. Biol. Pflanzen *49* : 251 - 271, 1973. [Ps.]

16828 - **KOLESNIKOV, P.A., ZORE, S.V., PSHENOVA, K.V., PETROCHENKO, E.I., PLETNIKOVA, N.K., MAKOVKINA, L.E., MUTUSKIN, A.A.** : Lokalizatsiya $NADH_2$ ($NADFH_2$) : p-benzokhinon-oksidoreduktaz v kletochnykh fraktsiyakh list'ev gorokha. [Localization of $NADH_2$ ($NADPH_2$) : p-benzoquinone oxidoreductases in cell fractions of pea leaves.] - Fiziol. Rast. *20* : 170 - 174, 1973. [Chl; in R, ab : E.]

16829 - **KOLESOVA, L.S.** : Fiziologicheskoe deĭstvie preparatov na osnove ditiokarbaminovykh kislot na vinogradnoe rastenie i primenenie ikh v kachestve zashchitnykh fungitsidov. [Physiological action of preparates based on dithiocarbamic acids in grapevine plants and their utilization as protective fungicides.] - Tr. VSKhIZO *68* (Voprosy Fiziologii Rasteniĭ) : 60 - 75, 1973. [Chl, Car; in R.]

16830 - **KOLLMAN, V.H., HANNERS, J.L., HUTSON, J.Y., WHALEY, T.W., OTT, D.G., GREGG, C.T.** : Large-scale photosynthetic production of carbon-13 labeled sugars : the tobacco leaf system. - Biochem. biophys. Res. Commun. *50* : 826 - 831, 1973.

16831 - **KOLMAKOV, P.V., LAVIN, P.I., TITLYANOV, E.A.** : Radiatsionnyĭ rezhim verkhne-

sublitoral'nogo grota i fiziologicheskie pokazateli adaptatsii rasteniĬ k us-
loviyam sil'nogo zateneniya. [Radiation regime of upper sublitoral and physiological characteristics of plant adaptation to intense shading.] - In : Upravlenie Skorost'yu i Napravlennost'yu Biosinteza u RasteniĬ. P. 20. Krasnoyarsk
1973. [Ps; in R.]

16832 - KOLODCHENKO, N.A. : Zapasy fitomassy i ênergii v bereznyakakh kolochnoĬ leso-
stepi severnogo Kazakhstana. [Reserves of phytomass and energy in birch groves
of the grassy forest steppe of northern Kazakhstan.] - Izv. Akad. Nauk kaz.
SSR, Ser. biol. *1973* (1) : 29 - 36, 1973. [In R, ab : Kaz.]

16833 - KOMÁREK, J., HINDÁK, F., JAVORNICKÝ, P. : Ecology of the green kryophilic algae from Belanské Tatry Mountains (Czechoslovakia). - Arch. Hydrobiol., Sup-
plementb. *41* : 427 - 449, 1973. [Ps, Chl.]

16834 - KOMÁRKOVÁ, J. : Primary production of phytoplankton and periphyton in Opato-
vický fishpond (South Bohemia) in 1972. - In : HEJNÝ, S. (ed.) : Ecosystem
Study on Wetland Biome in Czechoslovakia. Czechosl. IBP/PT-PP Report No. 3.
Pp. 197 - 212. Třeboň 1973.

16835 - KOMISSAROV, G.G. : Fiziko-khimicheskoe modelirovanie struktury i funktsii prirodnykh fotosinteticheskikh sistem. [Physico-chemical modelling of the structure and function of natural photosynthetic systems.] - Zh. fiz. Khim. *47* :
1633 - 1642, 1973. [In R.]

16836 - KOMOR, E. : Proton-coupled hexose transport in *Chlorella vulgaris*. - FEBS
Lett. *38* : 16 - 18, 1973.

16837 - KOMOR, E., LOOS, E., TANNER, W. : A confirmation of the proposed model for
the hexose uptake system of *Chlorella vulgaris*. Anaerobic studies in the light
and in the dark. - J. Membrane Biol. *12* : 89 - 99, 1973.

16838 - KOMOV, S.V., BOCHENIN, V.F. : RadiatsionnyĬ rezhim posadok kartofelya i ego
svyaz' s nekotorymi fitometricheskimi kharakteristikami. [Radiation regime of
potato fields and its relation to some phytometric characteristics.] - In :
Voprosy Regulyatsii Fotosinteza. Vol. 3. Pp. 131 - 139, 165. Ural'. gos. Univ.,
Sverdlovsk 1973. [In R.]

16839 - KOMPANIETS, V.V., PUTSEĬKO, E.K. : Zavisimost' fotochuvstvitel'nosti v siste-
me RNK-khlorofill ot kontsentratsii adsorbirovannogo pigmenta. [Relationship
between photosensitivity in the system RNA-chlorophyll and concentration of
adsorbed pigment.] - Biofizika *18* : 550 - 553, 1973. [In R, ab : E.]

16840 - KONDRAT'EVA, E.N. : Puti assimilyatsii ugleroda pri fotosinteze. [Pathways
of photosynthetic carbon assimilation.] - In : Sovremennye Problemy Fotosinteza. Pp. 196 - 212. Izd. mosk. Univ., Moskva 1973. [In R.]

16841 - KONDYLAKI, S., ARGYROUDI-AKONYUNOGLOU, J.H. : Studies on the formation of grana. - Biochem. Biophys. News Lett. *1973* (4) : 4 - 5, 1973.

16842 - KONINGS, A.W.T., GUILLORY, R.J. : Resolution of enzymes catalyzing energy-
-linked transhydrogenation. IV. Reconstitution of adenosine triphosphate-dri-
ven transhydrogenation in depleted chromatophores of *Rhodospirillum rubrum* by
the transhydrogenase factor and a soluble oligomycin-insensitive Mg^{++}-adenosi-
ne triphosphatase. - J. biol. Chem. *248* : 1045 - 1050, 1973.

16843 - KONISHI, K., OGAWA, T., INOUE, Y., SHIBATA, K. : *In vivo* chlorophyll forms in
higher plants and algae sensitive to treatment with lutein. - Plant Cell Physiol. *14* : 227 - 236, 1973.

16844 - KONONENKO, A.A., LUKASHEV, E.P., RUBIN, A.B., SAMUILOV, V.D., TIMOFEEV, K.N.,
VENEDIKTOV, P.S. : On the nature of the light-induced bacteriochlorophyll absorbance changes in chromatophores of *Rhodospirillum rubrum*. - FEBS Lett. *30*:
239 - 242, 1973.

16845 - KONONENKO, A.A., REMENNIKOV, S.M., RUBIN, A.B., RUBIN, L.B., VENEDIKTOV, P.S.,
LUKASHEV, E.P. : Kinetics of laser-induced oxidoreductions in the photosynthetic reaction centre of *Ectothiorhodospira shaposhnikovii*. - J. Photochem.
2 : 371 - 376, 1973/74.

16846 - KORBUT, V.L., KLESHNIN, A.F., MALINOVSKIĬ, A.V. : Optimizatsiya fotosinteticheskoĬ aktivnosti rasteniĬ regulirovaniem ikh obluchennosti. [Optimization

of photosynthetic activity of plants by regulation of their irradiance.] - In:
Upravlenie Skorost'yu i Napravlennost'yu Biosinteza u Rasteniǐ. Pp. 143 - 144.
Krasnoyarsk 1973. [In R.]

16847 - **KORDYUM, E.L., NEDUKHA, O.M.** : Dva typy dilennya khloroplastiv u chereshkakh
lystkiv akatsiǐ zhovtoǐ. [Two types of chloroplast division in leaf petioles
of *Caragana arborescens* LAM.] - Ukr. bot. Zh. *30* : 787 - 790, 812, 1973. [In
Ukr., ab : E, R.]

16848 - **KOREN'KOV, D.A., FILIMONOV, D.A., ZAKHAROV, V.N.** : Deǐstvie azotnykh udobreniǐ pri intensivnom ispol'zovanii lugovykh trav. III. Vliyanie azotnykh udobreniǐ na soderzhanie organicheskikh soedineniǐ v travakh. [Effect of nitrogen
fertilizers by intense usage of meadow grasses. III. Effect of nitrogen fertilizers on the content of organic compounds in grasses.] - Agrokhimiya *1973*
(12) : 3 - 12, 1973. [Car; in R.]

16849 - **KOROLEVA, O.Ya., SAPOZHNIKOV, D.I.** : O bystroǐ i medlennoǐ reaktsiyakh époksi-
datsii zeaksantina v violaksantinovom tsikle. [Fast and slow reactions of zeaxanthin epoxidation in the violaxanthin cycle.] - Dokl. Akad. Nauk SSSR *208* :
25,1 - 253, 1973. [In R.]

16850 - **KORONA, V.V., RUTKEVICH, N.M.** : K metodike opredeleniya produktov metabolizma
v svobodnom prostranstve list'ev. [Determining of metabolic products in apparent free space of leaves.] - In : Voprosy Regulyatsii Fotosinteza. Vol. 3.
Pp. 109 - 116, 163. Ural'. gos. Univ., Sverdlovsk 1973. [Photosynthates; in R.]

16851 - **KORTSCHAK, H.P., NICKELL, L.G.** : Photosynthetic carbon monoxide metabolism by
sugarcane leaves. - Plant Sci. Lett. *1* : 213 - 216, 1973.

16852 - **KORZH, B.V.** : Opredelenie skorosteǐ vozdushnogo potoka i ob"emov gazoprovodov.
[Determination of air flow rates and air tubing volumes.] - In : Metody Kompleksnogo Izucheniya Fotosinteza Vol. 2. Pp. 168 - 174. VASKHNIL, Leningrad
1973. [In R.]

16853 - **KORZHENEVSKAYA, T.G., GUSEV, M.V.** : Nekotorye kharakteristiki povedeniya sine-
-zelenoǐ vodorosli *Anabaena variabilis* v usloviyakh temnoty. [Some characteristics of behavior of the blue-green alga *Anabaena variabilis* in darkness.] -
Mikrobiologiya *42* : 963 - 968, 1973. [Ps, Chl, Car; in R, ab : E.]

16854 - **KOSHKIN, V.A., GALKIN, V.I., BYKOV, O.D.** : Sposob termostatirovaniya list'ev.
[Method of leaf thermoregulation.] - In : Metody Kompleksnogo Izucheniya Fotosinteza. Vol. 2. Pp. 175 - 180. VASKHNIL, Leningrad 1973. [Leaf chambers;
in R.]

16855 - **KOSINOVA, V.P., RUDIN, V.D.** : Vliyanie mikroélementov kobal'ta i molibdena na
urozhaǐ i kachestvo belokachannoǐ kapusty. [Effect of cobalt and molybdenum
microelements on yield and quality of white head cabbage.] - Nauch. Tr. stav-
ropol'. sel'skokhoz. Inst. *3* (36) (Primenenie Mikroélementov, Udobreniǐ i Sti-
mulyatorov v Sel'skom Khozyaǐstve) : 89 - 92, 1973. [In R.]

16856 - **KOSMAKOVA, V.E., ZVEREVA, E.G.** : Raspredelenie assimilyatov u rasteniǐ soi pri
zatoplenii pochvy. [Distribution of photosynthates in soybean plants on flood-
ed soil.] - Tr. biol.-pochv. Inst., nov. Ser. *20* [(123) Transport Assimilyatov
i Otlozhenie Veshchestv v Zapas u Rasteniǐ] : 204 - 208, 1973. [In R, ab : E.]

16857 - **KOSMAKOVA, V.E., ZVEREVA, E.G.** : Raspredelenie produktov fotosinteza v rasteniyakh soi v usloviyakh pereuvlazhneniya pochvy. [Distribution of photosynthates in soybean plants at surplus water content in soil.] - Uch. Zap. dal'-
nevost. gos. Univ. *61* (Ustoǐchivost' Rasteniǐ k Pereuvlazhneniyu Pochvy na
Dal'nem Vostoke i Deǐstvie v étikh Usloviyakh Mikroelementov) : 22 - 34, 1973.
[In R.]

16858 - **KOSTKA, W.** : Apparative Verbesserung der konduktometrischen Gaswechselbestim-
mung in Gelände - zugleich eine Anregung für den fakultativen Unterricht und
die "Wissenschaftlich-praktische Arbeit". - Wiss. Z. päd. Hochsch. "Liselotte
Herrmann" Günstrow, math.-naturwiss. Fak. *1973* (2) : 21 - 30, 1973.

16859 - **KOSTRIKOVA, L.N.** : Khloroplasty zarodysha i endosperma nekotorykh predstavite-
leǐ motyl'kovykh. [Chloroplasts of embryo and endosperm of some representatives of the *Viciaceae*.] - Byull. glav. bot. Sada *89* : 29 - 35, 1973. [In R.]

16860 - **KOSTYAEV, V.Ya.** : Deǐstvie fenola na vodorosli. [Effect of phenol on algae.]
         - Tr. Inst. Biol. vnutr. Vod Akad. Nauk SSSR *24* (27) - Vliyanie Fenola na Gid-
         robiontov : 98 - 113, 220, 1973. [Ps; in R.]

16861 - **KOTAŃSKA, M.** : Productivity of underground plant organs on some meadow and
         marsh communities of the Niepołomice Forest (southern Poland). - Bull. Acad.
         pol. Sci., Sér. Sci. biol. *21* : 555 - 560, 1973.

*16862 - **KOVALENKO, V.F.** : Vliyanie medi na fotosintez list'ev yabloni. [Effect of cop-
         per on apple leaf photosynthesis.] - Dokl. TSKhA *158* : 127 - 133, 1970. [In
         R.]

16863 - **KOVALENKO, V.F.** : Khlorofill v list'yakh yabloni pri mednom golodanii. [Chlo-
         rophyll in apple tree leaves under copper deficiency.] - Vestn. sel'skokhoz.
         Nauki *1973* (12) : 79 - 83, 1973. [In R, ab : E, F, G.]

16864 - **KOWALCZEWSKI, A., PREJS, K., SPODNIEWSKA, I.** : Seasonal changes of biomass of
         benthic algae in the littoral of Mikołajskie Lake. - Ekol. Pol. *21* : 209 -
         - 217, 1973. [Chl.]

16865   **KOYAMA, H., KAWANO, S.** : Biosystematic studies on *Maianthemum (Liliaceae - Po-*
         lygonatae). VII. Photosynthetic behaviour of *M. dilatatum* under changing tem-
         perate woodland environments and its biological implications. - Bot. Mag.
         (Tokyo) *86* : 89 - 101, 1973.

16866 - **KOZLOV, Yu.N., KISELEV, B.A., EVSTIGNEEV, V.B.** : Élektrokhimicheskoe issledo-
         vanie khlorofilla. II. Vosstanovlenie khlorofilla i dal'neǐshie prevrashcheni-
         ya anion-radikala. [Electrochemical study of chlorophyll. II. Reduction of
         chlorophyll and further transformations of anion-radical.] - Biofizika *18* :
         59 - 63, 1973. [In R, ab : E.]

16867 - **KOZLOVA, A.P.** : Intensivnost' fotosinteza i vodnyǐ rezhim grechikhi v svyazi
         s izbytkom ionov khlora v pochve. [Photosynthetic rate and water relations in
         buckwheat in relation to surplus of chlorine ions in soil.] - In : VECHER, A.
         S. (ed.) : Fotosintez i Ustoǐchivost' Rasteniǐ. Pp. 45 - 50. Nauka i Tekhnika,
         Minsk 1973. [In R.]

16868 - **KOZŁOWSKA, Z.** : Badanie deficytu wodnego u zbóż metodą Czerskiego. [Water de-
         ficit in cereals studied by Czerski's method.] - Hodowla Rośl. Aklim. Nasienn.
         17 : 433 - 439, 1973. [Ps; in Pol., ab : E, R.]

16869 - **KRAKKAI, I.** : Herbicides derived from chlortriazine as permanent stimulators
         of maize. - Proc. Res. Inst. Pomol. Skierniewice, Ser. E *1973* (3) : 351 - 357,
         1973. [Chl.]

*16870 - **KRÁL'OVIČ, J., KRÁL'OVÁ, V.** : O vplyve toxafénu na fotosyntézu a obsah chlo-
         rofylu lucerny. [The influence of toxaphene on the photosynthesis and the
         chlorophyll content of alfalfa.] - Agrochémia (Bratislava) *12* : 247 - 249,
         1972. [In Slovak.]

16871 - **KRANZ, A.R.** : Monogen kontrollierte Langzeit-Transformationen der Chlorophylle
         in Blättern von *Arabidopsis thaliana* (L.) HEYNH. - Z. Pflanzenphysiol. *70* :
         333 - 349, 1973.

16872 - **KRAPIVIN, V.F.** : Izuchenie vertikal'noǐ struktury ėkosistemy na osnove ee ma-
         tematicheskoǐ modeli. [Ecosystem vertical structure and its mathematical mo-
         del.] - Gidrobiol. Zh. *9* (2) : 5 - 10, 1973. [Primary production; in R, ab :
         E.]

16873 - **KRASNOBAJEW, V.** : X-ray inactivation studies of nonheme iron proteins. The
         action of X-rays on spinach ferredoxin in aqueous solutions. - Biophysik *10* :
         213 - 220, 1973.

16874 - **KRASNOVSKIǏ, A.A.** : Khlorofill i fotosintez. [Chlorophyll and photosynthesis.]
         - In : Sovremennye Problemy Fotosinteza. Pp. 64 - 84. Izd. mosk. Univ., Mosk-
         va 1973. [In R.]

16875 - **KRASNOVSKIǏ, A.A., BOKUCHAVA, E.M., DROZDOVA, N.N.** : Fotokhimicheskie okisli-
         tel'no-vosstanovitel'nye reaktsii bakteriokhlorofilla *b* fotosinteziruyushchikh
         bakteriǐ *Rhodopseudomonas viridis.* [Photochemical oxidation-reduction reactions
         of bacteriochlorophyll *b* of photosynthesizing bacteria *Rhodopseudomonas viri-*
         dis.] - Dokl. Akad. Nauk SSSR *211* : 981 - 984, 1973. [In R.]

16876 - **KRASNOVSKIĬ, A.A. ml.** : Issledovanie promezhutochnykh produktov fotoreaktsiĭ khlorofilla metodom termokhemilyuminestsentsii. [Study of intermediate products of chlorophyll photoreactions by thermochemiluminescence.] - Tr. mosk. Obshch. Ispyt. Prirody *49* (Problemy Biofotokhimii) : 59 - 67, 1973. [In R, ab : E.]

16877 - **KRASNOVSKIĬ, A.A. ml., ROMANYUK, V.A., LITVIN, F.F.** : O fosforestsentsii i zamedlennoĭ fluorestsentsii khlorofillov i feofitinov *a* i *b*. [Phosphorescence and delayed fluorescence of chlorophylls and pheophytins *a* and *b*.] - Dokl. Akad. Nauk SSSR *209* : 965 - 968, 1973. [In R.]

16878 - **KRASZNER-BERNDORFER, K., TELEGDY KOVÁTS, L.** : Plastochinone, Analytik und Vorkommen in Lebensmitteln. - Nahrung *17* : 693 - 701, 1973.

16879 - **KRAUSE, G.H.** : Einfluß des Sauerstoffs auf den Energiezustand der belichteten Blattzelle. - Ber. deut. bot. Ges. *86* : 197 - 202, 1973.

16880 - **KRAUSE, G.H.** : The high-energy state of the thylakoid system as indicated by chlorophyll fluorescence and chloroplast shrinkage. - Biochim. biophys. Acta *292* : 715 - 728, 1973.

16881 - **KRAUZ, V.O., TREUSHNIKOV, V.M., OPRITOV, V.A.** : Rol' potentsialov deĭstviya v osushchestvlenii regulyatornoĭ vzaimosvyazi osnovnykh biosintetĭcheskikh tsentrov vysshikh rastenii. [Role of action potentials in realizing the interrelation control of basic biosynthetic centres in higher plants.] - In : Upravlenie Skorost'yu i Napravlennost'yu Biosinteza u Rastenii. Pp. 21 - 22. Krasnoyarsk 1973. [Ps; in R.]

16882 - **KRAVCHUK, T.S.** : K metodam issledovaniya matematicheskikh modeleĭ regulirovaniya biotsenozov. [Methods of investigating mathematical models for biocenosis regulation.] - Gidrobiol. Zh. *9* (1) : 107 - 113, 1973. [In R.]

16883 - **KREMER, B.P.** : Isolation of mannitol from *Desmarestia viridis*. - Phytochemistry *12* : 609 - 610, 1973. [^{14}C-photosynthates.]

16884 - **KREMER, B.P.** : Untersuchungen zur Physiologie von Volemit in der marinen Braunalge *Pelvetia canaliculata*. - Mar. Biol. *22* : 31 - 35, 1973. [Ps.]

16885 - **KREMER, B.P., SCHMITZ, K.** : CO_2-Fixierung und Stofftransport in benthischen marinen Algen. IV. Zur ^{14}C-Assimilation einiger litoraler Braunalgen im submersen und emersen Zustand. - Z. Pflanzenphysiol. *68* : 357 - 363, 1973.

16886 - **KRENDELEVA, T.E.** : O vzaimosvyazi potoka élektronov v fotosinteze s protsessom fotofosforilirovaniya. [Relation of electron flow in photosynthesis and photophosphorylation processes.] - Tr. mosk. Obshch. Ispyt. Prirody *49* (Problemy Biofotokhimii) : 167 - 174, 1973. [In R, ab : E.]

16887 - **KRENDELEVA, T.E., ZAMAZOVA, L.M., TULBU, G.V.** : Fotosinteticheskaya aktivnost' khloroplastov gorokha, vyrashchennogo pri razlichnoĭ osveshchennosti. [Photosynthetic activity of chloroplasts of pea grown under various illuminance.] - Nauch. Dokl. vyssh. Shkoly, biol. Nauki *16* (9) : 76 - 82, 1973. [In R.]

16888 - **KRETZER, F.** : Molecular architecture of the chloroplast membranes of *Chlamydomonas reinhardi* as revealed by high resolution electron microscopy. - Amer. J. Bot. *60* (4 Suppl.) : 39, 1973.

16889 - **KRETZER, F.** : Molecular architecture of the chloroplast membranes of *Chlamydomonas reinhardi* as revealed by high resolution electron microscopy. - J. Ultrastruct. Res. *44* : 146 - 178, 1973.

16890 - **KRETZER, F.L.** : Molecular architecture of the chloroplast membranes of *Chlamydomonas reinhardi* as revealed by high resolution electron microscopy. - Diss. Abstr. Int. B *33* : 3482-B, 1973.

16891 - **KREY, J.** : Primary production in the Indian Ocean I. - In : ZEITSCHEL, B. (ed.) : The Biology of the Indian Ocean. Pp. 115 - 126. Springer-Verlag, Berlin - Heidelberg - New York 1973.

16892 - **KRIEDEMANN, P.E., TÖRÖKFALVY, E., SMART, R.E.** : Natural occurrence and photosynthetic utilisation of sunflecks by grapevine leaves. - Photosynthetica *7* : 18 - 27, 1973.

*16893 - **KRISHNAMURTHY, K., BOMMEGOWDA, A., VENUGOPAL, N., JAGANNATH, M.K., RAGHUNATHA,**

G., RAJASHEKARA, B.G. : Pattern of dry-matter accumulation and distribution in maize (*Zea mays* L.). - Indian J. Agron. *17* : 104 - 109, 1972. [Growth analysis.]

16894 - KRISHNAMURTHY, K., RAJASHEKARA, B.G., JAGANNATH, M.K., BOMMEGOWDA, A., RAGHU-NATHA, G., VENUGOPAL, N. : Photosynthetic efficiency of sorghum genotypes after head emergence. - Agron. J. *65* : 858 - 860, 1973.

16895 - KRISHNAMURTHY, K., SUNDARARAJ, V. : A survey of environmental features in a section of the Vellar-Coleroon estuarine system, South India. - Mar. Biol. *23* : 229 - 237, 1973. [Chl, Car.]

16896 - KRISTKALNE, S., KREICBERGS, O., VĪTOLA, Ā., GUBARE, G. : Apgaismojuma ietekme uz augu produktivitāti, slāpekļa, fosfora un kālija saturu un uzkrāšanos augā. [Effect of illuminance on plant productivity and content and accumulation of nitrogen, phosphorus and potassium.] - In : TautsaimniecĪbā DerĪgo Augu Agrotehnika un Selekcija. Pp. 89 - 104. Zinatne, Rīgā 1973. [In Latv., ab : R.]

16897 - KROGMANN, D.W. : Photosynthetic reactions and components of thylakoids. - In : CARR, N.G., WHITTON, B.A. (ed.) : The Biology of Blue-Green Algae. Pp. 80 - - 98. Blackwell sci. Publ., Oxford - London - Edinburgh - Melbourne 1973.

16898 - KROGMANN, D.W., BERG, S., DILLEY, R.A. : Macromolecular probes of chloroplast structure and function. - Fed. Proc. *32* : 670 Abs, 1973.

16899 - KRUEGER, A.P., KOTAKA, S., REED, E.J. : The effects of air ions plants. - In : The Sun in the Service of Mankind. Int. Congr. Pp. V.5-1 - V.5-7. Unesco House 1973. [Chloroplasts.]

16900 - KRUPINSKY, J.M., SCHAREN, A.L., SCHILLINGER, J.A. : Pathogenic variation in *Septoria nodorum* (BERK.) BERK. in relation to organ specificity, apparent photosynthetic rate and yield of wheat. - Physiol. Plant Pathol. *3* : 187 - - 194, 1973.

16901 - KU, H.S., McCURRY, S.D., TOLBERT, N.E. : Phosphogluconate phosphatase from bean leaves. - Plant Physiol. *51* (Suppl.) : 41, 1973. [Ps.]

16902 - KU, S.B., HUNT, L.A. : Effects of temperature on the morphology and photosynthetic activity of newly matured leaves of alfalfa. - Can. J. Bot. *51* : 1907- - 1916, 1973.

16903 - KUBICOVÁ, Z. : Using of clinical incubators for cultivating tissue cultures with possibilities of illumination. - Acta Fac. Rerum nat. Univ. Comenianae, Physiol. Plant. *6* : 51 - 59, 1973. [Photosynthesizing tissues.]

B16904 - KUBÍN, Š. : Zdroje Fotosyntheticky Účinného Záření a Metody Jeho Měření. [Photosynthetically Active Radiation - the Sources and the Methods of Measurement.] - Academia, Praha 1973. [In Czech.]

16905 - KUBOTA, F., AGATA, W., KAMATA, E. : [Dry matter production of forage plants. VIII. Influence of plant density on the dry matter production in forage plant populations.] - J. jap. Soc. Grassland Sci. *19* : 194 - 200, 1973. [In Jap., ab : E.]

16906 - KUBOTA, F., AGATA, W., KAMATA, E. : [Dry matter production of forage plants. IX. The effects of cutting frequency and manuring amount on the dry matter production of forage plants.] - J. jap. Soc. Grassland Sci. *19* : 201 - 207, 1973. [In Jap., ab : E.]

16907 - KUGRENS, P., WEST, J.A. : The ultrastructure of carpospore differentiation in the parasitic red alga *Levringiella gardneri* (SETCH.) KYLIN. - Phycologia *12*: 163 - 173, 1973. [Chloroplast.]

16908 - KUKIELSKA, C. : Primary productivity of crop fields. - Bull. Acad. Pol. Sci., Sér. Sci. biol. *21* : 109 - 115, 1973.

16909 - KUKUSHKIN, A.K., CHATORDAI, K. : K voprosu o mekhanizme medlennoĭ induktsii fluorestsentsii vysshikh rasteniĭ i vodorosleĭ. [Mechanism of slow induction of fluorescence in higher plants and algae.] - Biofizika *18* : 562 - 564, 1973. [In R, ab : E.]

16910 - KUKUSHKIN, A.K., TIKHONOV, A.N., BLYUMENFEL'D, L.A., RUUGE, E.K. : Teoreticheskoe issledovanie pervichnykh protsessov fotosinteza vysshikh rasteniĭ i vodorosleỹ. [Theoretical investigation of primary processes of photosynthesis of higher plants and algae.] - Dokl. Akad. Nauk SSSR *211* : 718 - 721, 1973. [Model; in R.]

16911 - **KULAEVA, O.N.** : Opredelenie tsitokininovoỹ aktivnosti veshchestv s pomoshch'-yu biotestov. [Determination of the cytokinin activity of substances by means of biological tests.] - In : Metody Opredeleniya Fitogormonov, Ingibitorov Rosta, Defoliantov i Gerbitsidov. Pp. 63 - 73. Nauka, Moskva 1973. [Chl; in R.]

16912 - **KULIEV, F.A., GASANOV, Z.M., RAMAZANOV, S.R., RZAEV, A.S.** : Vliyanie orosheniya na nekotorye biologicheskie i khozyaỹstvennye pokazateli subtropicheskoỹ khurmy. [Effect of irrigation on some biological and economical characteristics of subtropical *Diospyros*.] - Subtrop. Kul't. *1973* (5) : 45 - 49, 1973. [Ps productivity; in R.]

16913 - **KUL'TEBAEV, E.T.** : Soderzhanie organicheskogo ugleroda v list'yakh razlichnykh sortov yablon' v dnevnoỹ, sezonnoỹ dinamike i po yarusam krony. [Content of organic carbon in leaves of various apple tree cultivars in diurnal or seasonal dynamics and according to crown storey.] - Vestn. sel'skokhoz. Nauki Kazakhstana *16* (6) : 83 - 86, 1973. [In R, ab : Kazakh.]

16914 - **KUMAR, G.** : Vliyanie sistemy formirovaniya kustov na soderzhanie khlorofilla i intensivnost' fotosinteza v list'yakh vinograda sorta Rkatsiteli. [Effect of the system of shrub formation on chlorophyll content and photosynthetic rate in leaves of grapevine cv. Rkatsiteli.] - Tr. kishinev. sel'skokhoz. Inst. M.V. Frunze *118* (Vinogradar.) : 73 - 79, 105, 1973. [In R.]

16915 - **KUMAR, R., SILVA, L.** : Light ray tracing through a leaf cross section. - Appl. Opt. *12* : 2950 - 2954, 1973.

16916 - **KUNG, S.D., SAKANO, K., WILDMAN, S.G.** : Studies of the subunit structure of fraction I protein from tobacco leaves. - Plant Physiol. *51* (Suppl.) : 27, 1973.

16917 - **KÜNZLER, A., PFENNIG, N.** : Das Vorkommen von Bacteriochlorophyll a_P und a_{Gg} in Stämmen aller Arten der *Rhodospirillaceae*. - Arch. Mikrobiol. *91* : 83 - 86, 1973.

16918 - **KURGANOVA, L.N.** : Vliyanie γ-oblucheniya semyan na vosstanovitel'nuyu funktsiyu fotosinteticheskogo apparata. [Effect of γ-irradiation of seeds on the reducing function of the photosynthetic apparatus.] - Radiobiologiya *13* : 292 - 294, 1973. [In R, ab : E.]

16919 - **KUROIWA, S.** : Plant-community photosynthesis as related to insolation climate. - In : The Sun in the Service of Mankind. Pp. V.7-1 - V.7-10. Unesco, Paris 1973.

16920 - **KURSANOV, A.L.** : Fotosintez i transport assimilyatov v listovoỹ plastinke. [Photosynthesis and transport of photosynthates in the leaf blade.] - Tr. biol.-pochv. inst., dal'nevost. nauch. Tsentr Akad. Nauk SSSR, N.S. *20* [(123) Transport Assimilyatov i Otlozhenie Veshchestv v Zapas u Rasteniỹ] : 8 - 25, 1973. [In R, ab : E.]

16921 - **KURSANOV, A.L., PAVLINOVA, O.A.** : K voprosu o mekhanizme deỹstviya gidrazida maleinovoỹ kisloty na rost i sakharonakoplenie v sveklovichnom korne. [Mechanism of action of maleic hydrazide on growth and sugar accumulation in the bulb of sugar beet.] - In : Gidrazid Maleinovoỹ Kisloty kak Regulyator Rosta Rasteniỹ. Pp. 257 - 267, 365. Nauka, Moskva 1973. [Photosynthates; in R.]

16922 - **KURTH, H., KÜHNER, E.** : Untersuchungen am Phytoplankton des nördlichen Zentralatlantiks I. Das Phytoplankton der küstenfernen Region (offshore) auf 30° W in Relation zu einigen produktionsbestimmenden Umweltfaktoren. - Wiss. Z. Univ. Rostock, math.-naturwiss. Reihe *22* : 1159 - 1163, 1973.

16923 - **KURZ, W.G.W., GALLON, J.R., GAMBORG, O.L.** : Photosynthetic oxygen evolution and nitrogenase activity in the blue-green alga *Gleocapsa* sp. LB.795. - Plant Physiol. *51* (Suppl.) : 34, 1973.

16924 - **KUSAI, A., YAMANAKA, T.** : An NAD(P) reductase derived from *Chlorobium thiosulfatophilum* : purification and some properties. - Biochim. biophys. Acta *292* : 621 - 633, 1973.

16925 - **KUSAI, K., YAMANAKA, T.** : The oxidation mechanisms of thiosulphate and sulphide in *Chlorobium thiosulphatophilum* : roles of cytochrome *c*-551 and cytochrome *c*-553. - Biochim. biophys. Acta *325* : 304 - 314, 1973.

16926 - **KUSHNIRENKO, M.D., KRYUKOVA, E.V., PECHERSKAYA, S.N.** : Vodouderzhivayushchaya sposobnost' i belki list'ev i khloroplastov u rasteniĭ s razlichnoĭ ustoĭchivost'yu k zasukhe. [Water-retaining ability and proteins of leaves and chloroplasts in plants with different drought resistances.] - Fiziol. Rast. *20* : 582 - 589, 1973. [Chl; in R, ab : E.]

16927 - **KUTYURIN, V.M.** : Razlozhenie vody rasteniyami i mekhanizm fotosinteza. [Plant water splitting and mechanism of photosynthesis.] - In : Sovremennye Problemy Fotosinteza. Pp. 138 - 160. Izd. mosk. Univ., Moskva 1973. [In R.]

16928 - **KUTYURIN, V.M., ARTAMKINA, I.Yu., KORSUN, A.D., ANISIMOVA, I.N., MATVEEVA, I. V.** : O vzaimodeĭstvii okislennoĭ formy khlorofilla s vodoĭ. [Interaction between oxidized form of chlorophyll and water.] - Dokl. Akad. Nauk SSSR *212* : 243 - 245, 1973. [In R.]

16929   **KUTYURIN, V.M., SLAVNOVA, T.D., CHIBISOV, A.K.** : Vliyanie vody na vykhod kation-radikala khlorofilla v reaktsii fotookisleniya. [Effect of water on the yield of chlorophyll cation-radical during photooxidation.] - Biofizika *18* : 1004 - 1007, 1973. [In R, ab : E.]

16930 - **KUZ'MENKO, L.V.** : Pervichnaya produktsiya severnoĭ chasti Araviĭskogo morya. [Primary production of the Northern Arabian Sea.] - Okeanologiya *13* : 307 - 313, 1973. [In R, ab : E.]

16931 - **KUZ'MENKO, M.I.** : Neadekvatnost' pervichnoĭ produktsii i nakopleniya organicheskogo veshchestva u vodorosleĭ. [Inadequacy of algal primary production and accumulation of organic matter.] - Gidrobiol. Zh. *9* (4) : 73 - 75, 1973. [In R.]

16932 - **KUZNETSOVA, G.K., KHAZANOV, V.S.** : O spektral'noĭ ĕffektivnosti fotosinteza nekotorykh alkaloidonosnykh rasteniĭ pri adaptatsii k svetu raznogo kachestva. [Spectral effectiveness of photosynthesis of some alkaloid-bearing plants during adaptation to light of different quality.] - Fiziol. Rast. *20* : 554 - - 557, 1973. [In R, ab : E.]

16933 - **KVEDER, S., REVELANTE, N.** : Phytoplankton production in the North Adriatic (1967 - 1970). - Rapp. Commun. int. Mer médit. *21* : 441 - 443, 1973. [Chl.]

16934 - **KVĚT, J.** : Shoot biomass, leaf area index and mineral content in selected South Bohemian and South Moravian stands of common reed (*Phragmites communis* TRIN.). Results of 1968. - In : HEJNÝ, S. (ed.) : Ecosystem Study on Wetland Biome in Czechoslovakia. Czechosl. IBP/PT-PP Report No.3. Pp. 93 - 95. Třeboň 1973.

*16935 - **KVITKO, K.V.** : Mutatsionnyĭ analiz struktury genotipa zelenykh vodorosleĭ. [Mutation analysis of genotype structure in green algae.] - In : Geneticheskie Aspekty Fotosinteza. Tezisy Dokladov. Pp. 62 - 63. Donish, Dushanbe 1972. [In R.]

16936 - **KWIATKOWSKA, M., ANTOSZEWSKI, R.** : Cytoradioautographic studies on assimilation of $^{14}CO_2$ by strawberry leaves : Influence of abscisic acid. - Proc. Res. Inst. Pomol., Skierniewice, Ser. E *1973* (3) : 273, 1973.

16937 - **KYLIN, A., SUNDBERG, I., TILLBERG, J.-E.** : The number of sites for photophosphorylation *in vivo*. - In : Ninth International Congress of Biochemistry. Stockholm, 1 - 7 July, 1973. Abstract Book. P. 236. IUB, Stockholm 1973.

16938 - **LABORDE, J.A., SPURR, A.R.** : Chromoplast ultrastructure as affected by genes controlling grana retention and carotenoids in fruits of *Capsicum annuum*. - Amer. J. Bot. *60* : 736 - 744, 1973.

16939 - **LACH, H.-J., RUPPEL, H.G., BÖGER, P.** : Cytochrom 553 aus der Alge *Bumilleriopsis filiformis*. - Z. Pflanzenphysiol. *70* : 432 - 451, 1973.

16940 - **LADYGIN, V.G.** : Intensivnost' fotosinteza i nakoplenie pigmentov u iskhodnogo shtamma i mutantov *Chlamydomonas* v techenie vegetativnogo kletochnogo tsikla [Photosynthetic rate and accumulation of pigments in the parent strain and

mutants of *Chlamydomonas* during vegetative cellular cycle.] - Fiziol. Rast.
20 : 995 - 998, 1973. [In R, ab : E.]

*16941 - **LADYGIN, V.G., SADOVNIKOVA, L.G.** : Lokalizatsiya genov u pigmentnykh mutantov
Chlamydomonas reinhardi. [Localization of genes in pigment mutants of *Chlamy-domonas reinhardi*.] - In : Geneticheskie Aspekty Fotosinteza. Tezisy Dokladov.
Pp. 64 - 65. Donish, Dushanbe 1972. [In R.]

16942 - **LADYGIN, V.G., SADOVNIKOVA, L.G.** : Geneticheskiỹ analiz svetlo-zelenykh mu-
tantov *Chlamydomonas reinhardi*. [Genetic analysis of light-green mutants of
Chlamydomonas reinhardi.] - Genetika *9* (6) : 45 - 50, 1973. [Chl; in R, ab :
E.]

*16943 - **LADYGIN, V.G., SEMENOVA, G.A.** : Èlektronnomikroskopicheskoe izuchenie khloroplastov v zigotakh *Chlamydomonas reinhardi*. [Electron-microscopic study of
chloroplasts in zygotes of *Chlamydomonas reinhardi*.] - In : Geneticheskie Aspekty Fotosinteza. Tezisy Dokladov. Pp. 65 - 66. Donish, Dushanbe 1972. [In
R.]

16944 - **LADYGIN, V.G., SEMENOVA, G.A., TAGEEVA, S.V.** : Razvitie lamellyarnoỹ struktury
plastid u pigmentnykh mutantov *Chlamydomonas reinhardi*. [Plastid lamellar
structure development in the *Chlamydomonas reinhardi* pigment mutants.] - Tsitologiya *15* : 810 - 817, 1973. [In R, ab : E.]

16945 - **LAGUN, L.P.** : Vliyanie usloviỹ mineral'nogo pitaniya na fotosinteticheskuyu
deyatel'nost' yachmenya. [Effect of the conditions of mineral nutrition on
photosynthetic activity in barley.] - In : VECHER, A.S. (ed.) : Fotosintez i
Ustoỹchivost' Rasteniỹ. Pp. 51 - 57. Nauka i Tekhnika, Minsk 1973. [In R.]

16946 - **LAING, W.A., OGREN, W.L., HAGEMAN, R.H.** : Ribulose diphosphate carboxylase regulation of an oxygen response in soybean photosynthesis. - Plant Physiol. *51*
(Suppl.) : 40, 1973.

16947 - **LAĬSK, A.** : Matematicheskaya model' fotosinteza i fotodykhaniya. Obratimaya
fosforibulokinaznaya reaktsiya. [Mathematical model of photosynthesis and
photorespiration. Reversible phosphoribulokinase reaction.] - Biofizika *18* :
637 - 642, 1973. [In R, ab : E.]

16948 - **LANDE, A.** : Byglandsfjorden. Primary production and other limnological features in an oligotrophic Norwegian lake. - Hydrobiologia *42* : 335 - 344, 1973.

16949 - **LANDE, A.** : Studies on phytoplankton in relation to its production and some
physical-chemical factors in Lake Svinsjøen. - Arch. Hydrobiol. *72* : 71 - 86,
1973.

16950 - **LANDI, R., ANTONGIOVANNI, M.** : Contributo allo studio del carotene contenuto
nelle piante di sorgo : influenza di alcuni fattori biologici ed agronomici.
[Contribution to the study of carotene contents in sorghum plants : The influence of some biological and agronomic factors.] - Maydica *18* : 50 - 62,
1973. [In Span., ab : E.]

16951 - **LANDSBERG, J.J., JARVIS, P.G.** : A numerical investigation of the momentum balance of a spruce forest. - J. appl. Ecol. *10* : 645 - 655, 1973.

16952 - **LANDSBERG, J.J., JARVIS, P.G., SLATER, M.B.** : The radiation régime of a spruce forest. - In : SLATYER, R.O. (ed.) : Plant Response to Climatic Factors.
Pp. 411 - 418. Unesco, Paris 1973.

16953 - **LANDSBERG, J.J., POWELL, D.B.B., BUTLER, D.R.** : Microclimate in an apple or-
chard. - J. appl. Ecol. *10* : 881 - 896, 1973. [Energy balance.]

16954 - **LANG, N.J., WHITTON, B.A.** : Arrangement and structure of thylakoids. - In :
CARR, N.G., WHITTON, B.A. (ed.) : The Biology of Blue-Green Algae. Pp. 66 -
- 79. Blackwell sci. Publ., Oxford - London - Edinburgh - Melbourne 1973.

16955 - **LAPINA, L.P., BIKMUKHAMETOVA, S.A.** : Vliyanie NaCl i Na_2SO_4 na funktsional'-
nuyu aktivnost' fotosinteticheskogo apparata kukuruzy. [Effect of NaCl and
Na_2SO_4 on functional activity of photosynthetic apparatus of maize.] - Fiziol.
Rast. *20* : 798 - 805, 1973. [In R, ab : E.]

16956 - **LARCHER, W.** : Transformations and translocation of carbohydrates. - In :
PRECHT, H., CHRISTOPHERSEN, J., HENSEL, H., LARCHER, W. (ed.) : Temperature
and Life. Pp. 133 - 137. Springer-Verlag, Berlin - Heidelberg - New York 1973.
[Ps.]

16957 - **LARCHER, W., CERNUSCA, A., SCHMIDT, L.** : A. Stoffproduktion und Energiebilanz in Zwergstrauchbeständen auf dem Patscherkofel bei Innsbruck. - In : ELLEN-BERG, H. (ed.) : Ökosystemforschung. Pp. 175 - 194. Springer-Verlag, Berlin - Heidelberg - New York 1973.

16958 - **LARCHER, W., HEBER, U., SANTARIUS, K.A.** : Limiting temperatures for life functions. - In : PRECHT, H., CHRISTOPHERSEN, J., HENSEL, H., LARCHER, W. (ed.) : Temperature and Life. Pp. 195 - 292. Springer-Verlag, Berlin - Heidelberg - New York 1973. [Ps.]

16959 - **LARCHER, W., SCHMIDT, L., GRABHERR, G., CERNUSCA, A.** : Plant biomass and production of alpine shrubs heaths at Mt. Patscherkofel, Austria. - In : BLISS, L.C., WIELGOLASKI, F.E. (ed.) : Primary Production and Production Processes, Tundra Biome. Pp. 65 - 73. Tundra Biome Steering Comm., Edmonton 1973.

16960 - **LARPENT, J.-P., JACQUES, R., MONÉGER, R., ADABRA, Y.** : Sur les teneurs en caroténoïdes et en chlorophylles de thalles de *Draparnaldia mutabilis* (ROTH) CEDERG exposés à des radiations oligochromatiques. - Compt. rend. Acad. Sci. Paris, Sér. D *276* : 3417 - 3420, 1973.

16961 - **LARSON, G.L.** : A limnology study of a high mountain lake in Mount Rainier National Park, Washington State; USA. - Arch. Hydrobiol. *72* : 10 - 48, 1973. [Chl.]

16962 - **LARSON, P.R., DICKSON, R.E.** : Distribution of imported ^{14}C in developing leaves of eastern cottonwood according to phyllotaxy. - Planta *111* : 95 - 112, 1973.

16963 - **LARSSON, C., COLLIN, C., ALBERTSSON, P.-Å.** : The fine structure of chloroplast stroma crystals. - J. Ultrastruct. Res. *45* : 50 - 58, 1973.

16964 - **LASER, K.D.** : Plastids of sieve tube members in the stamen vascular bundle of *Sorghum bicolor (Gramineae)*. - Protoplasma *80* : 279 - 283, 1973.

16965 - **LASSIG, J., NIEMI, Å.** : Amounts of chlorophyll a in the Baltic during June and July 1969 and 1970. - Oikos *1973* (Suppl. 15) : 34 - 42, 1973.

16966 - **LATZKO, E., PAWLIZKI, K.** : Eigenschaften der photosynthetischen Glycerinaldehyd-3-Phosphat-Dehydrogenase. - Hoppe-Seyler's Z. physiol. Chem. *354* : 1216 - - 1217, 1973.

16967 - **LAUDI, G., MEDEGHINI BONATTI, P.** : Ultrastructure of chloroplasts of some *Chlamidospermae (Ephedra twediana, Gnetum montana, Welwitschia mirabilis)*. - Caryologia *26* : 107 - 114, 1973.

16968 - **LAVERGNE, D., BISMUTH, E.** : Simultaneous purification of two kinases from spinach leaves : Ribulose-5-phosphate kinase and phosphoglycerate kinase. - Plant Sci. Lett. *1* : 229 - 236, 1973.

16969 - **LAVOREL, J.** : Simulation par la méthode de Monte Carlo, d'un modèle d'unités photosynthétiques connectées. - Physiol. vég. *11* : 681 - 720, 1973.

16970 - **LAVOREL, J.** : Kinetics of luminescence in the $10^{-6} - 10^{-4}$ - s range in *Chlorella*. - Biochim. biophys. Acta *325* : 213 - 229, 1973.

16971 - **LAWANSON, A.O., ONWUEME, I.C.** : Effect of prior heat stress on protochlorophyll and chlorophyll formation in seedlings of *Colocynthis citrullus*. - Z. Pflanzenphysiol. *69* : 461 - 463, 1973.

16972 - **LAWLOR, D.W., MILFORD, G.F.J.** : The effect of sodium on growth of water-stressed sugar beet. - Ann. Bot. *37* : 597 - 604, 1973. [Ps.]

16973 - **LAWRIE, A.C., WHEELER, C.T.** : The supply of photosynthetic assimilates to nodules of *Pisum sativum* L. in relation to the fixation of nitrogen. - New Phytol. *72* : 1341 - 1348, 1973.

16974 - **LEBEDEV, S.I., ALEĬNIKOV, I.M.** : Pro uchast' fumarovoï i maleïnovoï kyslot u biosyntezi karotynoïdiv. [Participation of fumaric and maleic acids in carotenoid biosynthesis.] - Dopov. Akad. Nauk ukr. RSR, Ser. B : Geol., Geofiz., Khim., Biol. *1973* : 463 - 465, 1973. [In Ukr., ab : E, R.]

16975 - **LEBEDEV, S.I., SINGKH, P.** : Fotosinteticheskaya produktivnost' i kachestvo zerna ozimoĭ pshenitsy sortov Bezostaya 1 i Kavkaz v usloviyakh orosheniya na yuge USSR. [Photosynthetic productivity and grain quality of the winter

wheat cv. Bezostaya 1 and Kavkaz under irrigation in the south of the Ukrainian SSR.] - Fiziol. Biokhim. kul't. Rast. 5 : 451 - 455, 1973. [In R, ab :
E.]

16976 - LEBEDEVA, E.V., SIMONOV, V.M., NIKISHANOVA, T.I. : Osobennosti tekhnologii
bessubstratnogo kul'tivirovaniya i produktivnost' nekotorykh ovoshchnykh rastenii. [Technological features of cultivation without substrate and productivity of some vegetables.] - In : Upravlenie Skorost'yu i Napravlennost'yu
Biosinteza u Rastenii. Pp. 145 - 146. Krasnoyarsk 1973. [In R.]

16977 - LECHOWICZ, M.J., ADAMS, M.S. : Net photosynthesis of *Cladonia mitis* (SAND.)
from sun and shade sites on the Wisconsin Pine Barrens. - Ecology 54 : 413 -
- 419, 1973.

16978 - LECHOWSKI, Z. : The action spectrum in chloroplast translocation in multilay-
er leaf cells. - Acta Soc. Bot. Pol. 42 : 461 - 472, 1973.

16979 - LEE, J.H., OTA, Y. :[Interrelationship between the morphological and physiological characteristics of roots and shoots of rice plants.]- Bull. nat. Inst.
agr. Sci., Ser. D 24 : 61 - 105, 1973. [Chl; in Jap., ab : E.]

16980 - LEE, K.J., KOZLOWSKI, T.T. : Responses of woody plants to silicone antitrans-
pirant. - Amer. J. Bot. 60 (4 Suppl.) : 25, 1973. [Chl.]

16981 - LEE, R.E., THOMPSON, A. : The stromacentre of plastids of *Kalanchoë Pinnata*
PERSOON. - J. Ultrastruct. Res. 42 : 451 - 456, 1973.

16982 - LEE, S.S., WAUCHOPE, R.D., HAQUE, R., FANG, S.C. : Binding of mercury compounds
to the chloroplasts in relation to the inhibition of the Hill reaction. - Pestic. Biochem. Physiol. 3 : 225 - 229, 1973.

16983 - LEECH, R.M., RUMSBY, M.G., THOMSON, W.W. : Plastid differentiation, acyl lipid, and fatty acid changes in developing green maize leaves. - Plant Physiol.
52 : 240 - 245, 1973.

16984 - LEFORT-TRAN, M., COHEN-BAZIRE, G., POUPHILE, M. : Les membranes photosynthétiques des algues à biliproteines observées après cryodécapage. - J. Ultrastruct. Res. 44 : 199 - 209, 1973.

16985 - LEHMUSLUOTO, P.O., PESONEN, L. : Eutrophication in the Helsinki and Espoo sea
areas measured as phytoplankton primary production. - Oikos 1973 (Suppl. 15):
202 - 208, 1973.

16986 - LEIGH, J.S. Jr., DUTTON, P.L. : Identification of primary photosynthetic processes. - Ann. NY Acad. Sci. 222 : 838 - 845, 1973.

16987 - LEKESH, Ya., ZENISHCHEVA, L., BEZDEK, V. : Voprosy vysokoi produktivnosti ko-
rotkostebel'nykh sortov yachmenya. [Problems of high productivity in short-
-stalked barley cultivars.] - Rostlinná Výroba 19 : 549 - 558, 1973. [In R,
ab : Czech, E, G.]

16988 - LEMASSON, C., TANDEAU de MARSAC, N., COHEN-BAZIRE, G. : Role of allophycocyanin as a light-harvesting pigment in cyanobacteria. - Proc. nat. Acad. Sci.
USA 70 : 3130 - 3133, 1973.

16989 - LEMEUR, R. : A method for simulating the direct solar radiation regime in
sunflower, Jerusalem artichoke, corn and soybean canopies using actual stand
structure data. - Agr. Meteorol. 12 : 229 - 247, 1973.

16990 - LEMEUR, R. : Effects of spatial leaf distribution on penetration and interception of direct radiation. - In : SLATYER, R.O. (ed.) : Plant Response to
Climatic Factors. Pp. 349 - 356. Unesco, Paris 1973.

16991 - LEMOALLE, J. : L'énergie lumineuse et l'activité photosynthétique du phytoplancton dans le Lac Tchad. - Cah. ORSTOM, Sér. Hydrobiol. 7 : 95 - 116, 1973.

16992 - LEMON, E. : Predicting crop climate and net carbon dioxide exchange. - Photosynthetica 7 : 408 - 413, 1973.

16993 - LENZ, F., WILLIAMS, C.N. : Effect of fruit removal on net assimilation and
gaseous diffusive resistance of soybean leaves. - Angew. Bot. 47 : 57 - 63,
1973.

16994 - LEÓN, A.R. De, BRAUN, J.G. : Ciclo anual de la produccion primaria y su rela-

cion con los nutrientes en aguas canarias. [Annual cycle of primary production and its relation to the nutrients in Canary Island waters.] - Bol. Inst.
esp. Oceanogr. *167* : 1 - 24, 1973. [Chl; in Span., ab : E.]

16995 - **LEPPINK, G.J., THOMAS, J.B.** : On the identity of the 640-nm component observed in chlorophyll *b* absorption spectra *in vivo*. - Biochim. biophys. Acta *305*:
610 - 617, 1973.

16996 - **LESTER, J.N., GOLDSWORTHY, A.** : The occurrence of high CO_2-compensation points
in *Amaranthus* species. - J. exp. Bot. *24* : 1031 - 1034, 1973.

16997 - **LEVINE, R.P., DURHAM, H.A.** : The polypeptides of stacked and unstacked *Chlamydomonas reinhardi* chloroplast membranes and their relation to Photosystem
II activity. - Biochim. biophys. Acta *325* : 565 - 572, 1973.

16998 - **LEX, M., STEWART, W.D.P.** : Algal nitrogenase, reductant pools and photosystem
I activity. - Biochim. biophys. Acta *292* : 436 - 443, 1973.

16999 - **LHOSTE, J-M., GRIVET, J-P.** : Electron paramagnetic resonance in triplet states of photobiological interest : a phase sensitive method applied to chlorophylls and related compounds. - Advance. Rad. Res., Phys. Chem. *1* : 327 -
- 337, 1973.

17000 - **LI, B.D., TITLYANOV, E.A.** : Diapazon fenoticheskikh izmeneniĭ soderzhaniya
fotosinteticheskikh pigmentov u morskikh zelenykh vodorosleĭ v svyazi s ikh
ěkologicheskoĭ plastichnost'yu. [Range of phenotic changes in the content of
photosynthetic pigments in marine green algae in relation to their ecological
plasticity.] - In : Informatsionnye Materialy Akad. Nauk SSSR, Sibir. Otd.,
Sibir. Inst. Fiziol. Biokhim. Rast., Irkutsk *11* : 42 - 43, 1973. [In R.]

17001 - **LI, B.D., TITLYANOV, E.A.** : Rol' sveta v nakoplenii fotosinteticheskikh pigmentov khloroplastami morskikh prikreplennykh vodorosleĭ. [Role of light in
the accumulation of photosynthetic pigments by chloroplasts of marine attached algae.] - In : Upravlenie Skorost'yu i Napravlennost'yu Biosinteza u Rasteniĭ. P. 23. Krasnoyarsk 1973. [In R.]

17002 - **LIAAEN-JENSEN, S.** : Carotenoids anno 1972. - Planta med. *23* : 251 - 268, 1973.

17003 - **LIAAEN-JENSEN, S.** : Structural elucidation of carotenoids - a progress report.
- Pure appl. Chem. *35* : 81 - 112, 1973.

17004 - **LIBERA, W., ZIEGLER, H., ZIEGLER, I.** : Förderung der Hill-Reaktion und der
CO_2-Fixierung in isolierten Spinatchloroplasten durch niedere Sulfitkonzen-
trationen. - Planta *109* : 269 - 279, 1973.

17005 - **LICHTENTHALER, H.K.** : Hemmung der lichtinduzierten Bildung von Thylakoidlipi-
den durch Äthanol und durch Inhibitoren der Proteinsynthese. - Hoppe-Seyler's
Z. physiol. Chem. *354* : 1220, 1973.

17006 - **LICHTENTHALER, H.K.** : Regulation der Lipochinonsynthese in Chloroplasten. -
Ber. deut. bot. Ges. *86* : 313 - 329, 1973. [Chl, car.]

17007 - **LICHTENTHALER, H.K., VERBEEK, L.** : Die Hemmung der Carotinoidsynthese im
Stickstoffmangel. - Planta *112* : 265 - 271, 1973.

17008 - **LICHTLÉ, C.** : Ultrastructure du plaste de deux Rhodophycées : *Rhodochorton
purpureum* (LIGHFT.) ROSENVINGE et *Rhodothamniella floridula* (DILLWYN) J.
FELDMANN. - Compt. rend. Acad. Sci. Paris, Sér. D *277* : 1865 - 1868, 1973.

17009 - **LICHTLÉ, C.** : Dégénérescence du plaste du *Rhodochorton purpureum* (LIGHFT.)
ROSENVINGE. Rhodophycée. Acrochaétiale. - Compt. rend. Acad. Sci. Paris, Sér.
D *277* : 2341 - 2344, 1973.

17010 - **LIEN, S., GEST, H.** : On the identity of photo- and oxidative phosphorylation
coupling factors in *Rhodopseudomonas capsulata*. - Arch. Biochem. Biophys.
159 : 730 - 737, 1973.

17011 - **LIEN, S., GEST, H., SAN PIETRO, A.** : Regulation of chlorophyll synthesis in
photosynthetic bacteria. - J. Bioenerg. *4* : 423 - 434, 1973.

17012 - **LILLEY, R. McC., SCHWENN, J.D., WALKER, D.A.** : Inorganic pyrophosphatase and
photosynthesis by isolated chloroplasts. II. The controlling influence of
orthophosphate. - Biochim. biophys. Acta *325* : 596 - 604, 1973.

17013 - **LILLEY, R. McC., WALKER, D.A.** : The measurement of cyclic photophosphorylation in isolated chloroplasts by determination of hydrogen ion consumption.
An evaluation of the method using titration at constant pH. - Biochim. biophys. Acta *314* : 354 - 359, 1973.

17014 - **LIMAR', R.S., SAKHAROVA, O.V.** : Bystryĭ spektrofotometricheskiĭ metod opredeleniya pigmentov list'ev (po Nibomu). [Rapid spectrophotometric method for
determination of pigments in leaves (according to Nybom).] - In : Metody Kompleksnogo Izucheniya Fotosinteza. Vol. 2. Pp. 260 - 267. VASKHNIL, Leningrad
1973. [In R.]

17015 - **LIMBERGER, G.È., DZHANUMOV, D.A., VESELOVSKIĬ, V.A., TARUSOV, B.N.** : Dlitel'-
noe poslesvechenie list'ev i felodermy kory vetveĭ razlichnykh po zimostoĭ-
kosti form yabloni v svyazi s sostoyaniem pokoya. [Long-term afterglow of
leaves and phelloderm of the bark of shoots in apple-tree forms differing in
their winter hardiness in connection with the dormancy period.] - Sel'.-khoz.
Biol. *8* : 410 - 414, 1973. [In R, ab : E.]

17016 - **LINDEMAN, W.** : Emerson enhancement effect and the reactivation of photosynthesis in phosphate deficient *Lemna minor*. - Acta bot. neerl. *22* : 553 - 568,
1973.

17017 - **LIPSKAYA, G.A., MATVEENTSAVA, V.S., CHARKASKAYA, S.K.** : Uplyŭ roznykh spalu-
chénnyaŭ kobal'tu z inshymi mikraélementami na zmyanenne aktyŭnastsi réaktsyi
Khila. [Effect of various compounds of cobalt with other microelements on
changes in Hill reaction rate.] - Vestsi Akad. Navuk belarus. SSR, Ser. biyal.
Navuk *1973* (2) : 32 - 36, 138, 1973. [In Belorus., ab : R.]

17018 - **LIPSKAYA, G.A., SERGEĬCHIK, S.A., SERGEĬCHIK, A.A., MATVEENTSEVA, V.S.** :
Strukturnye i fiziologo-biokhimicheskie pokazateli fotosinteticheskogo apparata i urozhaĭ yachmenya pri vnesenii v pochvu raznykh doz kobal'to-
vykh mikroudobreniĭ. [Structural and physiological biochemical characteristics of photosynthetic apparatus and grain yield in barley under the soil application of various doses of cobalt microfertilizers.] - Agrokhimiya *1973*
(7) : 91 - 97, 1973. [In R.]

17019 - **LISOVSKIĬ, G.M., SID'KO, F.Ya., CHUCHALIN, A.I.** : Ispol'zovanie FAR posevom
pshenitsy v svetokul'ture. [Use of PhAR by wheat stand in light culture.] -
In : Upravlenie Skorost'yu i Napravlennost'yu Biosinteza u Rasteniĭ. Pp. 147 -
- 148. Krasnoyarsk 1973. [In R.]

*17020 - **LISOVSKIĬ, G.M., YAN, N.A.** : Produktivnost' i konkurentosposobnost' malokhlo-
rofil'nogo shtamma khlorelly. [Productivity and competing ability of a low-
-chlorophyll *Chlorella* strain.] - In : Geneticheskie Aspekty Fotosinteza. Tezisy Dokladov. Pp. 106 - 107. Donish, Dushanbe 1972. [In R.]

17021 - **LITTLE, C.H.A., LOACH, K.** : Effect of changes in carbohydrate concentration
on the rate of net photosynthesis in mature leaves of *Abies balsamea*. - Can.
J. Bot. *51* : 751 - 758, 1973.

17022 - **LITVIN, F.F.** : Sistema nativnykh form khlorofilla i ee funktsii v pervichnykh
protsessakh fotosinteza. [System of native forms of chlorophyll and its functions in primary processes of photosynthesis.] - In : Sovremennye Problemy
Fotosinteza. Pp. 175 - 195. Izd. mosk. Univ., Moskva 1973. [In R.]

17023 - **LITVIN, F.F., BELYAEVA, O.B., GULYAEV, B.A., SINESHCHEKOV, V.A.** : Organizat-
siya pigmentnoĭ sistemy fotosinteziruyushchikh organizmov i ee svyaz' s pervichnymi fotoprotsessami. [Organization of the pigment system of photosynthesizing organisms and its relation to primary photoprocesses.] - Tr. mosk.
Obshch. Ispyt. Prirody *49* : 132 - 147, 1973. [In R.]

17024 - **LITVINENKO, L.G., LEBEDEV, S.I.** : Izuchenie fotokhimicheskoĭ aktivnosti khloroplastov v svyazi s fluorestsentsieĭ khlorofilla u ryada sel'skokhozyaĭstvennykh rasteniĭ. [Photochemical activity of chloroplasts as connected with chlorophyll fluorescence in a number of crop plants.] - Nauch. Dokl. vyssh. Shkoly, biol. Nauki *16* (3) : 80 - 83, 1973. [In R.]

17025 - **LIU, P., WALLACE, D.H., OZBUN, J.L.** : Influence of translocation on photosynthetic efficiency of *Phaseolus vulgaris* L. - Plant Physiol. *52* : 412 - 415,
1973.

17026 - LÍVANSKÝ, K., PROKEŠ, B., DITTRT, F., BENEŠ, V. : Some problems of CO_2 absorption by algae suspensions. - Biotechnol. Bioeng. *1973* (Suppl. 4) : 513 - - 518, 1973.

17027 - LJUBEŠIĆ, N. : Transformations of plastids in white pumpkin fruits. - Acta bot. croat. *32* : 59 - 62, 1973.

17028 - LOACH, K., LITTLE, C.H.A. : Production, storage, and use of photosynthate during shoot elongation in balsam fir (*Abies balsamea*). - Can. J. Bot. *51* : 1161 - 1168, 1973.

17029 - LOACH, P.A., KATZ, J.J. : Primary photochemistry of photosynthesis : report on a conference held at Argonne National Laboratory, November 1971. - Photochem. Photobiol. *17* : 195 - 208, 1973.

17030 - LOBANOVA, A.S., POKROVSKAYA, M.Z. : Udlinenie srokov khraneniya kartofelya i ovoshcheĭ pri pomoshchi gidrazida maleinovoĭ kisloty. [Prolongation of the storage period of potato and vegetables by means of maleic hydrazide.] - In : Gidrazid Maleinovoĭ Kisloty kak Regulyator Rosta Rasteniĭ. Pp. 92 - 101, 362. Nauka, Moskva 1973. [Chl; in R.]

17031 - LODHI, M.A.K., NICKELL, G.L. : Effects of leaf extracts of *Celtis laevigata* on growth, water content, and carbon dioxide exchange rates of three grass species. - Bull. Torrey bot. Club *100* : 159 - 165, 1973.

17032 - LOEWENSCHUSS, H., WAKELYN, P.J. : Determination of chlorophyll *a* and *b* in plant extracts by combined sucrose thin-layer chromatography and atomic absorption spectrophotometry. - Anal. chim. Acta *63* : 230 - 235, 1973.

17033 - LOEWENSCHUSS, H., WAKELYN, P.J. : Occurrence of chlorophyll in dried bract of cotton plant. - J. agr. Food Chem. *21* : 319 - 321, 1973.

17034 - LÖFFELHARDT, W., LUDWIG, B., KINDL, H.: Thylakoid-gebundene L-Phenylalanin--Ammoniak-Lyase. - Hoppe-Seyler's Z. physiol. Chem. *354* : 1006 - 1012, 1973.

17035 - LOGAN, K.T. : Temporary stimulation of photosynthetic rate by a short photoperiod. - Bi-Monthly Res. Notes *29* : 1 - 2, 1973.

17036 - LOOS, J.A., van DOORN, R.Æ.H., ROOS, D. : A sensitive mechanized determination of ATP + ADP. - Anal. Biochem. *53* : 309 - 312, 1973.

17037 - LÖPPERT, H.G., BRODA, E. : ATP-Gehalt und ATP-Umsatz von *Chlorella* in Abhängigkeit von Belichtung und Belüftung. - Z. allgem. Mikrobiol. *13* : 499 - 506, 1973. [Ps.]

17038 - LORD, J.M., MERRETT, M.J. : The conversion of glycerate into pyruvate by *Chlorella* extracts. - New Phytol. *72* : 249 - 252, 1973.

17039 - LORENZEN, M., MITCHELL, R. : Theoretical effects of artificial destratification on algal production in impoundments. - Environm. Sci. Technol. *7* : 939 - - 944, 1973. [Productivity.]

17040 - LORIMER, G.H., ANDREWS,T.J. : Plant photorespiration - an inevitable consequence of the existence of atmospheric oxygen. - Nature *243* : 359 - 360, 1973.

17041 - LOSEV, A.P., ZEN'KEVICH, E.I., GURINOVICH, G.P. : Smeshannaya agregatsiya khlorofilla "*a*" s protokhlorofillom i feofitinom. [Mixed aggregation of chlorophyll *a* with protochlorophyll and pheophytin.] - Zh. prikl. Spektroskop. *19* : 262 - 268, 1973.

17042 - ĿOTOCKI, A., ŻELAWSKI, W. : Effect of ammonium and nitrate source of nitrogen on productivity of photosynthesis in Scots pine (*Pinus silvestris* L.) seedlings. - Acta Soc. Bot. Pol. *42* : 599 - 605, 1973.

17043 - LOTT, J.N.A., DARLEY, J.J. : Unusual thylakoid arrangements in plastids from *Cucurbita maxima* seed coat cells. - Cytobiologie *8* : 55 - 60, 1973.

*17044 - LOVELL, P.H., BARRATT, E., MOORE, K.G. : The rooted excised cotyledon as an experimental system. - New Phytol. *69* : 1185 - 1187, 1970. [Chl.]

17045 - LOVETT, J.V., CAMPBELL, D.A. : Effects of CCC and moisture stress on sunflower. - Exp. Agr. *9* : 329 - 336, 1973. [Stomatal resistance.]

17046 - LOVEYS, B.R., BIRD, A.F. : The influence of nematodes on photosynthesis in tomato plants. - Physiol. Plant Pathol. *3* : 525 - 529, 1973.

17047 - **LOVEYS, B.R., KRIEDEMANN, P.E.** : Rapid changes in abscisic acid-like inhibitors following alterations in vine leaf water potential. - Physiol. Plant. *28* : 476 - 479, 1973. [Ps.]

17048 - **LOVEYS, B.R., KRIEDEMANN, P.E., TÖRÖKFALVY, E.** : Is abscisic acid involved in stomatal response to carbon dioxide ? - Plant Sci. Lett. *1* : 335 - 338, 1973.

17049 - **LOZIER, R.H., BUTLER, W.L.** : Effects of photosystem II inhibitors on electron paramagnetic resonance signal II of spinach chloroplasts. - Photochem. Photobiol. *17* : 133 - 137, 1973.

17050 - **LOZOVA, G.I., SIDOROVA, I.A.** : Deyaki vlastyvosti fraktsiĭ pigmentvmishchuyuchykh kompleksiv, oderzhanykh gradientnym vysolyuvannyam. [Some properties of fractions of pigment containing complexes obtained by gradient salting out.] - Ukr. bot. Zh. *30* (1) : 15 - 20, 127, 1973. [In Ukr., ab : E, R.]

17051 - **LUEKING, D., TOKUHISA, D., SOJKA, G.** : Glycerol assimilation by a mutant of *Rhodopseudomonas capsulata.* - J. Bacteriol. *115* : 897 - 903, 1973. [Ps.]

17052 - **LUGO, A.E., RAMSEY, T., HOY, J.** : Chlorophyll studies in a southern hardwood forest in North Central Florida. - Florida Scientist *36* : 146 - 153, 1973.

17053 - **LUGOVAYA, E.Sh., ZAKHAROVA, N.I., KUTYURIN, V.M.** : Molekulyarnye vesa belkovykh sub"edinits lamell khloroplastov gorokha. [Molecular weights of protein subunits of pea chloroplast lamellae.] - Dokl. Akad. Nauk SSSR *212* : 1458 - - 1460, 1973. [In R.]

17054 - **LUKINA, G.A.** : Deĭstvie malykh doz fenola na fotosintez khlorelly. [Effect of small doses of phenol on photosynthesis in *Chlorella.*] - Tr. Inst. Biol. vnutr. Vod Akad. Nauk SSSR *24* (27) Vliyanie Fenola na Gidrobiontov : 114 - - 118, 220 - 221, 1973. [Ps; in R.]

17055 - **LUMPKIN, O., HILLEL, Z.** : Isolation of a fast decay in submillisecond *Chlorella* luminescence. - Biochim. biophys. Acta *305* : 281 - 291, 1973.

17056 - **LUNG, KHYONG VAN, LAPINA, L.P., STROGONOV, B.P.** : Izmenenie soderzhaniya osnovnykh élementov mineral'nogo pitaniya u rasteniĭ v usloviyakh zasoleniya. [Changes in the content of principle elements of mineral nutrition in plants in saline environment.] - Agrokhimiya *1973* (9) : 87 - 95, 1973. [Productivity; in R.]

17057 - **LÜNING, K., SCHMITZ, K., WILLENBRINK, J.** : CO_2 fixation and translocation in benthic marine algae. III. Rates and ecological significance of translocation in *Laminaria hyperborea* and *L. saccharina.* - Mar. Biol. *23* : 275 - 281, 1973.

17058 - **LUPATOV, V.M., KUTYURIN, V.M.** : Issledovanie kinetiki dezaktivatsii promezhutochnykh produktov, uchastvuyushchikh v protsesse obrazovaniya kisloroda pri fotosinteze. [Kinetics of deactivation of intermediates in oxygen formation in photosynthesis.] - Biofizika *18* : 930 - 931, 1973. [In R, ab : E.]

17059 - **LUPATOV, V.M., KUTYURIN, V.M., TRIFONOVA, I.A., AFANAS'EV, E.A.** : K voprosu o chisle fotokhimicheskikh reaktsiĭ, neobkhodimykh dlya osushchestvleniya protsessa razlozheniya vody i vydeleniya kisloroda rasteniyami pri fotosinteze. [Number of photochemical reactions necessary for the realization of the process of water decomposition and hydrogen liberation by plants during photosynthesis.] - Dokl. Akad. Nauk SSSR *210* : 1227 - 1229, 1973. [In R.]

17060 - **LUPATOV, V.M., TRIFONOVA, I.A., AFANAS'EV, E.A., KUTYURIN, V.M.** : Metod skorostnoĭ differentsial'noĭ kulonometrii v issledovaniyakh kinetiki vydeleniya kisloroda pri impul'snom osveshchenii rasteniĭ. [Method of high-speed differential coulombmetry applied for the investigation of kinetics of oxygen evolution during impulsive illumination of the plants.] - Fiziol. Rast. *20* : 1295 - 1299, 1973. [In R, ab : E.]

17061 - **LURIE, S.** : Delayed light and thermoluminescence studies of energy storage by the oxygen evolving photoreaction of photosynthesis. - Diss. Abstr. int. B *33* : 5152-B, 1973.

17062 - **LUST, N.** : La respiration et la photosynthèse de frênes qui poussent dans des conditions différentes. - Med. Fac. Landbouwwetensch. Rijksuniv. Gent *38* : 450 - 466, 1973. Sylva gandavensis *36* : 1 - 17, 1973.

17063 - **LUST, N.** : Étude sur la teneur en pigment de frênes qui croissent dans des conditions différentes. - Med. Fac. Landbouwwetensch. Rijksuniv. Gent *38* : 467 - 485, 1973. Sylva gandavensis *37* : 1 - 19, 1973.

17064 - **LUTSENKO, G.N., SAAKOV, V.S.** : Obnovlenie karotinoidov v zelenom rastenii. [Renovation of carotenoids in a green plant.] - Fiziol. Rast. *20* : 90 - 95, 1973. [In R, ab : E.]

17065 - **LÜTTGE, U.** : Photosynthetic O_2 evolution and apparent H^+ uptake by slices of greening barley and maize leaves in aerobic and anaerobic solutions. - Can. J. Bot. *51* : 1953 - 1957, 1973.

17066 - **LÜTTGE, U.** : Proton and chloride uptake in relation to the development of photosynthetic capacity in greening etiolated barley leaves. - In : ANDERSON, W.P. (ed.) : Ion Transport in Plants. Pp. 205 - 221. Academic Press, London - New York 1973.

17067 - **LÜTTGE, U., BALL, E.** : Ion uptake by slices from greening etiolated barley and maize leaves. - Plant Sci. Lett. *1* : 275 - 280, 1973. [Ps, Chl.]

17068 - **LUTZ, M., BRETON, J.** : Chlorophyll associations in the chloroplast : resonance Raman spectroscopy. - Biochem. biophys. Res. Commun. *53* : 413 - 418, 1973.

17069 - **LUUKKANEN, O.** : Havaintoja kuusen vapaapölytysjälkeläistöjen ja männyn metsikköalkuperien CO_2-aineen-vaihdunnasta. [CO_2 exchange in open-pollinated progenies of Norway spruce and provenances of Scots pine.] - Silva fennica *7* : 255 - 276, 1973. [In Fin., ab : E.]

17070 - **LUXMOORE, R.J., MILLINGTON, R.J., PETERS, D.B.** : Row-crop microclimate. - In: SLATYER, R.O. (ed.) : Plant Response to Climatic Factors. Pp. 377 - 388. Unesco, Paris 1973. [Radiation in canopy.]

17071 - **LYAKHNOVICH, Ya.P.** : O nekotorykh dlinnovolnovykh formakh khlorofilla u fotosinteziruyushchikh ob"ektov. [Some long-wave forms of chlorophyll in photosynthesizing objects.] - In : Formirovanie Pigmentnogo Apparata Fotosinteza. Pp. 88 - 104. Nauka i Tekhnika, Minsk 1973. [In R.]

17072 - **LYAKHNOVICH, Ya.P.** : Sostoyanie fotosinteziruyushchikh pigmentov pri dlitel'-nom funktsionirovanii kletok khlorelly v polnoĭ temnote. [State of photosynthesizing pigments under long-term functioning of *Chlorella* cells in darkness.] - In : VECHER, A.S. (ed.) : Fotosintez i Ustoĭchivost' Rasteniĭ. Pp. 58 - 68. Nauka i Tekhnika, Minsk 1973. [In R.]

*17073 - **LYASHCHENKO, I.F.** : Fenotipicheskaya izmenchivost' proyavleniya gena u razlichnykh tipov khlorofil'nykh mutatsiĭ podsolnechnika. [Phenotypic variability of gene display in different types of chlorophyll mutations in sunflower.] - Genetika *6* (5) : 169 - 172, 1970. [In R, ab : E.]

17074 - **LYNE, R.L., STEWART, W.D.P.** : Emerson enhancement of carbon fixation but not of acetylene reduction (nitrogenase activity) in *Anabaena cylindrica*. - Planta *109* : 27 - 38, 1973.

17075 - **LYTTLETON, J.W.** : Carbon dioxide incorporation by chloroplast extracts at high pH. - FEBS Lett. *38* : 4 - 6, 1973.

17076 - **MacCOLL, R., BERNS, D.S.** : Increased aggregation of *C*-phycocyanin produced by phenol and benzene. - Arch. Biochem. Biophys. *156* : 161 - 167, 1973.

17077 - **MacCOLL, R., HABIG, W., BERNS, D.S.** : Characterization of phycocyanin from *Chroomonas* species. - J. biol. Chem. *248* : 7080 - 7086, 1973.

17078 - **MACDOWALL, F.D.H.** : Growth kinetics of Marquis wheat. IV. Temperature dependence. - Can. J. Bot. *51* : 729 - 736, 1973. [Ps.]

17079 - **MACHE, R., LOISEAUX, S.** : Light saturation of growth and photosynthesis of the shade plant *Marchantia polymorpha*. - J. Cell Sci. *12* : 391 - 401, 1973.

17080 - **MACHE, R., ROZIER, C., LOISEAUX, S., VIAL, A.M.** : Synchronous division of plastids during the greening of cut leaves of maize. - Nature - new Biol. *242* : 158 - 160, 1973. [Chl.]

17081 - **MACNICOL, P.K.** : Metabolic regulation in the senescing tobacco leaf. II. Changes in glycolytic metabolite levels in the detached leaf. - Plant Physiol. *51* : 798 - 801, 1973. [Chl.]

17082 - **MACNICOL, P.K., YOUNG, R.E., BIALE, J.B.** : Metabolic regulation in the senescing tobacco leaf. I. Changes in pattern of ^{32}P incorporation into leaf disc metabolites. - Plant Physiol. *51* : 793 - 797, 1973. [Chl.]

17083 - **MADSEN, E.** : The effect of CO_2-concentration on development and dry matter production in young tomato plants. - Acta Agr. scand. *23* : 235 - 240, 1973. [Growth analysis.]

17084 - **MADSEN, E.** : Effect of CO_2-concentration on the morphological, histological and cytological changes in tomato plants. - Acta Agr. scand. *23* : 241 - 246, 1973. LChloroplasts in stomata.|

17085 - **MADZHAROVA, D., BENBASAM, E., BUBAROVA, M., CHAVDAROV, I.** : Novi formi listna i korenoplodna tselina. [New forms of curled celery and celeriac.] - Gradinar. lozar. Nauka *10* (3) : 85 - 93, 1973. [Chl; in Bulg., ab : E, R.]

17086 - **MAEDA, O., ICHIMURA, S.-E.** : On the high density of a phytoplankton population found in a lake under ice. - Int. Rev. ges. Hydrobiol. *58* : 673 - 689, 1973. [Ps.]

17087 - **MAEDA, O., ZAMMA, M., ICHIMURA, S.-E.** : Photosynthetic response of estuarine phytoplankton to salinity variations in their habitat. - Mer (Tokyo) *11* (3) : 137 - 140, 1973.

*17088 - **MAGAZZŮ, G.** : Risultati su un ciclo annuale di osservazioni sulla produzione primaria con il metodo del C^{14} nelle acque costiere del Basso Tirreno. [Results of the annual cycle of primary production observations with the ^{14}C method in coastal waters of Basso Tirreno.] - Pubbl. Staz. Zool. Napoli *38* (Suppl.) : 44, 1970. [In Ital.]

17089 - **MAGAZZŮ, G., ANDREOLI, C.** : Ciclo annuale della produzione primaria e del fitoplancton in una zona di avamporto (Milazzo). [Annual cycle of primary production of phytoplankton in the outer harbour of Milazzo.] - In : Atti del 5° Colloquio Internazionale di Oceanografia Medica. Pp. 379 - 398. Ed. Libraria Bonanzinga, Messina 1973. [In Ital., ab : E, F.]

17090 - **MAGAZZŮ, G., CRESCENTI, N., AINIS, L.** : II. Nota sulla distribuzione della biomassa planctonica al largo delle coste orientali Siciliane (estate 1972). [II. A note on the planktonic crops distribution off the eastern Sicily coast (summer 1972).] - Mem. Biol. mar. Oceanogr., N.S. *3* : 51 - 84, 1973. [Chl; in Ital., ab : E.]

17091 - **MAGOMEDOV, I.M., CHERNYAD'EV, I.I., KOVALEVA, L.B., DOMAN, N.G.** : O lokalizatsii ribulozodifosfatkarboksilazy v list'yakh rastenii s "C_4-putem fotosinteza". [Localization of ribulosediphosphate carboxylase in leaves of plants with the C_4-pathway of photosynthesis.] - Dokl. Akad. Nauk SSSR *213* : 737 - 738, 1973. [In R.]

17092 - **MAGOMEDOV, I.M., KOVALEVA, L.B.** : O lokalizatsii fosfoenolpiruvatkarboksilazy v list'yakh kukuruzy. [Localization of phosphoenolpyruvate carboxylase in maize leaves.] - Dokl. Akad. Nauk SSSR *209* : 1467 - 1469, 1973. [In R.]

17093 - **MAGOMEDOV, I.M., KOVALEVA, L.B., NAZYMOK, A.E.** : Obrazovanie i transport organicheskikh kislot pri fotosinteze v list'yakh kukuruzy. [Formation and transport of organic acids in photosynthesizing tissues of maize leaves.] - Tr. biol.-pochv. Inst., dal'nevost. nauch. Tsentr Akad. Nauk SSSR *20* (123) : 48 - - 52, 1973. [In R, ab : E.]

17094 - **MAGOMEDOV, I.M., SAAKOV, V.S.** : Vtoraya proizvodnaya spektrov pogloshcheniya dvukh tipov khloroplastov list'ev *Zea mays* L. [Second derivative of absorption spectra of two types of chloroplasts from *Zea mays* L. leaves.] - Bot. Zh. *58* : 1201 - 1204, 1973. [In R.]

17095 - **MAGYAROSY, A.C., BUCHANAN, B.B., SCHÜRMANN, P.** : Effect of a systemic virus infection on chloroplast function and structure. - Virology *55* : 426 - 438, 1973.

17096 - **MAIER, R.** : C. Produktions- und Pigmentanalysen an *Utricularia vulgaris* L. - In : ELLENBERG, H. (ed.) : Ökosystemforschung. Pp. 87 - 101. Springer-Verlag, Berlin - Heidelberg - New York 1973.

17097 - **MAKAROV, A.D., PROTASHCHIK, V.A.** : Primenenie fluorestsentnogo metoda dlya odnovremennogo opredeleniya ATF i NADF.H_2O v suspenzii izolirovannykh khloroplas-

tov. [Simultaneous determination of ATP and $NADPH_2$ by fluorescent technique
in suspension of isolated chloroplasts.] - Fiziol. Rast. *20* : 646 - 648, 1973.
[In R, ab : E.]

17098 - **MAKAROV, A.D., STAKHOV, L.F.** : O roli fosfodoksina v protsesse fotofosforilirovaniya. [Role of phosphodoxin in photophosphorylation.] - Biokhimiya *38* :
785 - 789, 1973. [In R, ab : E.]

*17099 - **MAKAROVA, N.N., VISHNYAKOVA, I.I.** : Intensivnost' fotosinteza i napravlennost'
ottoka assimilyatov u geneticheski raznokachestvennykh sortov ozimoĭ rzhi.
[Photosynthetic rate and the direction of assimilate flow in genetically different winter rye cultivars.] - In : Geticheskie Aspekty Fotosinteza. Tezisy Dokladov. Pp. 107 - 108. Donish, Dushanbe 1972. [In R.]

17100 - **MAKEDONSKA, Ts.** : Nyakoi osobenosti v pogl"shchaneto i razpredelenieto na C^{14}
s dvegodishni igli ot *Pinus silvestris* L. s razlichno vodno s"d"rzhanie. [Cer-
tain features of the incorporation and distribution of ^{14}C in two-year needles
of *Pinus silvestris* L. with different water content.] - In : Vtora Natsionalna Konferentsiya po Botanika (1969). Pp. 427 - 435. Bulg. Acad. Sci., Inst.
Bot., Sofia 1973. [In Bulg.]

17101 - **MAKHAEVA, L.V.** : O ploshchadi list'ev travostoya nagornoĭ lugovoĭ stepi Kryma.
[Leaf area of the grass stand of the highland meadow steppe of Crimea.] - Bot.
Zh. *58* : 676 - 680, 1973. [Biomass components.]

17102 - **MAKHMET, B.M.** : Dinamika vmistu khlorofilu v klena gostrolistogo (*Acer platanoides* L.) protyagom vegetatsiĭnogo periodu zalezhno vid umov zatinennya. [Dy-
namics of chlorophyll in *Acer platanoides* L. during vegetation period in dependence on shading conditions.] - Ukr. bot. Zh. *30* : 151 - 154, 265, 1973.
[In Ukr., ab : E, R.]

17103 - **MAKSIMOVA, I.V., KUZNETSOVA, A.Ch.** : Zavisimost' obrazovaniya organicheskikh
veshchestv, vydelyaemykh kletkami *Chlorella pyrenoidosa* ot protsessa fotosinteza. [Effect of photosynthesis on production of organic compounds liberated
by the cells of *Chlorella pyrenoidosa*.] - Mikrobiologiya *42* : 969 - 975, 1973.
[In R, ab : E.]

17104 - **MAKUS, D.J., PHARR, D.M., LOWER, R.L.** : Effect of applied gibberellins on leaf
area and leaf chlorophyll content in cucumber cultivars. - HortScience *8* :
494 - 496, 1973.

17105 - **MALKIN, R., BEARDEN, A.J.** : Detection of a free radical in the primary reaction of chloroplast photosystem II. - Proc. nat. Acad. Sci. USA *70* : 294 -
- 297, 1973.

17106 - **MALKIN, R., BEARDEN, A.J.** : Light-induced changes of bound chloroplasts plastocyanin as studied by EPR spectroscopy : The role of plastocyanin in noncyclic photosynthetic electron transport. - Biochim. biophys. Acta *292* : 169 -
- 185, 1973.

17107 - **MALKIN, R., BEARDEN, A.J.** : The role of plastocyanin in the photosynthetic
electron transport chain as studied by EPR spectroscopy. - Ann. New York Acad.
Sci. *222* : 846 - 857, 1973.

17108 - **MALKIN, R., KNAFF, D.B.** : Effect of oxidizing treatment on chloroplast Photosystem II reactions. - Biochim. biophys. Acta *325* : 336 - 340, 1973.

17109 - **MALKIN, R., KNAFF, D.B., BEARDEN, A.J.** : The oxidation-reduction potential of
membrane-bound chloroplast plastocyanin and cytochrome *f*. - Biochim. biophys.
Acta *305* : 675 - 678, 1973.

17110 - **MALKIN, S., HARDT, H.** : Kinetic characterization of T-jump thermoluminescence
in isolated chloroplasts. - Biochim. biophys. Acta *305* : 292 - 301, 1973.

17111 - **MALL, L.P., TUGNAWAT, R.K.** : Biomass estimation of producer in grassland ecosystem. - Curr. Sci. *42* : 868 - 869, 1973.

17112 - **MALL, L.P., TUGNAWAT, R.K.** : Ecopathological investigations (1) Influence of
plant pathogenic fungus on productivity of *Dichanthium annulatum* (FORSK)
STAPF. - Flora *162* : 437 - 441, 1973. [Chl, Car.]

17113 - **MALONE, T.C., GARSIDE, C., ANDERSON, R., ROELS, O.A.** : The possible occurrence

of photosynthetic microorganisms in deep-sea sediments of the North Atlantic.
- J. Phycol. *9* : 482 - 488, 1973. [Ps, Chl.]

17114 - **MAL'YAN, A.N., MAKAROV, A.D:, OPANASENKO, V.K., KARTASHOV, I.M.** : Ob uchastii
magniya v protsesse fotofosforilirovaniya. [Role of magnesium in photophosphorylation.] - Biokhimiya *38* : 815 - 822, 1973. [In R, ab : E.]

17115 - **MANETAS, J., MICHELINAKI, M., MICHALOPOULOS, G., AKOYUNOGLOU, G.** : Protochlorophyllide biosynthesis from δ-aminolevulinic acid in etiolated leaves of
Phaseolus vulgaris. - Biochem. Biophys. News Lett. *1973* (4) : 6 - 7, 1973.

*17116 - **MANGAN, J.L., PRYOR, M.J.** : Quantitative studies on the degradation of chloroplasts in the rumen. - J. Physiol. *200* : 18P - 19P, 1968.

17117 - **MANN, K.H.** : Seaweeds : their productivity and strategy for growth. - Science
182 : 975 - 981, 1973.

17118 - **MANSFIELD, T.A., MARTIN, F.S., MEIDNER, H.** : The sun and the stomatal apparatus. - In : The Sun in the Service of Mankind. Pp. V.12-1 - V.12-10. Unesco,
Paris 1973.

*17119 - **MANSUROV, N.I., MECHISLAVSKIĬ, Yu.A., MEDVEDOVSKAYA, L.E.** : Izmenenie fotosinteticheskoĭ deyatel'nosti khlopchatnika pri gibridizatsii. [Variations of the
photosynthetic activity of cotton during hybridization.] - In : Geneticheskie
Aspekty Fotosinteza. Tezisy Dokladov. Pp. 108 - 109. Donish, Dushanbe 1972.
[In R.]

17120 - **MARANI, A., AVIELI, E.** : Heterosis during the early phases of growth in intra-
specific and interspecific crosses of cotton. - Crop Sci. *13* : 15 - 18, 1973.
[Growth analysis.]

17121 - **MARČENKO, E.** : Plastids of the yellow Y-1 strain of *Euglena gracilis*. - Protoplasma *76* : 417 - 433, 1973.

17122 - **MAR'ENKO, V.A., SAAKOV, V.S.** : Proizvodnaya spektrofotometriya na baze regis-
triruyushchego spektrofotometra *SF-10*. [Derivative spectrophotometry on the
basis of recording spectrophotometer *SF-10*.] - Fiziol. Rast. *20* : 637 - 645,
1973. [Chl; in R, ab : E.]

17123 - **MARGALEF, R.** : Fitoplancton marino de la región de afloramiento del NW de Áf-
rica. II. Composición y distribución del fitoplancton (campaña "Sahara II"
del "Cornide de Saavedra"). [Marine phytoplankton of the upwelling area of NW
Africa. II. Composition and distribution of the phytoplankton collected during
the cruise "Sahara II" of the research ship "Cornide de Saavedra".] - Res. Ex-
ped. cient. Buque oceanogr. Cornide 2 (Invest. Pesquera, Suppl.) : 65 - 94,
1973. [Chl; in Span., ab : E.]

17124 - **MAROC, J., GARNIER, J.** : Recherche du cytochrome *b*-563 et du *P*700 chez trois
mutants non photosynthétiques de *Chlamydomonas reinhardti*. - Biochim. biophys.
Acta *292* : 477 - 490, 1973.

17125 - **MARRS, B., GEST, H.** : Genetic mutations affecting the respiratory electron-
-transport system of the photosynthetic bacterium *Rhodopseudomonas capsulata*.
- J. Bacteriol. *114* : 1045 - 1051, 1973. [Ps, Chl.]

17126 - **MARRS, B., GEST, H.** : Regulation of bacteriochlorophyll synthesis by oxygen
in respiratory mutants of *Rhodopseudomonas capsulata*. - J. Bacteriol. *114* :
1052 - 1057, 1973.

17127 - **MARSCHNER, H., MIX, G.** : Einfluβ von Natriumchlorid und Mycostatin auf den
Mineralstoffgehalt im Blattgewebe und die Feinstruktur der Chloroplasten. -
Z. Pflanzenernährung Bodenkunde *136* : 203 - 219, 1973.

17128 - **MARSHALL, C., WARDLAW, I.F.** : A comparative study of the distribution and
speed of movement of ^{14}C assimilates and foliar-applied ^{32}P-labelled phosphate in wheat. - Aust. J. biol. Sci. *26* : 1 - 13, 1973.

17129 - **MARTIN, F.W., TELEK, L., RUBERTÉ, R.M.** : Yellow pigments of *Dioscorea bulbi-
fera*. - J. agr. Food Chem. *22* : 335 - 337, 1973. [Chl, Car.]

17130 - **MARTY, D.** : Aspects particuliers de l'ontogenèse des thylakoïdes dans les
feuilles panachées de *Coleus blumei* BENTH. - Compt. rend. Acad. Sci. Paris,
Sér. D *277* : 45 - 48, 1973.

17131 - **MARWAN, M.A., SELIM, A.K.A., EL-SAYED, S.I.** : The coordination of chlorophylls and carotenoids in the leaves of plants produced from irradiated seeds in different tomato cultivars. - Egypt. J. Bot. *16* : 437 - 447, 1973.

17132 - **MARX, J.L.** : Photorespiration : Key to increasing plant productivity ? - Science *179* : 365 - 367, 1973.

17133 - **MAŠEK, V.** : Wpływ pyłu z zakładu hutniczego na rośliny. [The effect of metallurgical industry dust on plants.] - Gaz, Woda Tech. sanit. *47* : 183 - 187, 1973. [Ps, Chl; in Pol.]

17134 - **MASLOVA, N.F., NYUTIN, Yu.I.** : Opticheskie svoľstva list'ev tomatov, vyrash-chennykh iz obluchennykh semyan. [Optical properties in leaves of tomatoes grown from irradiated seeds.] - Fiziol. Biokhim. kul't. Rast. *5* : 407 - 410, 1973. [In R, ab : E.]

17135 - **MATHEW, C., RAMADASAN, A.** : Chlorophyll stability index (C.S.I.) in different varieties and hybrids of coconut. - Curr. Sci. *42* : 584 - 585, 1973.

17136 - **MATHIEU, Y., TZENOVA, M.** : Influence de ferricyanure sur la fixation de CO_2 par les chloroplastes isolés d'épinard (*Spinacia oleracea* L.). - Photosynthetica *7* : 395 - 401, 1973.

17137 - **MATHIS, P., KLEO, J.** : The triplet state of β-carotene and of analog polyenes of different length. - Photochem. Photobiol. *18* : 343 - 346, 1973.

17138 - **MATHIS, P., SAUER, K.** : Chlorophyll formation in greening bean leaves during the early stages. - Plant Physiol. *51* : 115 - 119, 1973.

17139 - **MATHIS, P., VERMEGLIO, A.** : Period 2 oscillation for photoreduction of cytochrome *f* in chloroplasts. - In : Ninth International Congress of Biochemistry. Stockholm, 1 - 7 July, 1973. Abstract Book. P. 229. IUB, Stockholm 1973.

17140 - **MATIENCO, B.T.** : The accumulation of different compounds in the carotenoidoplasts. - Proc. Res. Inst. Pomol., Skierniewice, Ser. E *1973* : 419, 1973.

17141 - **MATSUDA, Y.** : Studies on chloroplast development in *Chlamydomonas reinhardtii* II. Effects of interposed darkness on chlorophyll synthesis. - Plant Cell Physiol. *14* : 815 - 821, 1973.

17142 - **MATSUZAKI, A., MATSUSHIMA, S., TOMITA, T.** : [Analysis of yield-determining process and its application to yield-prediction and culture improvement of lowland rice. CXVI. The effects of the growth amount at the starting time of nitrogen depletion treatment and the length of its treatment period on the growth control of rice plants.] - Proc. Crop Sci. Soc. Jap. *42* : 362 - 369, 1973. [Chl; in Jap., ab : E.]

17143 - **MAUDINAS, B., OELZE, J., VILLOUTREIX, J., REISINGER, O.** : The influence of 2-hydroxybiphenyl on membranes of *Rhodospirillum rubrum*. - Arch. Mikrobiol. *93* : 219 - 228, 1973. [Chl.]

17144 - **MAUZERALL, D.** : Why chlorophyll ? - Ann.New York Acad. Sci. *206* : 483 - 494, 1973.

17145 - **MAUZERALL, D., CHIVVIS, A.** : A novel cyclical approach to the oxygen producing mechanism of photosynthesis. - J. theor. Biol. *42* : 387 - 395, 1973.

17146 - **MAXWELL, J.R., COX, R.E., EGLINTON, G., PILLINGER, C.T., ACKMAN, R.G., HOOPER, S.N.** : Stereochemical studies of acyclic isoprenoid compounds - II. The role of chlorophyll in the derivation of isoprenoid-type acids in a lacustrine sediment. - Geochim. cosmochim. Acta *37* : 297 - 313, 1973.

17147 - **MAYNE, B.C., HOBBS, L.J.** : Benzoate induced to luminescence of isolated chloroplasts and algae. - Plant Physiol. *51* (Suppl.) : 66, 1973.

17148 - **MAYO, J.M., DESPAIN, D.G., van ZINDEREN BAKKER, E.M. Jr.** : CO_2 assimilation by *Dryas integrifolia* on Devon Island, Northwest Territories. - Can. J. Bot. *51* : 581 - 588, 1973.

17149 - **McCARTY, R.E., FAGAN, J.** : Light-stimulated incorporation of *N*-ethylmaleimide into coupling factor 1 in spinach chloroplasts. - Biochemistry *12* : 1503 - - 1507, 1973.

17150 - **McCLOUD, D.E.** : Growth analysis of high yielding peanuts. - Soil Crop Sci. Soc. Florida Proc. *33* : 24 - 26, 1973.

17151 - **McCREE, K.J.** : Are we ready to abandon the footcandle ? - Plant Physiol. *51* (Suppl.) : 20, 1973. [Ps.]

17152 - **McCREE, K.J.** : The measurement of photosynthetically active radiation. - Solar Energy *15* : 83 - 87, 1973.

17153 - **McDANIEL, M.D., SILVEY, J.K.G.** : Phytoplankton community structure in selected southwestern reservoirs. - J. Phycol. *9* (Suppl.) : 10, 1973. [Chl, Car.]

17154 - **McDAVID, C.R., SAGAR, G.R., MARSHALL, C.** : The effect of root pruning and 6--benzylaminopurine on the chlorophyll content, $^{14}CO_2$ fixation and the shoot//root ratio in seedlings of *Pisum sativum* L. - New Phytol. *72* : 465 - 470, 1973.

17155 - **McEVOY, F.A., LYNN, W.S.** : Chloroplast membrane proteins II. Solubilization of the lipophilic components. - J. biol. Chem. *248* : 4568 - 4573, 1973.

17156 - **McEVOY, F.A., LYNN, W.S.** : The peptides of chloroplast membranes. I. The soluble coupling factor (Ca^{2+}-ATPase). - Arch. Biochem. Biophys. *156* : 335 - - 341, 1973.

17157 - **McFADDEN, B.A.** : Autotrophic CO_2 assimilation and the evolution of ribulose diphosphate carboxylase. - Bacteriol. Rev. *37* : 289 - 319, 1973.

17158 - **McLAUGHLIN, M., SHANNON, S., SWEET, R.D.** : Effect of linuron on carrot varieties. - Proc. northeast. Weed Sci. Soc. *27* : 208 - 217, 1973. [Chl.]

17159 - **McMAHON, J.W.** : Membrane filter retention - a source of error in the ^{14}C method of measuring primary production. - Limnol. Oceanogr. *18* : 319 - 324, 1973.

*17160 - **McMULLEN, M.** : The use of infrared photography in characterizing chlorophyll mutants. - Barley Genet. Newslett. *2* : 121 - 122, 1972.

17161 - **McNAUGHTON, S.J.** : Comparative photosynthesis of Quebec and California ecotypes of *Typha latifolia.* - Ecology *54* : 1260 - 1270, 1973.

17162 - **McPHEE, J., BRODY, S.S.** : Photophosphorylation by monomolecular films at an air-water interface. - Proc. nat. Acad. Sci. USA *70* : 50 - 53, 1973.

17163 - **McPHERSON, H.G., SLATYER, R.O.** : Mechanisms regulating photosynthesis in *Pennisetum typhoides.* - Aust. J. biol. Sci. *26* : 329 - 339, 1973.

17164 - **McWILLIAM, J.R., PHILLIPS, P.J., PARKES, R.R.** : Measurement of photosynthetic rate using labelled carbon dioxide. - Aust. CSIRO Div. Plant Ind. tech. Pap. *31* : 1 - 12, 1973.

17165 - **MEASURES, M., WEINBERGER, P., BAER, H.** : Variability of plant growth within controlled-environment chambers as related to temperature and light distribution. - Can. J. Plant Sci. *53* : 215 - 220, 1973.

17166 - **MEFFERT, M.-E.** : Einfluss von pH, CO_2-Konzentration und Bakterien auf das Wachstum von Blaualge *Oscillatoria redekei* van GOOR. - Arch. Hydrobiol. *72* : 186 - 201, 1973. [Ps.]

17167 - **MEINZER, F.C., RUNDEL, P.W.** : Crassulacean acid metabolism and water use efficiency in *Echeveria pumila.* - Photosynthetica *7* : 358 - 364, 1973.

17168 - **MEISTER, A.** : Resolution enhancement of absorption spectra of chlorophylls by Fourier transformation. - Stud. biophys. *35* : 231 - 236, 1973.

17169 - **MEISTER, A.** : Untersuchungen zur Zusammensetzung der Blattpigmente an Mutanten von *Lycopersicon esculentum, Lycopersicon pimpinellifolium* und *Antirrhinum majus.* - Kulturpflanze *21* : 295 - 311, 1973.

17170 - **MELESHKO, G.I., ANTONYAN, A.A., KAZAKOV, A.I., LEBEDEVA, E.K.** : Issledovanie vliyaniya povyshennykh kontsentratsiĭ kisloroda na metabolizm khlorelly. [Effect of increased oxygen concentrations on *Chlorella* metabolism.] - Kosmich. Biol. Med. *1973* (2) : 41 - 44, 1973. [Ps; in R, ab : E.]

17171 - **MENKE, W.** : Proteins of the thylakoid membrane, properties and functions. - Physiol. vég. *11* : 231 - 238, 1973.

17172 - **MENKE, W., HIRTZ, R.-D.** : The secondary structure of proteins in the thylakoid membrane. - Z. Naturforsch. *28c* : 128 - 130, 1973.

17173 - **MENKE, W., RADUNZ, A., KOENIG, F.** : Membrane and vesicle formation from fragments and proteins of thylakoids. - Z. Naturforsch. *28c* : 63 - 65, 1973.

17174 - **MENZEL, D.W., DUNSTAN, W.M.** : Growth measurements by analysis of carbon. - In: STEIN, J.R. (ed.) : Handbook of Phycological Methods. Culture Methods and Growth Measurements. Pp. 313 - 320. Cambridge Univ. Press, London 1973.

17175 - **MEREZHINSKIĬ, Yu.G., LAPINA, T.V.** : Fotofosforilirovanie i pervichnye produkty fotosinteza pri deĭstvii betanala. [Photophosphorylation and primary products of photosynthesis affected by betanal.] - Fiziol. Biokhim. kul't. Rast. . 5 : 293 - 297, 1973. [In R.]

17176 - **MERGENHAGEN, D., SCHWEIGER, H.G.** : Recording the oxygen production of a single *Acetabularia* cell for a prolonged period. - Exp. Cell Res. *81* : 360 - 364, 1973.

17177 - **MERRETT, M.J., LORD, J.M.** : Glycollate formation and metabolism by algae. - New Phytol. *72* : 751 - 767, 1973.

17178 - **MESSIKHA-KHANNA, R.G.** : Metodika issledovaniya rastitel'nykh pigmentov, prisutstvuyushchikh v morskoĭ vode. [Methods for studying plant pigments in sea water.] - Okeanologiya *13* : 704 - 705, 1973. [In R, ab : E.]

B17179 - Metody Kompleksnogo Izucheniya Fotosinteza. Vypusk 2. [Methods of Complex Research in Photosynthesis. Volume 2.] - Leningrad 1973. [In R.]

B17180 - Metody Vydeleniya i Issledovaniya Belkov - Kompontov Fotosinteticheskogo Apparata. [Methods of Isolation and Studying of Proteins - Components of Photosynthetic Apparatus.] - Akad. Nauk SSSR, Inst. Fotosinteza, Pushchino-na-Oke 1973. [In R.]

17181 - **MEYER, T.E., KENNEL, S.J., TEDRO, S.M., KAMEN, M.D.** : Iron protein content of *Thiocapsa pfennigii*, a purple sulfur bacterium of atypical chlorophyll composition. - Biochim. biophys. Acta *292* : 634 - 643, 1973.

*17182 - **MICHAELIS, P.** : Beiträge zum Problem der Plastiden-Abänderung IV. Über das Plasma- und Plastidenabänderungen auslösende, isotopen-(^{32}P)-induzierte Kerngen mp_1 von *Epilobium*. - Mol. gen. Genet. *101* : 257 - 306, 1968.

*17183 - **MICHAELIS, P.** : Beiträge zum Problem der Plastidenabänderung V. Über eine weitere isotopen-(^{35}S)-induzierte Kernmutante, die Plastidenabänderungen hervorruft. - Theor. appl. Genet. *38* : 314 - 320, 1968.

17184 - **MICHAELS, A., GIBOR, A.** : Ultrastructural changes in *Euglena* after ultraviolet irradiation. - J. Cell Sci. *13* : 799 - 809, 1973. [Chloroplast.]

17185 - **MICHEL, J.-M., SIRONVAL, C.** : Effet de la fréquence d'éclairs de faible intensité sur les caractères des changements spectraux produits dans une feuille étiolée. - Physiol. vég. *11* : 291 - 300, 1973.

17186 - **MICHEL, J.P., THIBAULT, P.** : Etude cinétique de la synthèse d'ATP *in vivo* en lumière mono- et bichromatique chez *Zea mays* : Effet antagoniste rouge-rouge lointain. - Biochim. biophys. Acta *305* : 390 - 396, 1973.

17187 - **MIKA, A., ANTOSZEWSKI, R.** : Photosynthesis and distribution of photosynthates in apple shoots treated by pinching and bark ringing. - Biol. Plant. *15* : 202- - 207, 1973.

17188 - **MIKHAILOVA, T.P., SHCHUKA, N.V., DAVIDENKO, R.G.** : Regulirovanie protsessa sozrevaniya list'ev tabaka s pomoshch'yu fiziologicheski aktivnykh veshchestv. [Regulation of maturation of tobacco leaves by means of physiologically active substances.] - Fiziol. Rast. *20* : 138 - 143, 1973. [Chl; in R, ab : E.]

17189 - **MIKHALEVA, E.N., SAZYKINA, N.A., KONOVALOV, I.N.** : Vliyanie zamorozkov na intensivnost' nekotorykh fiziologicheskikh protsessov u gorokha. [Effect of slight frosts on the rate of some physiological processes in pea.] - Byull. glav. bot. Sada *89* : 65 - 70, 1973. [Ps; in R.]

17190 - **MIKHEEVA, S.A., RYBIN, I.A.** : Zavisimost' perekhodnoĭ bioĕlektricheskoĭ reaktsii na vklyuchenie sveta ot intensivnosti osveshcheniya. [Transient bioelectric reaction after switching light on and off as dependent on illuminance.] - In : Voprosy Regulyatsii Fotosinteza. Vol. 3. Pp. 77 - 83, 162. Ural'. gos. Univ., Sverdlovsk 1973. [In R.]

17191 - **MILES, C.D., DANIEL, D.J.** : A rapid screening technique for photosynthetic mutants of higher plants. - Plant Sci. Lett. *1* : 237 - 240, 1973.

17192 - **MILES, D., BOLEN, P., FARAG, S., GOODIN, R., LUTZ, J., MOUSTAFA, A., RODRI-GUEZ, B., WEIL, C.** : Hg^{++} - a DCMU independent electron acceptor of photosystem II. - Biochem. biophys. Res. Commun. *50* : 1113 - 1119, 1973.

17193 - **MILLER, K.R., STAEHELIN, L.A.** : Fine structure of the chloroplast membranes of *Euglena gracilis* as revealed by freeze-cleaving and deep-etching techniques. - Protoplasma *77* : 55 - 78, 1973.

B17194 - **MILLER, L.P.** (ed.) : Phytochemistry. Vol. I. The Process and Products of Photosynthesis. - Van Nostrand Reinhold Company, New York - Cincinnati - Toronto - London - Melbourne 1973.

17195 - **MILLER, P.C.** : A model of temperature, transpiration rates and photosynthesis of sunlit and shaded leaves in vegetation canopies. - In : SLATYER, R.O. (ed.): Plant Response to Climatic Factors. Pp. 427 - 434. Unesco, Paris 1973.

17196 - **MINCHIN, F.R., PATE, J.S.** : The carbon balance of a legume and the functional economy of its root nodules. - J. exp. Bot. *24* : 259 - 271, 1973.

17197 - **MININBERG, S.Ya., LE ZU** : Vliyanie zasoleniya i podkormki mikroėlementami na sostoyanie pigmentov i aktivnost' khlorofillazy v list'yakh fasoli. [Effect of salting and dressing with trace elements on the state of pigments and activity of chlorophyllase in bean leaves.] - Fiziol. Biokhim. kul't. Rast. *5* : 187 - 190, 1973. [in R, ab : E.]

17198 - **MINX, L.** : Studium modelu organizace porostů. I. Simulační pokusy a vyvození vzorců. [Studies in modelling crop stand organization. I. Simulation experiments and derivation of formulae.] - Acta Univ. Agr., Fac. agron., Ser. A (Brno) *21* : 523 - 530, 1973. [in Czech.]

17199 - **MINX, L.** : Studium modelu organizace porostů. III. Některé možnosti využití modelu pro hodnocení a řízení organizace porostů cukrovky. [Studies in modelling the crop stand organization. III. Potential uses of the model for evaluating and managing the sugar beet stand organization.] - Acta Univ. Agr., Fac. agron., Ser. A (Brno) *21* : 717 - 727, 1973. [in Czech, ab : E, G, R.]

17200 - **MINX, L., POKORNÁ, Z.** : Studium modelu organizace porostů. II. Prověření platnosti simulačního modelu v provozních podmínkách. [Studies in modelling the crop stand organization. II. Testing of applicability of the simulation model in the field.] - Acta Univ. Agr., Fac. agron., Ser. A (Brno) *21* : 712 - - 716, 1973. [in Czech.]

17201 - **MIOVIČ, M.L., GIBSON, J.** : Nucleotide pools and adenylate energy charge in balanced and unbalanced growth of *Chromatium*. - J. Bacteriol. *114* : 86 - 95, 1973. [Chl.]

*17202 - **MIROSHNICHENKO, Yu.M., MIROSHNICHENKO, N.V.** : Influence of the Caspian sea level sinking on succession and vegetation productivity. - In : Eco-physiological Foundation of Ecosystems Productivity in Arid Zone. Pp. 124 - 128. Nauka, Leningrad 1972. [Biomass.]

17203 - **MIROSLAVOVA, S.A., TISHCHENKO, N.N., VASIL'EV, B.R.** : O vliyanii norm azotnogo pitaniya na formirovanie fotosinteticheskogo apparata v list'yakh *Bryophyllum daigremontianum*. [Effect of the norms of nitrogen nutrition on the formation of photosynthetic apparatus in *Bryophyllum daigremontianum*.] - In : Upravlenie Skorost'yu i Napravlennost'yu Biosinteza u Rastenii. Pp. 55 - 56. Krasnoyarsk 1973. [in R.]

17204 - **MISHRA, D., KAR, M.** : Effect of benzimidazole and nickel ions in the control of *Oryza sativa* leaf senescence by red and far-red light. - Phytochemistry *12* : 1521 - 1522, 1973. [Chl.]

17205 - **MISHRA, D., KAR, M., PRADHAN, P.K.** : Chemical regulation of acid inorganic pyrophosphatase activity during senescence of detached rice leaves. - Exp. Gerontol. *8* : 165 - 167, 1973. [Chl.]

17206 - **MISHRA, D., MISRA, B.** : Retardation of induced senescence of leaves from crop plants by benzimidazole and cytokinins. - Exp. Gerontol. *8* : 235 - 239, 1973. [Chl.]

17207 - **MISHRA, D., PRADHAN, P.** : Regulation of senescence in detached rice leaves
by light, benzimidazole and kinetin. - Exp. Geront. *8* : 153 - 155, 1973.
[Chl.]

17208 - **MITROFANOV, B.A., OKANENKO, A.S., POCHINOK, Kh.N., ZRAZHEVSKIĬ, M.N., MAKHOV-
SKAYA, M.A., GOLIK, K.N.** : Vliyanie vnekornevykh podkormok ozimoĭ pshenitsy
mochevinoĭ na intensivnost' fotosinteza i azotnyĭ obmen. [Effect of urea
spray application on winter wheat on photosynthetic rate and grain quality.]
- Fiziol. Biokhim. kul't. Rast. *5* : 232 - 238, 1973. [In R, ab : E.]

17209 - **MITSUI, A., KEISTER, D.** : A low potential chromatophore-bound cytochrome *C*
in *Rhodospirillum rubrum*. - Abstr. annu. Meet. amer. Soc. Microbiol. *73* :
161, 1973.

17210 - **MITSUI, A., SAN PIETRO, A.** : Large ferredoxin crystals from a blue-green alga, *Phormidium foveolarum*. - Plant Sci. Lett. *1* : 157 - 163, 1973.

17211 - **MIYATANI, D., TAKEUCHI, T.** : Study of the simultaneous measurement of ^3H and
^{14}C by means of a gas-proportional counter. - Int. J. appl. Rad. Isotopes
24 : 553 - 562, 1973.

172¹ - **MIZRAKH, S.A.** : Temnoustoĭchivost' pigmentnogo apparata pshenitsy v zavisimosti ot temperatury vozdukha i vozrasta rasteniĭ. [Dark resistance of pigment apparatus of wheat as dependent on air temperature and plant age.] - In:
Upravlenie Skorost'yu i Napravlennost'yu Biosinteza u Rasteniĭ. Pp. 24 - 25.
Krasnoyarsk 1973. [In R.]

17213 - **MŁODZIANOWSKI, F., KWINTKIEWICZ, M.** : The inhibition of kohlrabi chloroplast
degeneration by kinetin. - Protoplasma *76* : 211 - 226, 1973.

17214 - **MŁODZIANOWSKI, F., MŁODZIANOWSKA, L.** : Chloroplast degeneration and its inhibition by kinetin in detached leaves of *Cichorium intybus* L. - Acta Soc. Bot.
Pol. *42* : 649 - 656, 1973.

17215 - **MŁODZIANOWSKI, F., PONITKA, A.** : Ultrastructural changes of chloroplasts in
detached parsley leaves yellowing in darkness and the influence of kinetin on
that process. - Z. Pflanzenphysiol. *69* : 13 - 25, 1973.

17216 - **MŁODZIANOWSKI, F., WIECZOREK, W.** : DNA configuration in chloroplasts of *Brassica oleracea* var. *gongylodes* and *Petroselinum sativum*. - Biochem. Physiol.
Pflanzen *164* : 429 - 437, 1973.

17217 - **MO, Y., HARRIS, B.G., GRACY, R.W.** : Triosephosphate isomerases and aldolases
from light- and dark-grown *Euglena gracilis*. - Arch. Biochem. Biophys. *157* :
580 - 587, 1973.

17218 - **MOESTRUP, Ø., HOFFMAN, L.R.** : Ultrastructure of the green alga *Dichotomosip-
hon tuberosus* with special reference to the occurrence of striated tubules in
the chloroplast. - J. Phycol. *9* : 430 - 437, 1973.

17219 - **MOGILEVA, G.A., ZELENSKIĬ, M.I.** : O primenenii reaktsii Khilla dlya otsenki
fotosinteticheskoĭ deyatel'nosti kul'turnykh rasteniĭ. [Use of the Hill reaction for estimation of photosynthetic activity of cultivated plants.] - In :
Metody Kompleksnogo Izucheniya Fotosinteza. Vol. 2. Pp. 224 - 243. VASKHNIL,
Leningrad 1973. [In R.]

17220 - **MOHANTY, P., BRAUN, B.Z., GOVINDJEE** : Light-induced slow changes in chlorophyl
a fluorescence in isolated chloroplasts : Effects of magnesium and phenazine
methosulfate. - Biochim. biophys. Acta *292* : 459 - 476, 1973.

17221 - **MOHANTY, P., GOVINDJEE** : Light-induced changes in the fluorescence yield of
chlorophyll *a* in *Anacystis nidulans*. I. Relationship of slow fluorescence
changes with structural changes. - Biochim. biophys. Acta *305* : 95 - 104,
1973.

17222 - **MOHANTY, P., GOVINDJEE** : Light-induced changes in the fluorescence yield of
chlorophyll *a* in *Anacystis nidulans* II. The fast changes and the effect of
photosynthetic inhibitors on both the fast and slow fluorescence induction. -
- Plant Cell Physiol. *14* : 611 - 629, 1973.

17223 - **MOHANTY, P., GOVINDJEE** : Effect of phenazine methosulfate and uncouplers on
light induced chlorophyll *a* fluorescence yield changes in intact algal cells.
- Photosynthetica *7* : 146 - 160, 1973.

17224 - **MOKRONOSOV, A.T., BAGAUTDINOVA, R.I., BUBNOVA, E.A., KOBELEVA, I.V.** : Foto-sinteticheskiĭ metabolizm v palisadnoĭ i gubchatoĭ tkanyakh lista. [Photosynthetic metabolism in palisade and spongy leaf tissues.] - Fiziol. Rast. *20* : 1191 - 1197, 1973. [In R, ab : E.]

17225 - **MOKRONOSOV, A.T., BAGAUTDINOVA, R.I., FEDOSEEVA, G.P., NEKRASOVA, G.F., BOR-ZENKOVA, R.A., NAZAROV, S.K.** : Strukturnaya i funktsional'naya dinamika lista v ontogeneze. [Structural and functional dynamics of a leaf during ontogenesis.] - In : Voprosy Regulyatsii Fotosinteza. Vol. 3. Pp. 3 - 44, 161. Ural'. gos. Univ., Sverdlovsk 1973. [Ps; in R.]

17226 - **MOKRONOSOV, A.T., DAVYDOV, V.A.** : Indutsirovannyĭ sintez ureazy i metabolizm ékzogennoĭ mocheviny C^{14} v kletkakh khlorelly. [Induced urease synthesis and metabolism of exogenous ^{14}C-urea in *Chlorella* cells.] - Zap. sverdlov. Otd. vsesoyuz. bot. Obshch. *1973* (6) : 29 - 35, 1973. [Ps; in R.]

17227 - **MOKRONOSOV, A.T., DOBROV, A.V.** : Kamera dlya izucheniya fotosinteticheskogo metabolizma i opredeleniya potentsial'nogo fotosinteza na izolirovannykh list'-yakh. [Chamber for studying photosynthetic metabolism and determining potential photosynthesis in isolated leaves.] - In : Voprosy Regulyatsii Fotosinteza. Vol. 3. Pp. 149 - 152, 165. Ural'. gos. Univ., Sverdlovsk 1973. [In R.]

17228 - **MOKRONOSOV, A.T., NAZAROV, S.K., RAKHIMOVA, G.I.** : Metabolizm ékzogennogo alanina v list'yakh rasteniĭ. [Metabolism of exogenous alanine in plant leaves.] - Fiziol. Rast. *20* : 757 - 765, 1973. [In R, ab : E.]

17229 - **MOKRONOSOV, A.T., NEKRASOVA, G.F., POYARKOVA, N.M.** : Formirovanie fotosinteticheskogo apparata khlorelly pri raznykh kontsentratsiyakh CO_2. [Formation of photosynthetic apparatus in *Chlorella* in various CO_2 concentrations.] - In : Voprosy Regulyatsii Fotosinteza. Vol. 3. Pp. 84 - 92, 162. Ural'. gos. Univ., Sverdlovsk 1973. [In R.]

17230 - **MOLDAU, H.** : Effects of various water regimes on stomatal and mesophyll conductances of bean leaves. - Photosynthetica *7* : 1 - 7, 1973.

17231 - **MOLDAU, Kh.** : Vliyanie defitsita vody na soprotivlenie ust'its. (Matematiches-kaya model'.) [Effect of water deficit on stomatal resistance. (A mathematical model.)] - Eesti NSV Tead. Akad. Toim. Biol. [Izv. Akad. Nauk est. SSR] *22* : 348 - 357, 1973. [In R, ab : E, Est.]

17232 - **MOLOTKOVS'KYĬ, G. Kh., KOZHAKHMETOVA, A.É.** : Vmist zelenykh i zhovtykh pigmentiv kaliyu i natriyu v lystkakh dvodomno'i roslyny aktynidi'i. [Content of green and yellow pigments, potassium and sodium in leaves of dioecious plants of *Actinidia*.] - Dopov. Akad. Nauk ukr. RSR *35 B* : 557 - 558, 576, 1973. [In Ukr., ab : E, R.]

17233 - **MONSI, M., UCHIJIMA, Z., OIKAWA, T.** : Structure of foliage canopies and photosynthesis. - Annu. Rev. Ecol. Systematics *4* : 301 - 327, 1973.

17234 - **MONTALBINI, P.** : Effect of infection by *Uromyces phaseoli* (PERS.) WINT. on electron carrier quinones in bean leaves. - Physiol. Plant Pathol. *3* : 437 - - 441, 1973. [Ps, Chl.]

17235 - **MONTALBINI, P., CAPPELLI, C.** : Glycolic oxidase and NAD-glyoxylic reductase in rust susceptible bean leaves. - Phytopathol. Z. *77* : 348 - 355, 1973. [Chl.]

17236 - **MONTALBINI, P., RAGGI, V., CAPPELLI, C.** : Variazioni dell' attività fosfogli-colico fosfatasica in foglie di fagiolo affette da ruggine. [Phosphoglycolate phosphatase activity in rusted bean leaves.] - Ann. Fac. agraria Univ. Perugia *28* : 259 - 266, 1973. [Chl; in Ital., ab : E.]

17237 - **MONTENY, B.A.** : Anatomie et échange de CO_2 chez *Panicum maximum*. - Oecol. Plant. *8* : 125 - 140, 1973.

*17238 - **MONTIES, B.** : Biosynthèse, localisation et metabolisme de composes polypheno-liques dans les chloroplastes. - Bull. Liaison Groupe Polyphenols *3* : 1 - 2, 1972.

*17239 - **MONTIES, B.** : Mise au point faite dans le cadre de la R.C.P. 198 du C.N.R.S. (22 Juin 1972). Les composés polyphénoliques et leur métabolisme dans les chloroplastes. - Bull. Liaison Groupe Polyphenols *3* : 3 - 9, 1972.

17240 - **MONYO, J.H., WHITTINGTON, W.J.** : Genotypic differences in flag leaf area and their contribution to grain yield in wheat. - Euphytica *22* : 600 - 606, 1973.

17241 - **MOONEY, H.A., PARSONS, D.J.** : Structure and function of the California chaparral - an example from San Dimas. - In : di CASTRI, F., MOONEY, H.A. (ed.) : Mediterranean Type Ecosystems. Origin and Structure. (Ecol. Stud. Vol. 7.) Pp. 83 - 112. Springer Verlag, Berlin - Heidelberg - New York 1973. [Ps.]

17242 - **MOORE, R.T., EHLERINGER, J., MILLER, P.C., CALDWELL, M.M., TIESZEN, L.L.** : Gas exchange studies of four alpine tundra species at Niwot Ridge, Colorado. - In : BLISS, L.C., WIELGOLASKI, F.E. (ed.) : Primary Production and Production Processes. Tundra Biome. Pp. 211 - 217. Tundra Biome Steering Comm., Edmonton 1973. [Ps.]

17243 - **MOORE, R.T., MILLER, P.C., EHLERINGER, J., LAWRENCE, W.** : Seasonal trends in gas exchange characteristics of three mangrove species. - Photosynthetica *7* : 387 - 394, 1973.

17244 - **MORE, R.D., TROUGHTON, J.H.** : Production of $^{11}CO_2$ for use in plant translocation studies. - Photosynthetica *7* : 271 - 274, 1973.

17245 - **MORITA, K., KONO, M.** : Further evidence for the integrity of rice chloroplasts isolated in glutaraldehyde medium. - Soil Sci. Plant Nutr. *19* : 317 - 320, 1973.

17246 - **MORRIS, D.B., MARTIN-KAYE, P.H.A.** : Remote sensing of the environment. - Endeavour *32* : 117 - 121, 1973.

17247 - **MORRIS, I.** : The control of carbon dioxide fixation in algae and photosynthetic bacteria. - J. gen. Microbiol. *75* (2) : v, 1973.

17248 - **MORRIS, M.M., PARK, Y., MACKINNEY, G.** : On the photodecomposition of chlorophyll *in vitro*. - J. agr. Food Chem. *21* : 277 - 279, 1973.

17249 - **MOSER, W.** : C. Licht, Temperatur und Photosynthese an der Station "Hoher Nebelkogel" (3184 m). - In : ELLENBERG, H. (ed.) : Ökosystemforschung. Pp. 203- - 223. Springer-Verlag, Berlin - Heidelberg - New York 1973.

17250 - **MOSHKOV, B.S.** : Rol' luchístoĭ ėnergii v vyyavlenii potentsial'noĭ produktivnosti rasteniĭ. [Role of radiant energy in realizing potential productivity of plants.] - XXXII Timiryazevskoe Chtenie. 59 pp. Nauka, Moskva 1973. [Ps.]

17251 - **MOSHKOV, B.S., ODUMANOVA-DUNAEVA, G.A.** : Vliyanie fotosinteza na razvitie perilly maslichnoĭ (*Perilla ocymoides* L.) i abissinskoĭ kapusty (*Brassica carinata* A. BRAUN) v usloviyakh nepreryvnogo osveshcheniya. [The effect of photosynthesis on the development of *Perilla ocymoides* L. and *Brassica carinata* A. BRAUN under permanent illumination.] - Bot. Zh. *58* : 639 - 645, 1973. [In R, ab : E.]

17252 - **MOSS, B.** : Diversity of fresh-water phytoplankton. - Amer. Midland Naturalist *90* : 341 - 355, 1973. [Chl.]

17253 - **MOSS, G.P.** : Carotenoids and polyterpenoids. - In : OVERTON, K.H. (ed.) : Terpenoids and Steroids. Vol. 3. Pp. 230 - 244. Chem. Soc., London 1973.

17254 - **MOSZYŃSKA, B.** : Methods for assessing production of the upper parts of shrubs and certain perennial plants. - Ecol. pol. *21* (24) : 359 - 367, 1973.

17255 - **MOTKALYUK, O.B.** : Dinamika soderzhaniya karotinoidov i khlorofillov v verkhnikh list'yakh pshenitsy i yachmenya v kriticheskiĭ k nedostatku vlagi v pochve period. [Dynamics of the content of carotenoids and chlorophylls in the upper leaves of wheat and barley during critical period of water deficiency in soil.] - Fiziol. Rast. *20* : 1242 - 1247, 1973. [In R, ab : E.]

17256 - **MOUNTFORD, K.** : Parallel measurements of phytoplankton photosynthesis using dissolved oxygen and ^{14}C in the vicinity of a nuclear power plant. - Bull. N. Jersey Acad. Sci. *18* (2) : 26 - 29, 1973.

17257 - **MOURAVIEFF, I.** : Microphotométrie des fluctuations de la teneur en amidon des stomates éclairés par la lumiére de 436 nm et 665 nm en absence ou en présence de gaz carbonique. - Ann. Sci. nat. Bot. Biol. vég. *14* : 377 - 383, 1973.

17258 - **MUC, M.** : Primary production of plant communities of the Truelove Lowland, Devon Island, Canada - sedge meadows. - In : BLISS, L.C., WIELGOLASKI, F.E. (ed.) : Primary Production and Production Processes. Tundra Biome. Pp. 3 - 14. Tundra Biome Steering Comm., Edmonton 1973.

17259 - **MUES, R., EDELBLUTH, E., ZINSMEISTER, H.D.** : Das Carotinoidmuster von *Lophocolea bidentata* (L) DUM. - Österr. bot. Z. *122* : 177 - 184, 1973.

17260 - **MÜHLBACH, H.-P., WEGMANN, K.** : Photosynthetischer CO_2-Einbau und Photorespiration bei isolierten Zellen und Protoplasten aus Sonnenblumen und Mais. - Hoppe-Seyler's Z. physiol. Chem. *354* : 1226, 1973.

*17261 - **MUKHAMADIEV, B.T., KVITKO, K.V.** : O svyazi pigmentnogo sostava kletok mutantov s ikh ustoĭchivost'yu k ingibitoram fotofosforilirovaniya. [Relation of pigment composition in mutant cells to their resistance to inhibitors of photophosphorylation.] - In : Geneticheskie Aspekty Fotosinteza. Tezisy Dokladov. P. 90. Donish, Dushanbe 1972. [In R.]

17262 - **MUKHIN, E.N., GINS, V.K.** : Properties of ferredoxins from plants with C_3- and C_4-paths of carbon assimilation in photosynthesis. - In : Ninth International Congress of Biochemistry. Stockholm , 1 - 7 July, 1973. Abstract Book. P. 228. IUB, Stockholm 1973.

17263 - **MUKHIN, E.N., GINS, V.K., KULIKOV, A.V., LIKHTENSHTEĬN, G.I.** : Teploustoĭchivost' ferredoksinov vysshikh rasteniĭ v svyazi s aktivnost'yu v reaktsii fotovosstanovleniya NADF. [Heat resistance of ferredoxins of higher plants : their activity in NADP photoreduction.] - Fiziol. Rast. *20* : 1007 - 1012, 1973. [In R, ab : E.]

17264 - **MUKOHATA, Y.** : Thermal denaturation of thylakoids and inactivation of photophosphorylation in isolated spinach chloroplasts. - In : NAKAO, M. (ed.) : Organic Energy-Transducing Membranes. Pp. 219 - 237. Univ. Park Press, Baltimore, Md. 1973.

17265 - **MUKOHATA, Y., YAGI, T., HIGASHIDA, M., MATSUNO, A.** : Lipophilic interaction between thylakoid membranes and aliphatic compounds. - J. Bioenerg. *4* : 479 - -·490, 1973.

17266 - **MUKOHATA, Y., YAGI, T.,·HIGASHIDA, M., SHINOZAKI, K., MATSUNO, A.** : Biophysical studies on subcellular particles VI. Photosynthetic activities in isolated spinach chloroplasts after transient warming. - Plant Cell Physiol. *14* : 111- - 118, 1973.

17267 - **MUKOHATA, Y., YAGI, T., MATSUNO, A., HIGASHIDA, M.** : Biophysical studies on subcellular particles VII. Combined effects of alcohol and heat on photosynthetic activities in isolated spinach chloroplasts. - Plant Cell Physiol. *14*: 119 - 126, 1973.

17268 - **MÜLLER, D.** : Untersuchungen über die Photosynthesetätigkeit von Stark- und Schwachlichtpflanzen verschiedener *Pisum*-Mutanten. - Angew. Bot. *47* : 215 - - 226, 1973.

17269 - **MUMOLA, P.B., JARRETT, O. Jr., BROWN, C.A. Jr.** : Multiwavelength laser induced fluorescence of algae *in-vivo* : A new remote sensing technique. - In : Second Joint Conference on the Sensing of Environmental Pollutants. Pp. 53 - - 63. Pittsburgh, Pa. 1973. [Chl.]

17270 - **MUNRO, J.M.M., DAVIES, D.A.** : Potential pasture production in the uplands of Wales. 2. Climatic limitations on production. - J. brit. Grassland Soc. *28* : 161 - 169, 1973.

17271 - **MURAKAMI, S.** : Structural and energy states of photosynthetic membranes in relation to proton and cation gradients. - In : NAKAO, M. (ed.) : Organic Energy-Transducing Membranes. Pp. 291 - 313. Univ. Park Press, Baltimore, Md. 1973.

17272 - **MURAKAMI, T., TAKEDA, T.** : [An approach to the measurement of photosynthetic rate in single leaves by the aeration method. I. The effects of aeration speed, gradient of CO_2-concentration and propeller drive on the photosynthetic rate of mulberry leaf.] - Proc. Crop Sci. Soc. Jap. *42* : 170 - 177, 1973. [In Jap., ab : E.]

17273 - **MURAKAMI, T., TAKEDA, T.** : [The influence of leaf temperature on photosynthetic and respiratory rate in mulberry plants.] - J. Sericult. Sci. Jap. *42* : 157 - 163, 1973. [In Jap., ab : E.]

17274 - **MURAKAMI, T., TAKEDA, T.** : [Effect of light intensity on the photosynthetic rate in mulberry leaves.] - J. Sericult. Sci. Jap. *42* : 417 - 424, 1973. [In Jap., ab : E.]

17275 - **MURATA, N., FORK, D.C.** : The function of plastocyanin in electron transport of photosynthesis : The content of plastocyanin in spinach chloroplasts and subchloroplast particles prepared by different techniques. - Carnegie Inst. Year Book *72* : 376 - 384, 1973.

17276 - **MURATA, N., ITOH, S., OKADA, M.** : Induction of chlorophyll *a* fluorescence in isolated spinach chloroplasts at liquid nitrogen temperature. - Biochim. biophys. Acta *325* : 463 - 471, 1973.

17277 - **MURKOWSKI, A.** : Use of standard radiometric systems in the study of herbicides action on the photosynthetic luminescence of cultivable plants and the weeds. - Proc. Res. Inst. Pomol., Skierniewice, Ser. E *1973* : 595 - 603, 1973.

17278 - **MURR, S.M., SPURR, A.R.** : An albino mutant in *Plantago insularis* requiring thiamine pyrophosphate. - J. exp. Bot. *24* : 1271 - 1282, 1973.

17279 - **MURRAY, D.R., WARA-ASWAPATI, O., IRELAND, H.M.M., BRADBEER, J.W.** : The development of activities of some enzymes concerned with glycollate metabolism in greening bean leaves. - J. exp. Bot. *24* : 175 - 184, 1973. [Chl.]

17280 - **MURTY, K.S., NAYAK, S.K., SAHU, G.** : Photosynthetic efficiency in rice varieties. - ISNA Newslett. *2* (1) : 5 - 6, 1973.

17281 - **MUSCATINE, L., GREENE, R.W.** : Chloroplasts and algae as symbionts in molluscs. - Int. Rev. Cytol. *36* : 137 - 169, 1973. [Ps, Chl, Car.]

17282 - **MYNETT, A., WAIN, R.L.** : Herbicidal action of iodide : effect on chlorophyll content and photosynthesis in dwarf bean *Phaseolus vulgaris*. - Weed Res. *13* : 101 - 109, 1973.

17283 - **NADLER, K.D., SOVONICK, S.** : Effect of Δ-ALA on chloroplast membrane formation. - Plant Physiol. *51* (Suppl.) : 28, 1973.

17284 - **NAGY, A.H., BOKÁNY, A.J., ILLIK, M., BÁCS, B., DOMAN, N.G.** : Genetic properties of carboxylating enzyme capacity in plants with the C_4-dicarboxylic pathway of photosynthesis. - Ann. Univ. Sci. budap. Rolando Eötvös nomin., Sect. biol. *15* : 59 - 64, 1973.

17285 - **NAGY, A.H., GYURJÁN, I., SZÉKELY, S., DOMAN, N.G.** : Activities of enzymes related to photosynthesis in $^{14}CO_2$ fixation products in normal and carotenoid mutant maize leaves. - Photosynthetica *7* : 87 - 92, 1973.

17286 - **NAKAZAWA, F.** : [Studies on the dry matter production of sugar beet. Relationship between light condition and growing process, light intensity and photosynthesis.] - Bull. Fac. Agr. Meiji Univ. *30* : 25 - 30, 1973. [In Jap., ab : E.]

17287 - **NALBANDYAN, R.M., MUTUSKIN, A.A., PSHENOVA, K.V.** : Izmenenie opticheskikh i É.P.R. spektrov plastotsianinov pod vliyaniem organicheskikh rastvoritelei. [The alteration of optical and EPR spectra of plastocyanines under the effect of organic solvents.] - Dokl. Akad. Nauk SSSR *211* : 1217 - 1219, 1973. [In R.]

17288 - **NARBUT, N.A.** : Uchastie "svoikh" i "chuzhikh" assimilyatov v obrazovanii belkov v rastushchikh list'yakh soi. [Participation of "own" and "foreign" photosynthates in the formation of proteins in growing leaves of soybean plants.] - Tr. biol.-pochv. Inst., dal'nevost. nauch. Tsentr Akad. Nauk SSSR *20* (123) : 85 - 86, 1973. [In R, ab : E.]

17289 - **NASS, M.M.K., BEN-SHAUL, Y.** : Effects of ethidium bromide on growth, chlorophyll synthesis, ultrastructure and mitochondrial DNA in green and bleached mutant *Euglena gracilis*. - J. Cell Sci. *13* : 567 - 590, 1973.

*17290 - **NASYROV, Yu.S.** : Yaderno-plastidnyi kontrol' razvitiya i funktsional'noi aktivnosti khloroplastov. [Nuclear-plastid control of chloroplast development and functional activity.] - In : Geneticheskie Aspekty Fotosinteza. Tezisy Dokladov. Pp. 24 - 25. Donish, Dushanbe 1972. [In R.]

17291 - **NASYROV, Yu.S.** : Fotosintez i genetika khloroplastov. [Photosynthesis and genetics of chloroplasts.] - Priroda (Moskva) *1973* (1) : 56 - 61, 1973. [In R.]

17292 - **NÁTR, L.** : O možnosti regulace vodního provozu a intenzity fotosyntézy toxinem "fusicoccin". [The possibility of using the toxin "fusicoccin" for the control of water relations and photosynthetic rate.] - Rostlinná Výroba (Praha) *19* : 373 - 378, 1973. [In Czech, ab : E.]

17293 - **NÁTR, L.** : Využití analýzy difúzních odporů při studiu fotosyntetické charakteristiky. [Use of the analysis of diffusion resistances in the study of photosynthetic characteristics.] - Rostlinná Výroba (Praha) *19* : 653 - 661, 1973. [In Czech, ab : E, R.]

17294 - **NÁTR, L.** : The effect of plant density on grain yield and photosynthetic characteristics of spring barley varieties. - Rostlinná Výroba (Praha) *19* : 839 - 846, 1973.

17295 - **NAUMOVIČ, O., NAUMOVIČ, M.** : Sadržaj aneurina, karotina i tokoferola u raznim sórtama kukuruza. [Aneurine, tocopherol, and carotene contents in various maize types.] - Hrana Ishrana *14* : 130 - 136, 1973. [In Croat., ab : E.]

17296 - **NEAL, R.L. Jr.** : Remeasuring tree heights on permanent plots using rectangular coordinates and one angle per tree. - Forest Sci. *19* : 233 - 236, 1973.

17297 - **NEALES, T.F.** : Effect of night temperature on the assimilation of carbon dioxide by mature pineapple plants, *Ananas comosus* (L.) MERR. - Aust. J. biol. Sci. *26* : 539 - 546, 1973.

17298 - **NEALES, T.F.** : The effect of night temperature on CO_2 assimilation, transpiration, and water use efficiency in *Agave americana* L. - Aust. J. biol. Sci. *26* : 705 - 714, 1973.

17299 - **NEAMŢU, G., BODEA, C.** : Contribuţii la nomenclatura carotenoidelor. [Carotenoid nomenclature.] - Stud. Cercet. chim. *21* : 515 - 533, 1973. [In Roum.]

17300 - **NECHIPORENKO, G.A., IVANOV, V.P.** : Vydelenie i pogloshchenie kornyami kukuruzy i bobov mechennykh C^{14} organicheskikh veshchestv i uglekisloty. [Exudation and absorption of labelled ^{14}C organic substances and carbon dioxide by maize and broad bean roots.] - Fiziol. Rast. *20* : 577 - 581, 1973. [Ps; in R, ab : E.]

17301 - **NEDUKHA, E.M.** : Izuchenie ul'trastruktury karotinoplastov i soderzhanie pigmentov v parenkhimnykh kletkakh kornya morkovi. [Ultrastructure of carotenoplasts and pigment contents in parenchyma cells of the carrot root.] - Tsitologiya *15* : 218 - 220, 1973. [In R, ab : E.]

17302 - **NEDUKHA, O.M.** : Peretvorennya karotynoplastiv (khromoplastiv) u khloroplasty v klitynakh eksplantatu morkvi. [Transformation of carotenoplasts (chromoplasts) into chloroplasts in the carrot explantate cells.] - Ukr. bot. Zh. *30* : 655 - 658, 677, 1973. [In Ukr., ab : E, R.]

17303 - **NEKRASOV, Yu.I.** : Kompleks apparatury dlya registratsii skorosti peredvizheniya radioaktivnykh veshchestv v rasteniyakh. [Equipment for recording the rate of transport of radioactive substances in plants.] - Tr. biol.-pochv. Inst., dal'nevost. nauch. Tsentr Akad. Nauk SSSR, N.S. *20* (123) - Transport Assimilyatov i Otlozhenie Veshchestv v Zapas u Rastenii : 292 - 296, 1973. [In R, ab : E.]

17304 - **NEKRASOVA, G.F.** : Osobennosti geterotrofnoi fiksatsii CO_2 u khlorelly, adaptirovannoi k raznym kontsentratsiyam uglekisloty. [Characteristics of heterotrophic fixation of carbon dioxide in *Chlorella* adapted to different carbon dioxide concentrations.] - Ėkologiya *4* (5) : 103 - 105, 1973. [In R.]

17305 - **NEKRASOVA, G.F.** : Proyavlenie kislorodnogo ĕffekta Varburga u khlorelly, vyrashchennoi pri razlichnykh kontsentratsiyakh CO_2. [Warburg effect in *Chlorella* grown under various CO_2 concentrations.] - In : Voprosy Regulyatsii Fotosinteza. Vol. 3. Pp. 93 - 100, 163. Ural'. gos. Univ., Sverdlovsk 1973. [In R.]

17306 - **NELSON, N., DETERS, D.W., NELSON, H., RACKER, E.** : Partial resolution of the enzymes catalyzing photophosphorylation.|XIII. Properties of isolated subunits of coupling factor 1 from spinach chloroplasts. - J. biol. Chem. *248* : 2049 - 2055, 1973.

17307 - **NELSON, N., RACKER, E.** : Phosphate transfer from adenosine triphosphate in a model system. - Biochemistry *12* : 563 - 566, 1973. [Model of photophosphorylation.]

17308 - **NESTEROVA, E.S.** : Vplyv sirky ta fosforu na produktyvnist' riznykh vydiv syn'ozelenykh vodorostei. [Effect of sulphur and phosphorus on productivity of various genera of blue-green algae.] - Ukr. bot. Zh. *30* : 308 - 312, 401, 1973. [In Ukr., ab : E, R.]

17309 - **NETZEL, T.L., RENTZEPIS, P.M., LEIGH, J.** : Picosecond kinetics of reaction centers containing bacteriochlorophyll. - Science *182* : 238 - 241, 1973.

17310 - **NEUBERGER, A., SANDY, J.D., TAIT, G.H.** : Control of aminolaevulinate synthetase in *Rhodopseudomonas spheroides.* - Enzyme *16* : 79 - 85, 1973.

17311 - **NEUBERGER, A., SANDY, J.D., TAIT, G.H.** : Control of 5-aminolaevulinate synthetase activity in *Rhodopseudomonas spheroides.* The involvement of sulphur metabolism. - Biochem. J. *136* : 477 - 490, 1973.

17312 - **NEUBERGER, A., SANDY, J.D., TAIT, G.H.** : Control of 5-aminolaevulinate synthetase activity in *Rhodopseudomonas spheroides.* The purification and properties of an endogenous activator of the enzyme. - Biochem. J. *136* : 491 - 499, 1973.

17313 - **NEUMANN, D.** : Zur Ultrastruktur der Riesenplastiden aus dem Palisadenparenchym von *Peperomia metallica* LIND *et* RODIG. - Protoplasma *77* : 467 - 471, 1973.

17314 - **NEUMANN, D., PARTHIER, B.** : Effects of nalidixic acid, chloramphenicol, cycloheximide, and anisomycin on structure and development of plastids and mitochondria in greening *Euglena gracilis.* - Exp. Cell Res. *81* : 255 - 268, 1973.

17315 - **NEUMANN, J., BARBER, J., GREGORY, P.** : The relation between photophosphorylation and delayed light emission in chloroplasts. - Plant Physiol. *51* : 1069 - - 1073, 1973.

17316 - **NEUMANN, J., DRECHSLER, Z., YANNAI, Y.** : Evidence for stimulation of ATP formation by a membrane potential in chloroplasts. - In : Ninth International Congress of Biochemistry. Stockholm, 1 - 7 July, 1973. Abstract Book. P. 232. IUB, Stockholm 1973.

17317 - **NEUMANN, K.-H., RAAFAT, A.** : Further studies on the photosynthesis of carrot tissue cultures. - Plant Physiol. *51* : 685 - 690, 1973.

17318 - **NEUVILLE, D., DASTE, P.** : Sur la singularité de la production d'un pigment bleu-vert par la Diatomée *Navicula ostrearia* (GAILLON) BORY. - Compt. rend. Acad. Sci. Paris, Sér. D *276* : 3469 - 3472, 1973.

17319 - **NEWMAN, D.W., ROWELL, B.W., BYRD, K.** : Lipid transformations in greening and senescing leaf tissue. - Plant Physiol. *51* : 229 - 233, 1973. [Chl.]

17320 - **NGUEN TKHYU TKHYOK, AVDEEVA, T.A., ANDREEVA, T.F.** : Vliyanie fosfornogo pitaniya na aktivnost' fotosinteticheskogo apparata list'ev razlichnykh yarusov rastenii bobov. [Effect of phosphorus nutrition on the activity of photosynthetic apparatus in the leaves of broad bean plants from different tiers.] - Fiziol. Rast. *20* : 1024 - 1028, 1973. [In R, ab : E.]

*17321 - **NICHIPOROVICH, A.A.** : O raznoobrazii fotosinteticheskoi funktsii v rastitel'-nom mire kak osnove dlya vyvedeniya vysokoproduktivnykh form rasteni. [Diversity of the photosynthetic function in plant world as the basis for breeding of highly productive forms.] - In : Geneticheskie Aspekty Fotosinteza. Tezisy Dokladov. Pp. 109 - 111. Donish, Dushanbe 1972. [In R.]

17322 - **NICHIPOROVICH, A.A.** : Osnovy fotosinteticheskoi produktivnosti rastenii. [Basis of photosynthetic productivity of plants.] - In : Sovremennye Problemy Fotosinteza. Pp. 17 - 43. Izd. mosk. Univ., Moskva 1973. [In R.]

17323 - **NICHIPOROVICH, A.A., CHMORA, S.N., SLOBODSKAYA, G.A., AVDEEVA, T.A.** : Fotosinteticheskii CO_2-gazoobmen list'ev svekly i kukuruzy i ego svyaz' s tipami fitotsenozov. [Photosynthetic CO_2 exchange in leaves of beet and maize and its relation with the types of phytocenoses.] - Fiziol. Rast. *80* : 300 - 308, 1973. [In R, ab : E.]

17324 - **NICHIPOROVICH, A.A., CHMORA, S.N., SLOBODSKAYA, G.A., AVDEYEVA, T.A.** : Types of photosynthetic gas exchange of leaves of beet and maize in relation to the CO_2 regimes in their crop stands. - Plant Sci. Lett. *1* : 399 - 403, 1973.

17325 - **NICHOLS, M.S.** : Major cause of algae and weeds in Lake Mendota. - Trans. Wisc. Acad. Sci. Arts Lett. *61* : 229 - 234, 1973. [Ps.]

17326 - **NICHOLS, S.A.** : The effects of harvesting aquatic macrophytes on algae. - Trans. Wisc. Acad. Sci. Arts Lett. *61* : 165 - 172, 1973. [Production.]

17327 - **NICHOLSON-GUTHRIE, C.S.** : Isolation and characterization of a mutant of *Chlamydomonas reinhardi* lacking chlorophyll. - Diss. Abstr. Int. B *33* : 4137, 1973.

17328 - **NIELSEN, O.F.** : Protochlorophyll(ide) holochrome subunits from a mutant defective in the regulation of protochlorophyll(ide) synthesis. - FEBS Lett. *38* : 75 - 78, 1973.

17329 - **NIELSEN, O.F., KAHN, A.** : Kinetics and quantum yield of photoconversion of protochlorophyll(ide) to chlorophyll(ide) *a*. - Biochim. biophys. Acta *292* : 117 - 129, 1973.

17330 - **NIEMI, Å.** : Ecology of phytoplankton in the Tvärminne area, SW coast of Finland. I. Dynamics of hydrography, nutrients, chlorophyll *a* and phytoplankton. - Acta bot. fenn. *100* : 1 - 68, 1973. [Chl.]

17331 - **NIEUWHOF, M., de BRUYN, J.W., GARRETSEN, F.** : Methods to determine solidity and dry matter content of onions (*Allium cepa* L.). - Euphytica *22* : 39 - 47, 1973.

17332 - **NIIYA, I., KANEMATSU, H., IMAMURA, M., OKADA, M., MATSUMOTO, T.** : [Deterioration of oils and fats of hardened coconut oil series. IX. Green coloration of added β-carotene.] - J. Jap. Oil chem. Soc. [Yukagaku] *22* : 22 - 26, 1973. [In Jap.]

17333 - **NILOVSKAYA, N.T., BOKOVAYA, M.M.** : Dnevnoǐ khod fotosinteza posevov rasteniǐ v usloviyakh fitotrona. [Diurnal course of photosynthesis of plant stands in a phytotron.] - In : Upravlenie Skorost'yu i Napravlennost'yu Biosinteza u Rasteniǐ. Pp. 151 - 152. Krasnoyarsk 1973. [In R.]

17334 - **NILOVSKAYA, N.T., KORZHEVA, G.F., BOKOVAYA, M.M.** : O vzaimosvyazi gazoobmena i nakopleniya organicheskogo veshchestva u rasteniǐ. [Relationship between gas exchange and accumulation of organic matter in plants.] - Fiziol. Rast. *20* : 558 - 563, 1973. [In R, ab : E.]

17335 - **NIR, I., PEASE, D.C.** : Chloroplast organization and the ultrastructural localization of photosystems I and II. - J. Ultrastruct. Res. *42* : 534 - 550, 1973.

17336 - **NISHIKAWA, K., HOSOI, K., SUZUKI, J., YOSHIMURA, S., HORIO, T.** : Formation and decomposition of pyrophosphate related to bacterial photophosphorylation. - J. Biochem. (Tokyo) *73* : 537 - 553, 1973.

17337 - **NISHIMURA, M.** : [Interplay of molecular mechanism of biological reactions and the physical parameters of "reaction-fields" - a discussion concerning the primary processes of photosynthesis.] - Protein, nucl. Acid Enzyme [Tampaku-shitsu Kakusan Koso] *18* : 791 - 805, 1973. [In Jap.]

17338 - **NISHIMURA, M.** : [Photosynthesis and photodestruction : two actions of light on photosynthetic apparatus.] - Chemistry [Kagaku] *43* : 605 - 609, 1973. [In Jap.]

17339 - **NISHIMURA, M., AKAZAWA, T.** : Further proof for the catalytic role of the larger subunit in the spinach leaf ribulose-1,5-diphosphate carboxylase. - Biochem. biophys. Res. Commun. *54* : 842 - 848, 1973.

17340 - **NISHIMURA, M., TAKABE, T., SUGIYAMA, T., AKAZAWA, T.** : Structure and function of chloroplast proteins. XIX. Dissociation of spinach leaf ribulose-1,5-diphosphate carboxylase by *p*-mercuribenzoate. - J. Biochem. *74* : 945 - 954, 1973.

17341 - **NISHIZAKI, Y.** : Kinetics of ATP formation and proton efflux by acid-base transition in chloroplasts. - Biochim. biophys. Acta *314* : 312 - 319, 1973.

17342 - **NISHNIANIDZE, N.O., GOGIBERIDZE, G.S.** : [Physiological activity of fungicides in citrus plants.] - Tr. nauch.-issled. Inst. Zashchity Rast. gruz.SSR *25* : 50 - 54, 1973. [Ps; in Georg., ab : E, R.]

17343 - **NISIMOTO, Y., KAKUNO, T., YAMASHITA, J., HORIO, T.** : Two different NADH dehydrogenases in respiration of *Rhodospirillum rubrum* chromatophores. - J. Biochem. *74* : 1205 - 1216, 1973.

17344 - **NISIMOTO, Y., YAMASHITA, J., HORIO, T.** : Immunochemical studies on function of NADH : Hemeprotein oxidoreductase in electron transport system of chromatophores from *Rhodospirillum rubrum*. - J. Biochem. (Tokyo) *73* : 515 - 521, 1973.

17345 - **NISIMOTO, Y., YAMASHITA, J., HORIO, T.** : Immunological studies on function of NADH : Quinone oxidoreductase in electron transport system of chromatophores from *Rhodospirillum rubrum*. - J. Biochem. (Tokyo) *73* : 523 - 528, 1973.

17346 - **NITSCHE, H.** : Heteroxanthin in *Euglena gracilis*. - Arch. Mikrobiol. *90* : 151- - 155, 1973.

17347 - **NITSCHE, H.** : The structure of vaucheriaxanthin. - Z. Naturforsch. *28 c* : 641- - 645, 1973.

17348 - **NIXON, S.W., OVIATT, C.A.** : Ecology of a New England salt marsh. - Ecol. Monogr. *43* : 463 - 498, 1973. [Ps production.]

*17349 - **NIYAZMUKHAMEDOVA, M.B.** : Fotosintez i uglevodnyǐ obmen pigmentnykh mutantov *Arabidopsis thaliana* (L) HEYNH. [Photosynthesis and carbohydrate exchange of pigment mutants of *Arabidopsis thaliana* (L.) HEYNH.] - In : Geneticheskie Aspekty Fotosinteza. Tezisy Dokladov. P. 112. Donish, Dushanbe 1972. [In R.]

17350 - **NOBEL, P.S., WANG, C.-T.** : Ozone increases the permeability of isolated pea chloroplasts. - Arch. Biochem. Biophys. *157* : 388 - 394, 1973.

17351 - **NOBLE, R.D., CRANG, R.E.** : Photoduration effects on photosynthesis and photosynthetic apparatus in soybean. - Plant Physiol. *51* (Suppl.) : 28, 1973.

17352 - **NOBLE, R.D., LONG, D.W., BURLEY, J.W.** : Photosynthetic ^{14}C-labeling patterns in developing soybeans. - Can. J. Bot. *51* : 1505 - 1512, 1973.

17353 - **NORRIS, J.R., DRUYAN, M.E., KATZ, J.J.** : Electron nuclear double resonance of bacteriochlorophyll free radical *in vitro* and *in vivo*. - J. amer. chem. Soc. *95* : 1680 - 1682, 1973.

17354 - **van NOSTRAND, F.W. Jr.** : The *in vivo* orientation of chlorophyll : A spectroscopic study of magnetically oriented photosynthetic systems. - Diss. Abstr. Int. B *33* : 5167-B, 1973.

17355 - **NOVAK, V.A., IVANKINA, N.G.** : Svetoindutsirovannye izmeneniya vnutrikletochnogo ělektricheskogo potentsiala lista elodei kak otrazhenie vzaimootnosheniǐ mezhdu fotosintezom i dykhaniem. [Light-induced changes in the intracellular electrical potential reflecting interrelation between photosynthesis and respiration in an *Elodea* leaf.] - In : Upravlenie Skorost'yu i Napravlennost'yu Biosinteza u Rasteniǐ. Pp. 28 - 29. Krasnoyarsk 1973. [In R.]

17356 - **NOVOZHILOVA, M.I., BEREZINA, F.S.** : Pervichnaya produktsiya v Aral'skom more. [Primary production in the Aral sea.] - Gidrobiol. Zh. *9* (2) : 104 - 108, 1973. [In R.]

17357 - **NOY-MEIR, I.** : Desert ecosystems : environment and producers. - Annu. Rev. Ecol. Systematics *4* : 25 - 51, 1973. [Primary production.]

17358 - **NULTSCH, W.** : Relation between photomotion and photosynthesis. - In : **CHECCUCCI, A., WEALE, R.A.** (ed.) : Primary Molecular Events in Photobiology. Pp. 245- - 273. Elsevier, Amsterdam - London - New York 1973.

17359 - **NUÑES, M^A.A., BRUMBY, D., DAVIES, D.D.** : Estudo comparativo do metabolismo fotossintético em folhas de cafeeiro, beterraba e cana-de-açúcar. [Comparison of photosynthetic mechanism in leaves of coffee plant, sugar beet and sugar cane.] - Garcia Orta, Sér. Estud. agron. (Lisboa) *1* (1) : 1 - 13, 1973. [In Port., ab : E.]

17360 - **NYUNT, U.T., WISKICH, J.T.** : The effects of 2,5-dibromo-3-methyl-6-isopropyl-
-*p*-benzoquinone on isolated pea chloroplasts. - Plant Cell Physiol. *14* : 1099-
- 1105, 1973.

17361 - **NYUPPIEVA, K.A.** : Izmenenie fotosinteticheskogo apparata u razlichnykh po
ustoĭchivosti ̩vidov kartofelya v rezul'tate deĭstviya zamorozkov. [Changes in
photosynthetic apparatus of some potato species differing in frost resistance
induced by morning frosts.] - Thesis. Pp. 1 - 33. Petrozavod. gos. Univ. O.V.
Kuusinena, Petrozavodsk 1973. [In R.]

17362 - **NYUPPIEVA, K.A., OSIPOVA, O.P.** : Kratkovremennoe deĭstvie otritsatel'nykh
temperatur na soderzhanie i sostoyanie pigmentov v list'yakh razlichnykh po
ustoĭchivosti vidov kartofelya. [Short-term effect of negative temperatures
on the content and state of pigments in leaves of potato plants with different resistance to low temperatures.] - Fiziol. Rast. *20* : 17 - 23, 1973. [In
R, ab : E.]

17363 - **OAKLEY, B.R., DODGE, J.D.** : The ultrastructure and cytochemistry of microbodies in *Porphyridium*. - Protoplasma *80* : 233 - 244, 1973.

17364 - **OBOLONSKII, V.V., RYBIN, I.A.** : Rol' spektral'nogo sostava i intensivnosti
sveta v formirovanii svetostimuliruemoĭ bioélektricheskoĭ reaktsii rasteniya.
[Role of spectral composition and intensity of light in the formation of
light-stimulated bioelectrical reaction of the plant.] - In : Upravlenie Skorost'yu i Napravlennost'yu Biosinteza u Rastenĭ. Pp. 30 - 31. Krasnoyarsk
1973. [In R.]

17365 - **OBRAZTSOV, A.S.** : Fiziologicheskie aspekty selektsii rastenĭ na skorospelost'
i produktivnost'. [Physiological aspects of plant selection with respect to
early ripeness and productivity.] - Fiziol. Rast. *20* : 175 - 182, 1973.
[Growth analysis; in R, ab : E.]

17366 - **OCHIAI-YANAGI, S., MATSUKA, M., HASE, E.** : Studies on chlorophyll formation
in *Chlorella prototheooides*. III. Effects of chloramphenicol, cycloheximide,
and ethionine on chlorophyll formation. - Plant Cell Physiol. *14* : 299 - 305,
1973.

*17367 - **ODINTSOVA, M.S.** : K voprosu o proiskhozhdenii plastid. [Origin of plastids.]
- In : Geneticheskie Aspekty Fotosinteza. Tezisy Dokladov. Pp. 13 - 14. Donish, Dushanbe 1972. [In R.]

17368 - **OECHEL, W.C., COLLINS, N.J.** : Seasonal patterns of CO_2 exchange in bryophytes
at Barrow, Alaska. - In : Proceedings of the Conference on Primary Production
Processes, Tundra Biome. Pp. 197 - 203. IBP, Dublin 1973.

17369 - **OESTERHELT, D., STOECKENIUS, W.** : Functions of a new photoreceptor membrane.
- Proc. nat. Acad. Sci. USA *70* : 2853 - 2857, 1973. [Bacteriorhodopsin.]

17370 - **OETTMEIER, W., LOCKAU, W.** : Cytochrome *c* reducing substances in photosynthetic
electron transport. - Z. Naturforsch. *28 C* : 717 - 721, 1973.

17371 - **OGAWA, T., INOUE, Y., KITAJIMA, M., SHIBATA, K.** : Action spectra for biosynthesis of chlorophylls *a* and *b* and β-carotene. - Photochem. Photobiol. *18* :
229 - 235, 1973.

17372 - **OGAWA, T., SHIBATA, K.** : A simple porometer for precise recording of leaf resistance. - Plant Cell Physiol. *14* : 1039 - 1043, 1973.

17373 - **OGAWA, T., SHIBATA, K.** : Selective formation of photosystem 1 under illumination at low intensity. - Physiol. Plant. *29* : 112 - 117, 1973.

17374 - **OGREN, W.L., RINNE, R.W.** : Photosynthesis and seed metabolism. - In : CALD-
WELL, B.E. (ed.) : Soybeans : Improvement, Production, and Uses. Pp. 391 -
- 416. Amer. Soc. Agron., Madison 1973.

17375 - **OHAD, I.** : The role of membrane proteins of cytoplasmic origin in the assembly
of the chloroplast membranes. - In : Ninth International Congress of Biochem-
stry. Stockholm, 1 - 7 July, 1973. Abstract Book. P. 250. IUB, Stockholm 1973.

17376 - **OKALI, D.U.U., DODOO, G.** : Seedling growth and transpiration of two West Afri-
can mahogany species in relation to water stress in the root medium. - J. Ecol.
61 : 421 - 472, 1973. [Growth analysis.]

17377 - **OKAYAMA, S.** : [Primary reaction and electron transport in photosystem II.] - Protein, nucl. Acid Enzyme [Tampakushitsu, Kakusan Koso] *18* : 103 - 113, 1973. [In Jap.]

17378 - **OKINO, T.** : Studies on the blooming of *Microcystis aeruginosa*. - Jap. J. Bot. *20* : 381 - 402, 1973. [Chi.]

17379 - **OKSENYUK, Yu.F.** : Primenenie gerbitsidov v plodovodstve Primorskogo kraya. [Use of herbicides in fruit growing in Primor'e.] - Tr. dal'nevost. nauch.- -issled. Inst. sel'. Khoz. *13* (Part 2) : 311 - 313, 1973. [Chi; in R.]

17380 - **OKU, T.**, **SUGAHARA, K.**, **TOMITA, G.** : Electron transfer and energy dependent reactions in glutaraldehyde-fixed chloroplasts. - Plant Cell Physiol. *14* : 385 - 396, 1973.

17381 - **OKU, T.**, **TOMITA, G.** : Protochlorophyllide holochrome. 3. Optical and photo-chemical properties of a constituted chlorophyll *a*-holochromic protein. - Photosynthetica *7* : 67 - 72, 1973.

17382 - **OLSON, J.M.**, **PHILIPSON, K.D.**, **SAUER, K.** : Circular dichroism and absorption spectra of bacteriochlorophyll-protein and reaction center complexes from *Chlorobium thiosulfatophilum*. - Biochim. biophys. Acta *292* : 206 - 217, 1973.

17383 - **ONDOK, J.P.** : Interception of photosynthetically active radiation by a *Phragmites* stand. - In : HEJNÝ, S. (ed.) : Ecosystem Study on Wetland Biome in Czechoslovakia. Czechosl. IBP/PT-PP Report No. 3. Pp. 133 - 142. Třeboň 1973.

17384 - **ONDOK, J.P.** : Photosynthetically active radiation in a stand of *Phragmites communis* TRIN. I. Distribution of irradiance and foliage structure. - Photosynthetica *7* : 8 - 17, 1973.

17385 - **ONDOK, J.P.** : Photosynthetically active radiation in a stand of *Phragmites communis* TRIN. II. Model of light extinction in the stand. - Photosynthetica *7* : 50 - 57, 1973.

17386 - **ONDOK, J.P.** : Photosynthetically active radiation in a stand of *Phragmites communis* TRIN. III. Distribution of irradiance on sunlit foliage area. - Photosynthetica *7* : 311 - 319, 1973.

17387 - **ONDOK, J.P.** : Some basic concepts of modelling freshwater littoral ecosystems with respect to radiation regime of a pure *Phragmites* stand. - Pol. Arch. Hydrobiol. *20* : 101 - 109, 1973. [Ps.]

17388 - **ONDOK, P.**, **PŘIBÁŇ, K.** : PP/IBP initial level experiments in South Bohemia. - Annu. Rep. algol. Lab. Třeboň *1970* : 177 - 182, 1973. [Production.]

17389 - **ONWUEME, I.C.**, **LAWANSON, A.O.** : Effect of heat stress on subsequent chlorophyll accumulation in seedlings of *Colocynthis citrullus*. - Planta *110* : 81 - - 84, 1973.

*B17390 - **OPARIN, A.I.** (ed.) : Fiziologiya Sel'skokhozyaĭstvennykh Rasteniĭ. Tom I. Fiziologiya Rastitel'noĭ Kletki. Fotosintez. Dykhanie. [Physiology of Agricultural Plants. Vol. I. Physiology of Plant Cell. Photosynthesis. Respiration.] - Izdat. mosk. Univ., Moskva 1967. [In R.]

17391 - **OPATRNÝ, Z.** : Androgenese *in vitro* v prašníkových kulturách chlorofylových mutantů *Nicotiana tabacum* L. [Androgenesis *in vitro* in anther cultures of chlorophyll mutants of *Nicotiana tabacum* L.] - In : LANDA, Z., NOVÁK, F.J., OPATRNÝ, Z. (ed.) : I. Kolokvium\Využití Kultur Rostlinných Explantátů *in Vitro* v Genetice a Šlechtění. Pp. 271 - 278. Ústav Experimentální Botaniky ČSAV, Praha - Olomouc 1973. [In Czech, ab : E.]

17392 - **OPATRNÝ, Z.** : Androgenesis *in vitro* in anther cultures of chlorophyll mutants of *Nicotiana tabacum*. - Biol. Plant. *15* : 286 - 289, 1973.

17393 - **OPATRNÝ, Z.**, **LANDA, Z.** : Regenerace chlorofylových chimer z listových explantátů *Nicotiana tabacum* L. [Regeneration of chlorophyll chimerae from leaf explants of *Nicotiana tabacum* L.] - In : LANDA, Z., NOVÁK, F.J., OPATRNÝ, Z. (ed.) : I. Kolokvium Využití Kultur Rostlinných Explantátů *in Vitro* v Genetice a Šlechtění. Pp. 287 - 294. Ústav Experimentální Botaniky ČSAV, Praha - Olomouc 1973. [In Czech, ab : E.]

17394 - **OPHIR, I.**, **BEN-SHAUL, Y.** : Separation and ultrastructure of proplastids from dark-grown *Euglena* cells. - Plant Physiol. *51* : 1109 - 1116, 1973.

17395 - **OPHIR, I., BEN-SHAUL, Y.** : Structural organization of developing chloroplasts in *Euglena*. - Protoplasma *80* : 109 - 127, 1973.

17396 - **OPRITOV, V.A., MICHURIN, S.V.** : Bioělektricheskie potentsialy kak zveno v mekhanizme sopryazheniya aktivnogo transporta assimilyatov s metabolizmom. [Bioelectric potentials as a link in the mechanism of correlation of active transport of photosynthates with metabolism.] - Tr. biol.-pochv. Inst. dal'nevost. nauch. Tsentra Akad. Nauk SSSR *20* (123) : 101 - 106, 1973. [In R, ab : E.]

17397 - **O'REILLY, J.** : Oxidation-reduction potential of the ferro-ferricyanide system in buffer solutions. - Biochim. biophys. Acta *292* : 509 - 515, 1973.

17398 - **ORITANI, T., YOSHIDA, R.** : Studies on nitrogen metabolism in crop plants. XII. Cytokinins and abscisic acid-like substances levels in rice and soybean leaves during their growth and senescence. - Proc. Crop Sci. Soc. Jap. *42* : 280- - 287, 1973. [Chl.]

17399 - **ORT, D.R., IZAWA, S.** : Studies on the energy-coupling sites of photophosphorylation. II. Treatment of chloroplasts with NH_2OH plus ethylenediaminetetraacetate to inhibit water oxidation while maintaining energy-coupling efficiencies. - Plant Physiol. *52* : 595 - 600, 1973.

17400 - **ORT, D.R., IZAWA, S., GOOD, N.E., KROGMANN, D.W.** : Effects of the plastocyanin antagonists KCN and poly-*L*-lysine on partial reactions in isolated chloroplasts. - FEBS Lett. *31* : 119 - 122, 1973.

17401 - **ORTHOEFER, F.T., DUGAN, L.R. Jr.** : The coupled oxidation of chlorophyll *a* with linoleic acid catalysed by lipoxidase. - J. Sci. Food Agr. *24* : 357 - - 365, 1973.

17402 - **OSHCHEPKOV, V.P., NIKITINA, K.A., GUSEV, M.V., KRASNOVSKIĬ, A.A.** : Vydelenie molekulyarnogo vodoroda kul'turami sinezelenykh vodoroslеĭ. [Molecular hydrogen liberation by blue-green algae cultures.] - Dokl. Akad. Nauk SSSR *213* : 739 - 742, 1973. [In R.]

17403 - **OSMOND, C.B., ALLAWAY, W.G., SUTTON, B.G., TROUGHTON, J.H., QUEIROZ, O., LÜTTGE, U., WINTER, K.** : Carbon isotope discrimination in photosynthesis of CAM plants. - Nature *246* : 41 - 42, 1973.

17404 - **OSTROUMOV, S.A.** : O chem govorit skhodstvo mitokhondriĭ, khloroplastov i prokariotov. [Similarity of mitochondria, chloroplasts and prokaryots.] - Priroda (Moskva) *1973* (3) : 21 - 29, 1973. [In R.]

17405 - **OSTROUMOV, S.A., SAMUILOV, V.D., SKULACHEV, V.P.** : Transhydrogenase-induced responses of carotenoids, bacteriochlorophyll and penetrating anions in *Rhodospirillum rubrum* chromatophores. - FEBS Lett. *31* : 27 - 30, 1973.

17406 - **OSTROVSKAYA, L.K., GAMAYUNOVA, M.S., KOCHUBEĬ, S.M., GRIGORA, M.Yu., SILAEVA, A.M., YAKOVENKO, A.M.** : Nekotorye sravnitel'nye kharakteristiki fragmentov, soderzhashchikh fotosistemu I., iz lamell stromy i lamell gran khloroplastov. [Some comparative characteristics of the photosystem I-containing fragments from stroma lamellae and chloroplast granum lamellae.] - Dokl. Akad. Nauk SSSR *209* : 1457 - 1460, 1973. [In R.]

17407 - **OSWALD, W.J.** : Productivity of algae in sewage disposal. - Solar Energy *15* : 107 - 117, 1973. [Ps, Chl.]

17408 - **OUITRAKUL, R., IZAWA, S.** : Electron transport and photophosphorylation in chloroplasts as a function of the electron acceptor. II. Acceptor-specific inhibition by KCN. - Biochim. biophys. Acta *305* : 105 - 118, 1973.

17409 - **OUTLAW, W.H., FISHER, D.B.** : Sucrose compartmentation in *Vicia faba* leaflets. - Plant Physiol. *51* (Suppl.) : 62, 1973. [Photosynthates.]

17410 - **OWEN, J.M.** : An analogue device for correcting densitometric measurements of coloured spots on chromatograms for the effect of light scattering. - J. Chromatogr. *79* : 165 - 171, 1973. [Car.]

17411 - **OWENS, O.v.HARTZ** : The kinetics and quantum yield of photophosphorylation in *Anacystis nidulans* (RICHT.) DROUET. - Diss. Abstr. Int. B *33* : 4657, 1973.

17412 - **PACOLD, I., ANDERSON, L.E.** : Energy charge control of the Calvin cycle enzyme 3-phosphoglyceric acid kinase. - Biochem. biophys. Res. Commun. *51* : 139 - - 143, 1973.

17413 - **PAILLOTIN, G.** : Etats excités et processus photochimiques primaires dans la photosynthèse des plantes supérieures. - In : Ecole de Roscoff. La Photobiologie. Pp. 87 - 95. CNRS, Paris 1973.

17414 - **PAKARINEN, P., VITT, D.H.** : Primary production of plant communities of the Truelove Lowland, Devon Island, Canada - moss communities. - In : BLISS, L.C., WIELGOLASKI, F.E. (ed.) : Primary Production and Production Processes, Tundra Biome. Pp. 37 - 46. Tundra Biome Steering Comm., Edmonton 1973.

*17415 - **PALIS, R.K., PAULSEN, G.M., POWERS, W.L.** : Effect of some antitranspiration chemicals on photosynthesis and respiration of corn and soybean seedlings. - Trans. Kansas Acad. Sci. *75* : 218 - 225, 1972.

17416 - **PALLAS, J.E. Jr.** : Diurnal changes in transpiration and daily photosynthetic rate of several crop plants. - Crop Sci. *13* : 82 - 84, 1973.

1741 - **PALLAS, J.E. Jr., SAMISH, Y.B.** : Photosynthesis and light response of the cultivated peanut, *Arachis hypogaea* L. - Plant Physiol. *51* (Suppl.) : 68, 1973.

-18   - **PALMER, A.F.E., HEICHEL, G.H., MUSGRAVE, R.B.** : Patterns of translocation, respiratory loss, and redistribution of ^{14}C in maize labeled after flowering. - Crop Sci. *13* : 371 - 376, 1973.

17419 - **PALMER, B.J.** : A review paper on the effect of carbon dioxide and aerosols on climate modification. - Environ. Lett. *5* : 249 - 265, 1973. [Atmos. CO_2 conc., also prediction; source and sink.]

17420 - **PALTRIDGE, G.W.** : On the shape of trees. - J. theor. Biol. *38* : 111 - 137, 1973. [Ps.]

17421 - **PANITSKIĬ, V.V., PANITSKAYA, M.P.** : Energeticheskiĭ balans i vlagoobespechennost' pshenitsy v vostochnoĭ Sibiri. [Energy balance and water availability for wheat in eastern Siberia.] - Sel'sko-khoz. Biol. *8* : 617 - 618, 1973. [In R.]

17422 - **PANTER, R.A., BOARDMAN, N.K.** : Lipid biosynthesis by isolated plastids from greening pea, *Pisum sativum.* - J. Lipid Res. *14* : 664 - 671, 1973.

17423 - **PAPAGEORGIOU, G., ARGOUDELIS, C.** : Cation-dependent quenching of the fluorescence of chlorophyll *a in vivo* by nitroaromatic compounds. - Arch. Biochem. Biophys. *156* : 134 - 142, 1973.

17424 - **PAPASTEPHANOU, C., BARNES, F.J., BRIEDIS, A.V., PORTER, J.W.** : Enzymic synthesis of carotenes by cell-free preparations of fruit of several genetic selections of tomatoes. - Arch. Biochem. Biophys. *157* : 415 - 425, 1973.

17425 - **PARCEVAUX, S. de** : Importance des échanges gazeux au niveau des feuilles dans l'écophysiologie de diverses plantes. - Oecol. Plant. *8* : 41 - 62, 1973. [Ps.]

17426 - **PARCEVAUX, S. de** : Influence de la régulation stomatique sur la transformation de l'énergie d'origine solaire les feuilles. - In : The Sun in the Service of Mankind. Pp. V.15-1 - V.15-8. Unesco, Paris 1973. [Model.]

17427 - **PARCEVAUX, S. de, PERRIER, A.** : Bilan énergétique de la feuille. Application de l'étude des cinétiques de température à la détermination des résistances aux flux gazeux. - In : SLATYER, R.O. (ed.) : Plant Response to Climatic Factors. Pp. 127 - 135. Unesco, Paris 1973.

17428 - **PARK, K.E.Y., ANDERSON, L.E.** : Appearance of three chloroplast isoenzymes in dark-grown pea plants and pea seeds. - Plant Physiol. *51* : 259 - 262, 1973.

17429 - **PARK, R.B.** : Structure of photoreceptive units. - In : CHECCUCCI, A., WEALE, R.A. (ed.) : Primary Molecular Events in Photobiology. Pp. 1 - 20. Elsevier, Amsterdam - London - New York 1973. [Chloroplast.]

17430 - **PARK, Y., MORRIS, M.M., MACKINNEY, G.** : On chlorophyll breakdown in senescent leaves. - J. agr. Food Chem. *21* : 279 - 281, 1973.

17431 - **PARKER, B.C., SAMSEL, G.L. Jr., PRESCOTT, G.W.** : Comparison of microhabitats of macroscopic subalpine stream algae. - Amer. Midland Naturalist *90* : 143 - - 153, 1973. [Ps, Chl.]

17432 - PARKER, R.R., SIBERT, J. : Effect of pulpmill effluent on dissolved oxygen in a stratified estuary - I. Empirical observations. - Water Res. *7* : 503 - 514, 1973. [Primary production.]

17433 - PARSONS, T.R. : Coulter Counter for phytoplankton. - In : STEIN, J.R. (ed.) : Handbook of Phycological Methods. Culture Methods and Growth Measurements. Pp. 345 - 358. Cambridge Univ. Press, London 1973. [Production.]

*17434 - PARSONS, T.R., STEPHENS, K., TAKAHASHI, M. : The fertilization of Great Central Lake. I. Effect of primary production. - Fish. Bull. *70* : 13 - 23, 1972. [Chl.]

17435 - PASHCHENKO, P.N. : Fiziologicheskiĭ podkhod k issledovaniyu mekhanizma obrazovaniya svobodnogo i svyazannogo NADF H$_2$ v khloroplastakh. [Physiological approach to the study of the mechanism of the formation of free and bound NADPH$_2$ in chloroplasts.] - In : Upravlenie Skorost'yu i Napravlennost'yu Biosinteza u Rasteniĭ. P. 34. Krasnoyarsk 1973. [In R.]

17436 - PASHCHENKO, V.N. : Izmenchivost' fotosinteticheskoĭ edinitsy v tsenoze ,rasteniĭ. [Variability of photosynthetic unit in a plant stand.] - In : Upravlenie Skorost'yu i Napravlennost'yu Biosinteza u Rasteniĭ. Pp. 32 - 33. Krasnoyarsk 1973. [In R.]

17437 - PASSERA, C., FERRARI, G. : Evaluation of UV-induced mutants of *Chlorella vulgaris* for single cell protein I. The screening effect of cycloheximide and 6-methylpurine. - Physiol. Plant. *29* : 6 - 9, 1973. [Ps, Chl.]

17438 - PASTERNAK, D., WILSON, G.L. : Illuminance, stomatal opening, and photosynthesis in sorghum and cotton. - Aust. J. agr. Res. *24* : 527 - 532, 1973.

17439 - PATRICK, J.W., WAREING, P.F. : Auxin-promoted transport of metabolites in stems of *Phaseolus vulgaris* L. - J. exp. Bot. *24* : 1158 - 1171, 1973. [Photosynthates.]

17440 - PATTERSON, C.O.P., MYERS, J. : Photosynthetic production of hydrogen peroxide by *Anacystis nidulans*. - Plant Physiol. *51* : 104 - 109, 1973.

17441 - PAULET, A., LECHEVALLIER, D., BAZIER, R., COSTES, C., MONÉGER, R. : Extraction progressive des stérols de chloroplastes lyophilisés de Blé par des solvants de polarité croissante. - Physiol. vég. *11* : 655 - 661, 1973.

17442 - PEARMAN, G.I., GARRATT, J.R. : Carbon dioxide measurements above a wheat crop, 1. Observations of vertical gradients and concentrations. - Agr. Meteorol. *12* : 13 - 25, 1973.

17443 - PÉAUD-LENOËL, C., HANSON, A.D. : Une exigence en kinétine pour le développement de chloroplastes fonctionnels dans les cals de tissus de *Nicotiana tabacum* croissants par photosynthèse. - Compt. rend. Acad. Sci. Paris, Sér. D *277* : 1853 - 1855, 1973. [Chl.]

17444 - PEAVEY, D.G., GIBBS, M. : Emerson enhancement in the isolated chloroplast. - Plant Physiol. *51* (Suppl.) : 67, 1973.

17445 - PECHT, I. : Pulse radiolysis study of the reduction of spinach ferredoxin. - FEBS Lett. *33* : 259 - 262, 1973.

17446 - PEISKER, M. : CO$_2$-Aufnahme, Transpiration und Blattemperatur unter dem Einfluß von Änderungen der Stomataweite. - Kulturpflanze *21* : 97 - 109, 1973.

17447 - PEL, B. Van, BRONCHART, R., KEBERS, F., COCITO, C. : Structure and function of cytoplasmic organelles in transiently and permanently bleached *Euglena*. - Exp. Cell Res. *78* : 103 - 110, 1973. [Chl.]

17448 - PELKONEN, P., HARI, P., LUUKKANEN, O., SMOLANDER, H. : Notes on CO$_2$ exchange during accelerated seasonal development in birch. - In : Proceedings of IUFRO Working Party S2.01.4. Symposium on Dormancy in Trees. Pp. 1 - 6. Kórnik 1973.

17449 - PELL, E.J., BRENNAN, E. : Changes in respiration, photosynthesis, adenosine 5'-triphosphate, and total adenylate content of ozonated pinto bean foliage as they relate to symptom expression. - Plant Physiol. *51* : 378 - 381, 1973.

17450 - PELROY, R.A., BASSHAM, J.A. : Kinetics of glucose incorporation by *Aphanocapsa* 6714. - J. Bacteriol. *115* : 943 - 948, 1973.

17451 - PENDLAND, J.C., ALDRICH, H.C. : Ultrastructural organization of chloroplast thylakoids of the green alga *Oocystis marssonii*. - J. Cell Biol. *57* : 306 - - 314, 1973.

17452 - PERRIER, E.R., ASTON, A., ARKIN, G.F. : Wind flow characteristics on a soybean leaf compared with a leaf model. - Physiol. Plant. *28* : 106 - 112, 1973.

17453 - PERSKAYA, E.B., GEĬKO, N.S., RACHINSKIĬ, V.V., KRETOVICH, V.L. : Vklyuchenie C^{14} v ketokislotnuyu fraktsiyu list'ev fasoli pri kratkovremennom fotosinteze. [^{14}C incorporation in the ketoacid fraction of bean leaves during short--term photosynthesis.] - Izv. Akad. Nauk SSSR, Ser. biol. *1973* : 423 - 425, 1973. [In R, ab : E.]

17454 - PETERS, J.C., BOLDT, P.E., deLOACH, C.J. : Photometric and geometric estimates of leaf surface area of cabbage, mustard, and turnip plants as an aid to studying insect population dynamics. - J. econ. Entomol. *66* : 104 - 107, 1973.

17455 - PETERSON, L.W., KLEINKOPF, G.E., HUFFAKER, R.C. : Evidence of lack of turnover of ribulose 1,5-diphosphate carboxylase in barley leaves. - Plant Physiol. *51* : 1042 - 1045, 1973.

17456 - PETKOVA, R., ZEĬNALOV, Yu., DILOVA, S. : State of the pigment-protein complex in higher plants. I. Influence of temperatures of 25 - 70 °C. - Photosynthetica *7* : 226 - 231, 1973.

17457 - PETROV, A., MANOLOV, P. : Effect of leaf/fruit ratio on the translocation and distribution of ^{14}C assimilates in young peach trees. - Proc. Res. Inst. Pomol., Skierniewice, Ser. E *1973* (3) : 111 - 120, 1973.

*17458 - PEYRIÈRE, M. : Les problèmes cytologiques de *Rytiphlea tinctoria* Algue rouge à floridorubine. - Compt. rend. Acad. Sci. Paris, Sér. D *266* : 2253 - 2255, 1968. [Chloroplast.]

17459 - PFLÜGER, R. : Investigations on ion fluxes of chloroplasts with an intact envelope. - Z. Naturforsch. *28 c* : 779 - 780, 1973.

17460 - PFLÜGER, R., MENGEL, K. :The photochemical activity of chloroplasts from various plants with different potassium nutrition. - Büntehof. Abs. *3* : 37 - - 38, 1972-1973.

17461 - PHAN, C.T. : Chloroplasts of the peel and the internal tissues of apple-fruits. - Experientia *29* : 1555 - 1557, 1973.

17462 - PHAN, C.T. : The photosynthetic apparatus of internal tissues of apple-fruits. - Plant Physiol. *51* (Suppl.) : 68, 1973.

17463 - PHILIP, T. : Nature of xanthophyll esterification in grapefruit. - J. agr. Food Chem. *21* : 963 - 964, 1973.

17464 - PHILIP, T. : The nature of carotenoid esterification in citrus fruits. - J. agr. Food Chem. *21* : 964 - 966, 1973.

17465 - PHILIPPOVICH, I.I., BEZSMERTNAYA, I.N., OPARIN, A.I. : On the localization of polyribosomes in the system of chloroplast lamellae. - Exp. Cell Res. *79* : 159 - 168, 1973.

17466 - PHILIPSON, K.D., SAUER, K. : Comparative study of the circular dichroism spectra of reaction centers from several photosynthetic bacteria. - Biochemistry *12* : 535 - 539, 1973.

17467 - PHILIPSON, K.D., SAUER, K. : Light-scattering effects on the circular dichroism of chloroplasts. - Biochemistry *12* : 3454 - 3458, 1973.

17468 - PICK, U., AVRON, M. : Inorganic sulfate and selenate as energy transfer inhibitors of photophosphorylation. - Biochim. biophys. Acta *325* : 297 - 303, 1973.

17469 - PICK, U., ROTTENBERG, H., AVRON, M. : Effect of phosphorylation on the size of the proton gradient across chloroplast membranes. - FEBS Lett. *32* : 91 - - 94, 1973.

17470 - PIJANOWSKI, B.S. : Salinity corrections for dissolved oxygen measurements. - Environ. Sci. Technol. *7* : 957 - 958, 1973.

17471 - PIKE, J.D. : Effects of temperature and light intensity on the pigmentation and photosynthesis of *Chlorella pyrenoidosa*. - J. Phycol. *9* (Suppl.) : 5 - 6, 1973.

17472 - PILSON, M.E.Q., BETZER, S.B. : Phosphorus flux across a coral reef. - Ecology *54* : 581 - 588, 1973. [Ps.]

17473 - PINCUS, M., WINDREICH, S., MILLER, I.R. : The preparation of stable mixed monolayers of β-carotene and their transfer to glass slides. - Biochim. biophys. Acta *311* : 317 - 319, 1973.

17474 - PINEWITSCH, W., WASILJEWA, W. : Untersuchungen der Carotinoide von *Anacystis nidulans*. - Arch. Hydrobiol. *41* (Suppl.) : 373 - 391, 1973.

17475 - PINFIELD, N.J., STOBART, A.K., CRAWFORD, R.M., BECKETT, A. : Carbon assimilation by sycamore cotyledons during early seedling development. - J. exp. Bot. *24* : 1203 - 1207, 1973.

17476 - PISEK, A. : Photosynthesis. - In : PRECHT, H., CHRISTOPHERSEN, J., HENSEL, H., LARCHER, W. (ed.) : Temperature and Life. Pp. 102 - 127. Springer-Verlag, Berlin - Heidelberg - New York 1973.

17477 - PJON, C.-J., FUJITA, Y. : Immunological identification of pigment component of a photochemically active chromoprotein (ACP) isolated from the blue-green alga *Anabaena cylindrica*. - Plant Cell Physiol. *14* : 201 - 205, 1973.

17478 - PLANTE-CUNY, M.-R. : Recherches sur la production primaire benthique en milieu marin tropical. I. Variations de la production primaire et des teneurs en pigments photosynthétiques sur quelques fonds sableux. Valeur des résultats obtenus par la méthode du 14 C. - Cah. ORSTOM, Sér. Océanogr. *11* : 317 - 348, 1973.

*B17479 - Plastidnyĭ Apparat i Zhiznedeyatel'nost' Rasteniĭ. [Plastid Apparatus and Plant Viability.] - Nauka i Tekhnika, Minsk 1971. [In R.]

17480 - PLATT, T., IRWIN, B. : Caloric content of phytoplankton. - Limnol. Oceanogr. *18* : 306 - 310, 1973.

17481 - PLATT, T., IRWIN, B., SUBBA RAO, D.V. : Primary productivity and nutrient measurements on the spring phytoplankton bloom in Bedford Basin, 1971. - Fisheries Res. Board Can., tech. Rep. *423* : 1 - 45, 1973.

17482 - PLATT, T., SUBBA RAO, D.V. : Some current problems in marine phytoplankton productivity. - Fisheries Res. Board Can., tech. Rep. *370* : 1 - 90, 1973.

17483 - PLAUT, Z., BRAVDO, B. : Response of carbon dioxide fixation to water stress. Parallel measurements on isolated chloroplasts and intact spinach leaves. - Plant Physiol. *52* : 28 - 32, 1973.

17484 - PLESNIČAR, M., BENDALL, D.S. : Some evidence on the site of action of plastocyanin in the photosynthetic electron-transport chain between cytochromes f and $P700$. - Europe. J. Biochem. *34* : 483 - 488, 1973.

17485 - PLESNIČAR, M., BENDALL, D.S. : The photochemical activities and electron carriers of developing barley leaves. - Biochem. J. *136* : 803 - 812, 1973.

17486 - PLOTNIKOV, V.A. : Mutagennyĭ éffekt dimetilsul'fata u podsolnechnika (inbrednaya liniya). [Mutagenic effect of dimethyl sulfate in sunflower (inbred line).] - Genetika *9* (5) : 15 - 22, 1973. [Chl; in R, ab : E.]

17487 - POCKER, Y., NG, J.S.Y. : Plant carbonic anhydrase. Properties and carbon dioxide hydration kinetics. - Biochemistry *12* : 5127 - 5134, 1973.

17488 - POINCELOT, R.P. : Differences in lipid composition between undifferentiated and mature maize chloroplasts. - Plant Physiol. *51* : 802 - 804, 1973.

17489 - POINCELOT, R.P. : Isolation and characterization of envelope membranes from spinach chloroplasts. - Plant Physiol. *51* (Suppl.) : 17, 1973.

17490 - POINCELOT, R.P. : Isolation and lipid composition of spinach chloroplast envelope membranes. - Arch. Biochem. Biophys. *159* : 134 - 142, 1973.

17491 - **POLEVAYA, V.S.** : Regulyatsiya rosta i sinteza pigmentov v kul'ture tkani ru-
ty. [Regulation of growth and pigment synthesis in tissue culture of rue.] -
In : Upravlenie Skorost'yu i Napravlennost'yu Biosinteza u Rasteniĭ. Pp. 119-
- 120. Krasnoyarsk 1973. [In R.]

17492 - **POLING, S.M., HSU, W.-J., YOKOYAMA, H.** : New chemical inducers of carotenoid
biosynthesis. - Phytochemistry *12* : 2665 - 2667, 1973.

*17493 - **PONOMAREVA, M.M.** : Fotosintez. [Photosynthesis.] - In : Biokompleksnaya Kharakteristika Osnovnykh Tsenozoobrazovatelei Rastitel'nogo Pokrova Tsentral'-
nogo Kazakhstana. Pp. 44 - 47; 77 - 78; 102 - 103; 124 - 125; 145 - 146;
165 - 167; 188 - 189; 215; 232 - 233; 246 - 247. Nauka, Leningrad 1969. [In
R.]

17494 - **PONYATOVSKAYA, V.M., MAKAREVICH, V.N.** : Ob izuchenii produktsionnogo protsessa v lugovykh soobshchestvakh. [Production process in meadow communities.] -
Bot. Zh. *58* : 997 - 1004, 1973. [In R.]

17495 - **POOVAIAH, B.W., LEOPOLD, A.C.** : Deferral of leaf senescence with calcium. -
Plant Physiol. *52* : 236 - 239, 1973. [Chl.]

1749ʳ - **POPE, D.H., BERGER, L.R.** : Algal photosynthesis at increased hydrostatic pressure and constant pO_2. - Arch.Mikrobiol. *89* : 321 - 325, 1973.

1⁷497 - **POPE, D.H., BERGER, L.R.** : An apparatus to measure the rate of oxygen evolution while maintaining pO_2 constant during photosynthetic growth in closed
culture vessels capable of operation at increased hydrostatic pressures. -
Biotechnol. Bioeng. *15* : 505 - 518, 1973.

17498 - **POPOVA, O.F., SAPOZHNIKOV, D.I.** : Deĭstvie sveta razlichnoĭ intensivnosti na
reaktsii violaksantinovogo tsikla v zeleneyushchikh prorostkakh kukuruzy.
[Effect of light of different intensity on reactions of violaxanthin cycle in
greening maize seedlings.] - Fiziol. Rast. *20* : 628 - 631, 1973. [In R.]

17499 - **POPOVICI, N., ZOLYNEAK, C., SIMA-DĘLIU, E.** : Influenţe fiziologice ale teofi-
linei în perioada postgerminatorie la porumb. [Physiological effects of theo-
phylline on maize in the post-germination period.] - An. şti. Univ. "Al.I. Cu-
za" Iaşi, Secţ. IIa, *19* : 31 - 42, 1973. [Ps, Chl; in Roum., ab : F.]

17500 - **PORATH, D., BEN SHAUL, Y.** : Growth, greening, and phytochrome in etiolated
Spirodela (Lemnaceae). - Plant Physiol. *51* : 474 - 477, 1973. [Chl.]

17501 - **POROKHNEVICH, N.V.** : Issledovanie proyavleniya deĭstviya tsinka i medi na fo-
tosinteticheskiĭ apparat l'na u rasteniĭ vtorogo semennogo pokoleniya. [Effect
of zinc and copper on the photosynthetic apparatus in flax plants of the second seed generation.] - In : Materialy Dokladov 3-go Delegatskogo Şobraniya
Belorusskogo Botanicheskogo Obshchestva. Pp. 184 - 186. Izd. Akad. Nauk BSSR,
Minsk 1973. [In R.]

17502 - **POROKHNEVICH, N.V.** : Izmenenie sootnosheniya mezhdu khlorofillami *a* i *b* v
list'yakh l'na pod vliyaniem tsinka. [Effect of zinc on the ratio between
chlorophylls *a* and *b* in flax leaves.] - Fiziol. Rast. *20* : 1029 - 1035, 1973.
[In R, ab : E.]

17503 - **POROKHNYA, A.D., KURILOVA, G.A.** : Fotosinteticheskaya produktivnost' risa pri
razlichnom urovne fosfornogo pitaniya. [Photosynthetic productivity of rice
at various levels of phosphorus nutrition.] - Tr. vses. nauch.-issled. Inst.
Risa *1973* (3) : 57 - 61, 1973. [In R.]

17504 - **PORRA, R.J.** : A new method for the determination of chlorophylls. - Proc.
aust. biochem. Soc. *6* : 77, 1973.

17505 - **PORRA, R.J.** : The regulation of chlorophyll formation in *Chlorella fusca* and
a new assay for chlorophylls *a* and *b*. - Hoppe Seyler's Z. physiol. Chem. *354*:
845, 1973.

17506 - **PORRA, R.J., GRIMME, L.H.** : The regulation of chlorophyll formation in *Chlorella fusca* and a new assay for chlorophylls *a* and *b*. - Enzyme *16* : 115 - 123,
1973.

17507 - **PORTER, J., JOST, M.** : Light shielding by gas-vacuoles in *Microcystis aeruginosa*. - J. gen. Microbiol. *75* : XXII, 1973.

17508 - PORTIS, A.R. Jr., McCARTY, R.E. : On the pH-dependence of the light-induced hydrogen ion gradient in spinach chloroplasts. - Arch. Biochem. Biophys. *156*: 621 - 625, 1973.

17509 - POSSINGHAM, J.V. : Chloroplast growth and division during the greening of spinach leaf discs. - Nature, new Biol. *245* : 93 - 94, 1973.

17510 - POSSINGHAM, J.V. : Effect of light quality on chloroplasts replication in spinach. - J. exp. Bot. *24* : 1247 - 1260, 1973.

17511 - POTRYKUS, I. : Transplantation of chloroplasts into protoplasts of *Petunia*. - Z. Pflanzenphysiol. *70* : 364 - 366, 1973.

17512 - POTTER, J.R., BOYER, J.S. : Chloroplast response to low leaf water potentials II. Role of osmotic potential. - Plant Physiol. *51* : 993 - 997, 1973.

17513 - POYARKOVA, N.M., DROZDOVA, I.S., VOSKRESENSKAYA, N.P. : Effect of blue light on the activity of carboxylating enzymes and $NADP^+$-dependent glyceraldehyde--3-phosphate dehydrogenase in bean and maize plants. - Photosynthetica *7* : 58 - 66, 1973.

17514 - PRAKASH, A., RASHID, M.A., JENSEN, A., SUBBA RAO, D.V. : Influence of humic substances on the growth of marine phytoplankton : Diatoms. - Limnol. Oceanogr. *18* : 516 - 524, 1973. [Chl]

17515 - PREDKEL', K.I., VECHER, A.S. : Soderzhanie razlichnykh po rastvorimosti form zheleza v rastitel'nykh tkanyakh i vydelennykh iz nikh plastidakh. [Content of different in solubility forms of iron in plant tissues and plastids isolated from them.] - In : VECHER, A.S. (ed.) : Fotosintez i Ustoĭchivost' Rastenĭ. Pp. 69 - 75. Nauka i Tekhnika, Minsk 1973. [Chl; in R.]

17516 - PŘIBIL, S., DYKYJOVÁ, D. : Seasonal differences in caloric contents of some emergent macrophytes. (A preliminary report.). - In : HEJNÝ, S. (ed.) : Ecosystem Study on Wetland Biome in Czechoslovakia. Czechosl. IBP/PT-PP Report No. 3. Pp. 97 - 99. Třeboň 1973.

17517 - PRICE, C.A., BREDEN, E.N., VASCONCELOS, A.C. : Isolation of intact spinach chloroplasts in the CF-6 continuous-flow zonal rotor; implications for membrane-bound organelles. - Anal. Biochem. *54* : 239 - 246, 1973.

17518 - PRIESTLEY, C.A. : The use of apple leaf discs in studies of [14]carbon translocation. - Proc. Res. Inst. Pomol., Skierniewice, Ser. E *1973* (3) : 121 - 128, 1973. [Photosynthates.]

17519 - PRINCE, R.C., CROFTS, A.R. : Photochemical reaction centres from *Rhodopseudomonas capsulata Ala pho+*. - FEBS Lett. *35* : 213 - 216, 1973.

17520 - PRINS, H.B.A. : The action spectrum of photosynthesis and the rubidiumchloride uptake by leaves of *Vallisneria spiralis*. - Proc. koninkl. nederl. Akad. Wetenschappen (Amsterdam), Ser. C *76* : 495 - 499, 1973.

17521 - PRIOUL, J.L. : Eclairement de croissance et infrastructure des chloroplastes de *Lolium multiflorum* LAM. Relation avec les résistances au transfert de CO_2. - Photosynthetica *7* : 373 - 381, 1973.

17522 - PRIOUL, J.L., BOURDU, R. : Graphical display of photosynthetic adaptability to irradiance. - Photosynthetica *7* : 405 - 407, 1973.

17523 - PRISTAVU, N., WEGMANN, K. : Influenţa NO_3^- şi a NH_4^+ asupra metabolismului carbonului la alga *Dunaliella tertiolecta*. [Effect of NO_3^- and NH_4^+ on carbon metabolism in the alga *Dunaliella tertiolecta*.] - Stud. Cercet. Biol., Ser. bot. *25* : 425 - 429, 1973. [Chl; in Roum., ab : G.]

17524 - PRISTUPA, N.A. : O transporte assimilyatov v tkanyakh lista kukuruzy. [Transport of radioactive photosynthates in the tissues of maize leaves.] - Tr. biol.-pochv. Inst., dal'nevost. nauch. Tsentr Akad. Nauk SSSR *20* (123) : 39 - 43, 1973. [in R, ab : E.]

17525 - PROCHÁZKA, S. : Study on the distribution of assimilates in spring barley during the period of grain formation. - Acta Univ. Agr., Ser. A, Fac. agron. (Brno) *21* : 705 - 710, 1973.

17526 - PROCHÁZKA, S., PEŠKA, J., KAPLANOVÁ, M. : Studium distribuce asimilátů vytvořených klasem pšenice ozimé. [Study on the distribution of photosynthates produced by the ear of winter wheat.] - Acta Univ. Agr., Ser. A, Fac. agron. (Brno) *21* : 531 - 537, 1973. [In Czech, ab : E, R.]

17527 - **PROCTOR, J.T.A.** : Developmental changes in radish caused by brief end-of-day exposures to far-red radiation. - Can. J. Bot. *51* : 1075 - 1077, 1973. [Chl.]

*B17528 - Produktsiya Morskogo Fitoplanktona. Retrospektivnaya Bibliografiya. [Production of Marine Phytoplankton. Retrospective Bibliography.] - Tsentral'nyĭ Nauchno-Issledovatel'skiĭ Institut Informatsii i Tekhniko-Ekonomicheskikh Issledovaniĭ Rybnogo Khozyaĭstva, Moskva 1972. [In R.]

17529 - **PROKHORCHIK, R.A.** : Vliyanie atrazina i fenazona na fotovosstanovlenie NADF khloroplastami rasteniĭ lyupina. [Effect of atrazine and phenazone on the photoreduction of NADP by chloroplasts of lupine.] - In : **VECHER, A.S.** (ed.): Fotosintez i Ustoĭchivost' Rasteniĭ. Pp. 93 - 97. Nauka i Tekhnika, Minsk 1973. [In R.]

17530 - **PROKHORCHIK, R.A., VOLYNETS, A.P.** : Deĭstvie flavonoidov na aktivnost' khloroplastov lyupina i gorokha. [Effect of flavonoids on the activity of lupine and pea chloroplasts.] - Fiziol. Biokhim. kul't. Rast. *5* : 623 - 628, 1973. [In R, ab : E.]

17531 - **PROMNITZ, L.C.** : A photosynthate allocation model for tree growth. - Diss. Abstr. int. B *33* : 4608-B, 1973.

17532 - **PRONINA, N.B.** : Aktivnost' adenilatkinazy v khloroplastakh gorokha. [Adenylate kinase activity in pea chloroplasts.] - Nauch. Dokl. vyssh. Shkoly, biol. Nauki *16* (11) : 78 - 82, 1973. [In R.]

17533 - **PRONINA, N.B., MUKHIN, E.N., KHOLODENKO, N.Ya.** : Fotokhimicheskiĭ sintez ATF v khloroplastakh gorokha pod vliyaniem fosfodoksina i pteridinov. [Photochemical synthesis of ATP in pea chloroplasts as affected by phosphodoxine and pteridines.] - Sel'sko-khoz. Biol. *8* : 415 - 419, 1973. [In R, ab : E.]

17534 - **PRONINA, N.B., MUKHIN, E.N., KHOLODENKO, N.Ya.** : Vliyanie fosfodoksina na fotovosstanovlenie ferritsianida i sopryazhennoe fosforilirovanie v khloroplastakh gorokha. [Effect of phosphodoxine on photoreduction of ferricyanide and coupled phosphorylation in pea chloroplasts.] - Fiziol. Biokhim. kul't. Rast. *5* : 64 - 67, 1973. [In R, ab : E.]

17535 - **PROSKURYAKOV, M.A.** : Metodika analiza razmeshcheniya elovogo drevostoya po elementam mikrorel'efa v gorakh Tyan'-shanya. [A method of analyzing spruce stand distribution to elements of microtopography in the Tien Shan mountains.] - Ekologiya *4* (2) : 90 - 91, 1973. [In R.]

17536 - **PUCHEU, N.L., KERBER, N.L., GARCÍA, A.F.** : Some environmental factors influencing the state of the membranes isolated by gradient centrifugation from cell-free extracts of *Rps. viridis*. - FEBS Lett. *33* : 119 - 124, 1973.

17537 - **PUCKETT, K.J., NIEBOER, E., FLORA, W.P., RICHARDSON, D.H.S.** : Sulphur dioxide: its effect on photosynthetic ^{14}C fixation in lichens and suggested mechanisms of phytotoxicity. - New Phytol. *72* : 141 - 154, 1973.

17538 - **PUCKRIDGE, D.W.** : A quantitative account of the influence of solar radiation, water and nitrogen supply on the photosynthesis of wheat communities in the field. - In : **SLATYER, R.O.** (ed.) : Plant Response to Climatic Factors. Pp. 519 - 525. Unesco, Paris 1973.

17539 - **PUGH, P.R.** : An evaluation of liquid scintillation counting techniques for use in aquatic primary production studies. - Limnol. Oceanogr. *18* : 310 - 319, 1973.

17540 - **dePUIT, E.J., CALDWELL, M.M.** : Seasonal pattern of net photosynthesis of *Artemisia tridentata*. - Amer. J. Bot. *60* : 426 - 435, 1973.

17541 - **PULICH, W.M., WARD, C.H.** : Physiology and ultrastructure of an oxygen-resistant *Chlorella* mutant under heterotrophic conditions. - Plant Physiol. *51* : 337 - 344, 1973. [Chloroplast.]

17542 - **PUMPYANSKAYA, S.L., RADOMYSL'SKAYA, T.M.** : Zavisimost' ul'trafioletovykh spektrov propuskaniya list'ev ot fotoperiodicheskikh vozdeĭstviĭ. [Dependence of

UV transmission spectra of leaves on photoperiodical effect.] - Dokl. Akad. Nauk SSSR *212* : 1250 - 1253, 1973. [In R.]

17543 - **PUPILLO, P., PICCARI, G.G.** : The effect of NADP on the subunit structure and activity of spinach chloroplast glyceraldehyde-3-phosphate dehydrogenase. - Arch. Biochem. Biophys. *154* : 324 - 331, 1973.

17544 - **PURITCH, G.S.** : Effect of water stress on photosynthesis, respiration, and transpiration of four *Abies* species. - Can. J. Forest Res. *3*: 293 - 298, 1973.

17545 - **PUROHIT, A.N., TREGUNNA, E.B.** : Carbon dioxide exchange in vegetative and reproductive plants. - Plant Physiol. *51* (Suppl.) : 11, 1973.

17546 - **PYT'EVA, N.F., RIZNICHENKO, G.Yu., RATYNI, A.I., RUBIN, A.B.** : Teoreticheskoe obosnovanie skhemy pervichnykh èlektron-transportnykh reaktsiĭ fotosinteza bakterial'nogo tipa. [Theoretical analysis of the scheme of the primary electron transfer reactions in bacterial-type photosynthesis.] - Stud. biophys. *38* : 139 - 154, 1973. [In R, ab : E.]

17547 - **PYT'EVA, N.F., RIZNICHENKO, G.Yu., RUBIN, A.B.** : Teoreticheskoe i èksperimental'noe issledovanie perekhodnykh protsessov v èlektron-transportnoĭ tsepi bakterial'nogo fotosinteza. [Theoretical and experimental study of the transient processes in the bacterial photosynthetic electron transport chain.] - Stud. biophys. *35* : 173 - 180, 1973. [In R, ab : E.]

17548 - **PYT'EVA, N.F., RUBIN, A.B.** : Matematicheskoe modelirovanie protsessov èlektronnogo transporta pri fotosinteze bakteriĭ. [Mathematical modelling of the electron transport in bacterial photosynthesis.] - Stud. biophys. *35* : 165 - - 172, 1973. [In R, ab : E.]

17549 - **QASIM, S.Z.** : Productivity of backwaters and estuaries. - In : ZEITSCHEL, B. (ed.) : The Biology of the Indian Ocean. Pp. 143 - 154. Springer-Verlag, Berlin - Heidelberg - New York 1973. [Ps, Chl.]

17550 - **QUEBEDEAUX, B., HARDY, R.W.F.** : Oxygen as a new factor controlling reproductive growth. - Nature *243* : 477 - 479, 1973. [CO$_2$ compensation point.]

17551 - **QURESHI, F.A., SPANNER, D.C.** : Movement of [^{14}C] sucrose along the stolon of *Saxifraga sarmentosa*. - Planta *110* : 145 - 152, 1973.

17552 - **RABINOWITCH, E.** : Photosynthesis : an unfolding discovery. - In : The Sun in the Service of Mankind. Pp. V.1-1 - V.1-5. Unesco, Paris 1973.

17553 - **RADI, A.F., AHMED, A.M., KHODARY, S.A.** : Influence of mineral nutrients on photosynthesis and some related metabolic processes. II. - Potassium. - Bull. Fac. Sci. (Assiut) *2* : 13 - 23, 1973.

17554 - **RADMER, R., KOK, B.** : A kinetic analysis of the oxidizing and reducing sides of the O$_2$-evolving system of photosynthesis. - Biochim. biophys. Acta *314* : 28 - 41, 1973.

17555 - **RADOMYSL'SKAYA, T.M.** : Ob izmerenii ul'trafioletovykh spektrov propuskaniya list'ev rasteniĭ. [Measurement of ultraviolet transmission spectra of plant leaves.] - Zh. prikl. Spektroskop. *18* : 269 - 274, 1973. [In R.]

17556 - **RADOMYSL'SKAYA, T.M., PUMPYANSKAYA, S.L.** : O svyazi ul'trafioletovykh spektrov propuskaniya zhivykh list'ev s protsessami fotosinteza i dykhaniya. [Correlation between ultra-violet spectra of living leaves and processes of photosynthesis and respiration.] - Fiziol. Rast. *20* : 109 - 113, 1973. [In R, ab : E.]

17557 - **RADOSEVICH, S.R., APPLEBY, A.P.** : Studies on the mechanism of resistance to simazine in common groundsel. - Weed Sci. *21* : 497 - 500, 1973. [Ps.]

17558 - **RADUNZ, A., SCHMID, G.H.** : Reactions of antisera to lutein and plastoquinone with chloroplast preparations and their effects on photosynthetic electron transport. - Z. Naturforsch. *28c* : 36 - 44, 1973.

17559 - **RAJAN, A.K., BETTERIDGE, B., BLACKMAN, G.E.** : Differences in the interacting effects of light and temperature on growth of four species in the vegetative phase. - Ann. Bot. *37* : 287 - 313, 1973. [Growth analysis.]

*17560 - **RAKHMANKULOVA, M.E.** : Osobennosti organizatsii i funktsional'noĭ aktivnosti
khloroplastov khlopchatnika pri gibridizatsii. [Specificity of organization
and activity of chloroplasts from cotton plants during hybridization.] - In :
Geneticheskie Aspekty Fotosinteza. Tezisy Dokladov. Pp. 66 - 67. Donish, Dushanbe 1972. [In R.]

17561 - **RAKHMANKULOVA, M.E., TUKHTAEVA, G.M., RODIMTSEVA, N.E., KHODZHAEV, A.S.** :
Vliyanie griba *V. dahliae* na strukturu i funktsii khloroplastov khlopchatnika.
[Effect of the fungus *V. dahliae* on the structure and function of cotton chlo-
roplasts.] - Uz. biol. Zh. *1973* (3) : 21 - 23, 1973. [In R, ab : Uz.]

17562 - **RAKHTEENKO, I.N., MARTINOVICH, B.S., KAUROV, I.A., MIN'KO, I.F., KROT, L.A.** :
Vodorastvorimye metabolity i ikh rol' vo vzaimootnosheniyakh rasteniĭ v fitotsenozakh. [Water-soluble metabolites and their role in plant interrelations
in phytocenoses.] - Fiziol. Rast. *20* : 385 - 391, 1973. [Ps; in R, ab : E.]

17563 - **RAKHTSEENKA, I.N., MARTSINOVICH, B.S., KROT, L.A., KABASHNIKAVA, G.I.** : Uza-
emadzeyanne drévavykh parod praz ikh karanёvyya vydzyalenni. [Interaction of
trees *via* their root excretions.] - Vestsi Akad. Navuk belarus.SSR, Ser. biyal.
Navuk *1973* (1) : 10 - 16, 137, 1973. [Ps; in Belorus., ab : R.]

17564 - **RAKITIN, Yu.V.** : Gidrazid maleinovoĭ kisloty, priroda ego deĭstviya i prakti-
cheskoe primenenie. [Malein hydrazide, the nature of its action, and its practical use.] - In : Gidrazid Maleinovoĭ Kisloty kak Regulyator Rosta Rasteniĭ.
Pp. 5 - 39, 361. Nauka, Moskva 1973. [Ps; in R.]

17565 - **RAKITIN, Yu.V., STREL'NIKOVA, B.D.** : Izuchenie deĭstviya gidrazida maleinovoĭ
kisloty kak ingibitora prorastaniya klubneĭ kartofelya. [Action of maleic hydrazide as an inhibitor of germination of potato tubers.] - In : Gidrazid Maleinovoĭ Kisloty kak Regulyator Rosta Rasteniĭ. Pp. 102 - 128, 362. Nauka,
Moskva 1973. [Ps; in R.]

17566 - **RAKITIN, Yu.V., VAKULENKO, V.V.** : Gidrazid maleinovoĭ kisloty kak sredstvo
tormozheniya rosta kustarnikov v zhivykh izgorodyakh. [Maleic hydrazid as a
tool for inhibiting growth of shrubs in hedges.] - In : Gidrazid Maleinovoĭ
Kisloty kak Regulyator Rosta Rasteniĭ. Pp. 219 - 229, 364. Nauka, Moskva 1973.
[Ps; in R.]

17567 - **RAMAKRISHNAN, T.V., FRANCIS, F.J.** : Color and carotenoid changes in heated
paprika. - J. Food Sci. *38* : 25 - 28, 1973.

17568 - **RAMASWAMY, D.K., NAIR, P.M.** : Temperature and light dependent formation of
δ-aminolevulinic acid synthetase in potatoes. - In : Proceedings of the Symposium on Control Mechanisms in Cellular Processes. Pp. 567 - 579. Bhabha at.
Res. Centre, Bombay 1973.

17569 - **RAMATI, A., ESHEL, A., LIPHSCHITZ, N., WAISEL, Y.** : Localization of ions in
cells of *Potamogeton lucens* L. - Experientia *29* : 497 - 501, 1973. [Chloro-
plast.]

17570 - **RAMSHAW, J.A.M., BROWN, R.H., SCAWEN, M.D., BOULTER, D.** : Higher plant plastocyanin. - Biochim. biophys. Acta *303* : 269 - 273, 1973.

17571 - **RANDALL, P.J., BOUMA, D.** : Zinc deficiency, carbonic anhydrase, and photosynthesis in leaves of spinach. - Plant Physiol. *52* : 229 - 232, 1973.

17572 - **RANGELOVA, E.** : Sravnitelni izsledvaniya v"rkhu aktivnostta na reaktsiyata na
Khil v tseli i puknati khloroplasti. [Hill reaction activity in intact
and cracked chloroplasts.] - Izv. Inst. Fiziol. Rast. "M. Popov" b"lg. Akad.
Nauk *18* : 265 - 275, 1973. [In Bulg., ab : E, R.]

17573 - **RANGNEKAR, P.V.** : Translocation of 14C-photosynthate under calcium stress. -
Plant Physiol. *51* (Suppl.) : 63, 1973.

17574 - **RANGNEKAR, P.V., FORWARD, D.F.** : Foliar nutrition and wood growth in red pine :
effects of darkening and defoliation on the distribution of ^{14}C-photosynthate
in young trees. - Can. J. Bot. *51* : 103 - 108, 1973.

17575 - **RAPER, C.D. Jr., DOWNS, R.J.** : Factors affecting the development of flue-cured
tobacco grown in artificial environments : IV. Effects of carbon dioxide depletion and light intensity. - Agron. J. *65* : 247 - 252, 1973. [Growth analysis.]

17576 - RAPER, C.D. Jr., WEEKS, W.W., DOWNS, R.J., JOHNSON, W.H. : Chemical properties of tobacco leaves as affected by carbon dioxide depletion and light intensity. - Agron. J. *65* : 988 - 992, 1973. [Photosynthates.]

17577 - RAPOPORT, V.L., ZHADIN, N.N. : Izmenenie fotokhimicheskoĭ aktivnosti ftalotsianina magniya v reaktsii s kislorodom pri agregatsii. [Change in the photochemical activity of magnesium phtalocyanine in the reaction with oxygen during aggregation.] - Dokl. Akad. Nauk SSSR *212* : 1155 - 1158, 1973. [In R.]

17578 - RAPP, J., HIND, G. : The role of cytochromes in photosynthesis. - Fed. Proc. *32* : 632, 1973.

17579 - RAPS, S. : Reversion of *Scenedesmus* photosynthetic mutants. - Biochim. biophys. Acta *305* : 384 - 389, 1973.

17580 - RASKIN, V.I. : Effektivnost' reaktsii fotovosstanovleniya protokhlorofillida. [Efficiency of the reaction of photoreduction of protochlorophyllide.] - In : Formirovanie Pigmentnogo Apparata Fotosinteza. Pp. 30 - 49. Nauka i Tekhnika, Minsk 1973. [In R.]

17581 - RAVEN, C.W. : Chlorophyll formation and phytochrome. - Med. Landbouwhogesch. Wageningen *73-9* : 1 - 100, 1973.

17582 - RAVEN, C.W., SPRUIT, C.J.P. : Induction of rapid chlorophyll accumulation in dark grown seedlings. III. Transport model for phytochrome action. - Acta bot. neerl. *22* : 135 - 143, 1973.

17583 - RAW, I., HOLLEMAN, G.W. : Water - energy for life. - Chemistry *46* (5) : 6 - - 11, 1973. [H_2O in Ps.]

17584 - REBEIZ, C.A., CASTELFRANCO, P.A. : Protochlorophyll and chlorophyll biosynthesis in cell-free systems from higher plants. - Annu. Rev. Plant Physiol. *24* : 129 - 172, 1973.

17585 - REBEIZ, C.A., CRANE, J.C., NISHIJIMA, C., REBEIZ, C.C. : Biosynthesis and accumulation of microgram quantities of chlorophyll by developing chloroplasts *in vitro*. - Plant Physiol. *51* : 660 - 666, 1973.

17586 - REBEIZ, C.A., LARSON, S., WEIER, T.E., CASTELFRANCO, P.A. : Chloroplast maintenance and partial differentiation *in vitro*. - Plant Physiol. *51* : 651 - 659, 1973.

17587 - RÉDEI, G.P. : Extra-chromosomal mutability determined by a nuclear gene locus in *Arabidopsis*. - Mutation Res. *18* : 149 - 162, 1973. [Chloroplast.]

17588 - RÉDEI, G.P., PLURAD, S.B. : Hereditary structural alterations of plastids induced by a nuclear mutator gene in *Arabidopsis*. - Protoplasma *77* : 361 - 380, 1973.

17589 - REED, D.W., KE, B. : Spectral properties of reaction center preparations from *Rhodopseudomonas spheroides*. - J. biol. Chem. *248* : 3041 - 3045, 1973.

17590 - REED, J.R., SAMSEL, G.L. Jr., BLOOD, F.B. : Preliminary comparison of two oxidation ponds with different trophic states in central Virginia. - Virginia J. Sci. *24* : 75 - 80, 1973. [Chl.]

17591 - REES, A.R., THORNLEY, J.H.M. : A simulation of tulip growth in the field. - Ann. Bot. *37* : 121 - 131, 1973. [Ps.]

17592 - REEVES, A.F., WICKLIFF, J.L., HARRIS, W.M. : Characterization of the citrine phenotype of the tomato. - J. Hered. *64* : 9 - 11, 1973. [Chl.]

17593 - REEVES, S.G., HALL, D.O. : The stoichiometry (ATP/$2e^-$ ratio) of non-cyclic photophosphorylation in isolated spinach chloroplasts. - Biochim. biophys. Acta *314* : 66 - 78, 1973.

17594 - REHFELD, D.W., JENSEN, R.G. : Metabolism of separated leaf cells. III. Effects of calcium and ammonium on product distribution during photosynthesis with cotton cells. - Plant Physiol. *52* : 17 - 22, 1973.

17595 - REIMANN, K. : Die Erfassung toxischer Hemmungen währen der Algenassimilation mittels direkter Sauerstoffmessung im Sapromat. - Z. Wasser- Abwasser-Forsch. *6* : 129 - 132, 1973.

17596 - **REIMER, A., DESMARAIS, R.** : Micrometeorological energy budget methods and apparent diffusivity for boreal forest and grass sites at Pinawa, Manitoba, Canada. - Agr. Meteorol. *11* : 419 - 436, 1973.

17597 - **REINACH, P., AUBREY, B.B., BRODY, S.S.** : Monomolecular films of bacteriochlorophyll and derivatives at an air-water interface : Surface and spectral properties. - Biochim. biophys. Acta *314* : 360 - 371, 1973.

17598 - **REJMÁNKOVÁ, E.** : Biomass, production and growth rate of duckweeds (*Lemna gibba* and *L. minor*). - In : HEJNÝ, S. (ed.) : Ecosystem Study on Wetland Biome in Czechoslovakia. Czechosl. IBP/PT-PP Report No. 3. Pp. 101 - 106. Třeboň 1973.

17599 - **REJMÁNKOVÁ, E.** : Chlorophyll content in leaves of *Phragmites communis* TRIN. - In : HEJNÝ, S. (ed.) : Ecosystem Study on Wetland Biome in Czechoslovakia. Czechosl. IBP/PT-PP Report No. 3. Pp. 143 - 145. Třeboň 1973.

17600 - **REJMÁNKOVÁ, E.** : Seasonal changes in the growth rate of duckweeds (*Lemna gibba* L.) in the littoral of the Nesyt fishpond. - In : KVĚT, J. (ed.) : Littoral of the Nesyt Fishpond. Studie ČSAV 15. Pp. 103 - 106. Academia, Praha 1973.

17601 - **REMY, R.** : Appearance and development of photosynthetic activities in wheat etioplasts greened under continuous or intermittent light - evidence for water-side photosystem II deficiency after greening under intermittent light. - Photochem. Photobiol. *18* : 409 - 416, 1973.

17602 - **REMY, R.** : Pre-existence of chloroplast lamellar protein in wheat etioplasts. Functional and protein changes during greening under continuous or intermittent light. - FEBS Lett. *31* : 308 - 312, 1973.

17603 - **RENGER, G.** : Die Primärprozesse der Photosynthese. - Angew. Chem. *85* : 515 - - 516, 1973.

17604 - **RENGER, G.** : Studies on the reactivity of the positive charges trapped in the photosynthetic watersplitting enzyme system! - In : Ninth International Congress of Biochemistry, Stockholm, 1 - 7 July, 1973. Abstract Book. P. 241. IUB, Stockholm 1973.

17605 - **RENGER, G.** : Studies on the mechanism of destabilization of the positive charges trapped in the photosynthetic water-splitting enzyme system Y by a deactivation-accelerating agent. - Biochim. biophys. Acta *314* : 390 - 402, 1973.

17606 - **RENGER, G.** : The action of 3-(3,4-dichlorophenyl)-1,1-dimethylurea on the water-splitting enzyme system Y of photosynthesis. - Biochim. biophys. Acta *314* : 113 - 116, 1973.

17607 - **RENGER, G., BOUGES-BOCQUET, B., BÜCHEL, K.-H.** : The modification of the trapping properties within the photosynthetic watersplitting enzyme system Y. - J. Bioenerg. *4* : 491 - 505, 1973.

17608 - **RENGER, G., BOUGES-BOCQUET, B., DELOSME, R.** : Studies on the ADRY agent-induced mechanism of the discharge of the holes trapped in the photosynthetic water-splitting enzyme system Y. - Biochim. biophys. Acta *292* : 796 - 807, 1973.

17609 - **RENK, H.** : Produkcja Pierwotna Toni Wodnej Poludniowego Bułtyku. [Primary Production of Water in South Baltic Sea.] - Studia i Materialy, Ser. A. Vol. 12. Pp. 1 - 126. Morski Inst. Rybacki, Ośrodek Wydawniczy, Gdynia 1973. [In Pol.]

17610 - **RENK, H.** : Primary production and chlorophyll content in the Baltic Sea. Part II. Chlorophyll-*a* distribution. - Pol. Arch. Hydrobiol. *20* : 237 - 255, 1973.

17611 - **REPKA, Ĭ., SARICH, M., MAREK, Ĭ.** : Vliyanie nedostatka makroėlementov na strukturu khloroplastov parenkhimnoĭ obkladki provodyashchikh puchkov kukuruzy. [Effect of deficiency of macroelements on structure of chloroplasts in parenchyma coating of conducting bundles of maize.] - Fiziol. Rast. *20* : 766 - 768, 1973. [In R, ab : E.]

17612 - **REVELANTE, N., GILMARTIN, M.** : Some observations on the chlorophyll maximum and primary production in the eastern North Pacific. - Int. Rev. ges. Hydrobiol. *58* : 819 - 834, 1973.

17613 - **REY, L.** : Ultrastructure des chloroplastes au cours de leur évolution pathologique dans le tissu central de la jeune galle de *Pontania proxima*. - Compt. rend. Acad. Sci. Paris, Sér. D *276* : 1157 - 1160, 1973.

17614 - REYNOLDS, C.S. : The phytoplankton of Crose Mere, Shropshire. - Brit.phycol. J. *8* : 153 - 162, 1973. [Chl.]

17615 - REYSSAC, J. : Aspect quantitatif du phytoplancton de la baie du Lévrier (Mauritanie). - Bull. Mus. nat. Hist. natur., 3e Sér. , No. 149, Écol. gén. *5* : 101 - 112, 1973. [Chl.]

17616 - RICHARD, F., NIGON, V. : La synthèse de l'acide δ-aminolévulinique et de la chlorophylle lors de l'éclairement d'*Euglena gracilis* étiolées. - Biochim. biophys. Acta *313* : 130 - 149, 1973.

17617 - RICHARDSON, D.H.S. : Photosynthesis and carbohydrate movement. - In : AHMADJIAN, V., HALE, M.E. (ed.) : The Lichens. Pp. 249 - 288. Academic Press, New York - London 1973.

17618 - RICHARDSON, D.H.S., FINEGAN, E.J. : Primary production of plant communities of the Truelove Lowland, Devon Island, Canada - lichen communities. - In : BLISS, L.C., WIELGOLASKI, F.E. (ed.) : Primary Production and Production Processes, Tundra Biome. Pp. 47 - 55. Tundra Biome Steering Comm., Edmonton 1973.

17619 - RICHARDSON, D.H.S., PUCKETT, K.J. : Sulphur dioxide and photosynthesis in lichens. - In : Air Pollution and Lichens. Pp. 283 - 298. Athlone Press, Univ. London, London 1973.

17620 - RICHARDSON, M. : Microbodies (glyoxysomes and peroxisomes) in plants. - Sci. Progr. *61* : 41 - 61, 1973.

17621 - RICHMOND, B.J., BIDWELL, R.G.S. : CO_2 fixation in a mutant of corn, white-3, which lacks carotenoids. - Can. J. Bot. *51* : 1927 - 1929, 1973.

17622 - RIED, A., ŠETLÍK, I., BOSSERT, U., BERKOVÁ, E. : The effect of low irradiances on oxygen exchange in green and blue-green algae. 1. Analysis based on transients and adenylate equilibria. - Photosynthetica *7* : 161 - 176, 1973.

17623 - RIENITS, K.G., HARDT, H., AVRON, M. : ATP driven reverse electron transport in chloroplasts. - FEBS Lett. *33* : 28 - 32, 1973.

17624 - RIJTEMA, P.E. : The effect of light and water potential on dry matter production of field crops. - In : SLATYER, R.O. (ed.) : Plant Response to Climatic Factors. Pp. 513 - 518. Unesco, Paris 1973.

17625 - RIMAI, L., HEYDE, M.E., GILL, D. : Vibrational spectra of some carotenoids and related linear polyenes. A Raman spectroscopic study. - J. amer. chem. Soc. *95* : 4493 - 4501, 1973.

17626 - RISSER, P.G., JOHNSON, F.L. : Carbon dioxide exchange characteristics of some prairie grass seedlings. - Southwest. Natur. *18* : 85 - 91, 1973.

17627 - RIVERA, A.M., HERAS, L. : Efecto de distintos niveles de salinidad sobre el contenido de clorofila, composición mineral y crecimiento en centeno (*Secale Cereale*) tetraploide. [Effect of different salinity levels on chlorophyll content, mineral composition and growth of tetraploid rye.] - An. Estac. exp. Aula Dei *12* : 100 - 108, 1973. [In Span., ab : E.]

17628 - ROBERTSON, E.E. : Solar energy and biomass. - Agrologist *2* (6) : 8 - 10, 1973.

17629 - ROBSON, M.J. : The growth and development of simulated swards of perennial ryegrass. I. Leaf growth and dry weight change as related to the ceiling yield of a seedling sward. - Ann. Bot. *37* : 487 - 500, 1973.

17630 - ROBSON, M.J. : The growth and development of simulated swards of perennial ryegrass. II. Carbon assimilation and respiration in a seedling sward. - Ann. Bot. *37* : 501 - 518, 1973.

*17631 - RODIN, L.E., BAZILEVICH, N.I., MIROSHNICHENKO, Yu.M. : Productivity and biogeochemistry of *Artemisieta* in the mediterranean area. - In : Eco-physiological Foundation of Ecosystems Productivity in Arid Zone. Pp. 193 - 198. Nauka, Leningrad 1972. [Biomass.]

16632 - RODIONOV, V.S. : Izmenenie kontsentratsii galakto- i fosfolipidov v list'yakh kartofelya v zavisimosti ot osveshchennosti. [Effect of illuminance on concentration of galactolipids and phospholipids in potato leaves.] - Fiziol. Rast. *20* : 753 - 756, 1973. [Chl; in R, ab : E.]

17633 - **ROGERS, L.J., KERSLEY, J., LEES, D.N.** : Physicochemical properties of membrane proteins of photosynthetic organelles. - Physiol. vég. *11* : 327 - 360, 1973.

17634 - **ROKOS, J.A.S.** : Determination of ubiquinone in subnanomole quantities by spectrofluorometry of its product with alkaline ethylcyanoacetate. - Anal. Biochem. *56* : 26 - 33, 1973.

17635 - **ROMANENKO, V.I., DOBRYNIN, E.G.** : Potreblenie kisloroda, temnovaya assimilyatsiya CO_2 i intensivnost' fotosinteza v natural'nykh i profil'trovannykh probakh vody. [Oxygen utilization, dark assimilation of CO_2 and the rate of photosynthesis in natural and filtered water samples.] - Mikrobiologiya *42* : 573 - - 575, 1973. [In R, ab : E.]

17636 - **ROMANOVA, A.K., VEDENINA, I.Ya.** : Ingibirovanie ribulozodifosfatkarboksilazy adenozintrifosfatom u avtotrofnykh organizmov. [Inhibition of ribulosediphosphate carboxylase by adenosine triphosphate in autotrophic organisms.] - Dokl. Akad. Nauk SSSR *211* : 241 - 244, 1973. [In R.]

*17637 - **ROMANOVA, A.V., LYASHCHENKO, I.F.** : Belkovyĭ obmen v list'yakh khlorofil'nykh mutantov podsolnechnika. [Protein metabolism in the leaves of the sunflower chlorophyll mutants.] - Genetika *6* (9) : 55 - 60, 1970. [In R, ab : E.]

17638 - **ROSA, N.** : The "leaf plastochron index" : a non-destructive measure of vegetative plant growth. - Plant Physiol. *51* (Suppl.) : 57, 1973.

17639 - **ROSENBERG, A.** : Chloroplast lipids of photosynthesizing eukaryotic protists. - In : ERWIN, J.A. (ed.) : Lipids and Biomembranes of Eukaryotic Microorganisms. Pp. 233 - 257. Academic Press, New York - London 1973.

17640 - **ROSENBERG, N.J., BROWN, K.W.** : Measured and modelled effects of microclimate modification on evapotranspiration by irrigated crops in a region of strong sensible heat advection. - In : SLATYER, R.O. (ed.) : Plant Response to Climatic Factors. Pp. 539 - 546. Unesco, Paris 1973. [Resistances.]

17641 - **ROUGÉ, P.** : Identification et localisation de la ferrédoxine dans les électrophorégrammes en gel de polyacrylamide des protéines solubles foliaires. - Compt. rend. Acad. Sci. Paris, Sér. D *277* : 1241 - 1244, 1973.

17642 - **ROUHANI, I., VINES, H.M., BLACK, C.C. Jr.** : Isolation of mesophyll cells from *Sedum telephium* leaves. - Plant Physiol. *51* : 97 - 103, 1973. [Ps, Chl.]

17643 - **ROUSAR, D.C.** : Seasonal and spatial changes in primary production and nutrients in Lake Michigan. - Water Air Soil Pollut. *2* : 497 - 514, 1973.

17644 - **ROVIRA, A.D., BOWEN, G.D.** : The influence of root temperature on ^{14}C assimilate profiles in wheat roots. - Planta *114* : 101 - 107, 1973.

17645 - **ROY, R.N., WRIGHT, B.C.** : Sorghum growth and nutrient uptake in relation to soil fertility : I. Dry matter accumulation patterns, yield, and N content of grain. - Agron. J. *65* : 709 - 711, 1973. [Growth analysis.]

17646 - **ROY, S.K.** : A simple and rapid method for estimation of total carotenoid pigments in mango. A direct colorimetric method developed. - J. Food Sci. Technol. (Mysore) *10* : 45, 1973.

17647 - **RUBIN, A.B.** : Élektronnyĭ transport v fotosinteze i sopryazhennye s nim protsessy. [Photosynthetic electron transport and related processes.] - In : Sovremennye Problemy Fotosinteza. Pp. 126 - 137. Izd. mosk. Univ., Moskva 1973. [In R.]

17648 - **RUBIN, A.B.** : O regulyatsii pervichnykh protsessov transporta élektronov v fotosinteze. [Regulation of primary processes of electron transfer in photosynthesis.] - Tr. mosk. Obshch. Ispyt. Prirody *49* (Problemy Biofotokhimii) : 161-166, 1973. [In R, ab : E.]

17649 - **RUBIN, B.A.** : Fotosintez i évolyutsiya bioénergeticheskikh protsessov. [Photosynthesis and evolution of bioenergetic processes.] - In : Sovremennye Problemy Fotosinteza. Pp. 44 - 63. Izd. mosk. Univ., Moskva 1973. [In R.]

17650 - **RUBIN, B.A.** : Sovremennye predstavleniya ob évolyutsii sistem énergoobmena v

zhivoĭ prirode. [Current concepts of the evolution of energy exchange systems
in living nature.] - Sel'skokhoz. Biol. *8* : 483 - 502, 1973. [Ps; in R, ab :
E.]

17651 - **RUBIN, L.B., KHOKHLOV, R.V., PASHCHENKO, V.Z.** : Primenenie lazerov v biofoto-
fizicheskikh issledovaniyakh. [The application of lasers in biophotophysical
investigations.] - Tr. mosk. Obshch. Ispyt. Prirody *49* (Problemy Biofotokhimii) : 258 - 264, 1973. [Ps; in R, ab : E.]

17652 - **RUBIN, L.B., SHVINKA, Yu.È.** : Regulyatornoe deĭstvie blizhnego UF-sveta na
razvitie *Ectothiorhodospira shaposhnikovii*. Karotinoidy kak vozmozhnyĭ pig-
ment-sensibilizator. [The regulatory role of UV-light on the development of
Ectothiorhodospira shaposhnikovii cell. Carotenoids as possible pigment-sen-
sitizers.] - Mol. Biol. (Moskva) *7* : 817 - 824, 1973. [In R, ab : E.]

17653 - **RUCKENBAUER, P.** : Yielding ability and translocation pattern of $^{14}CO_2$-labelled
assimilates in contrasting wheat varieties. - Proc. Res. Inst. Pomol., Skierniewice, Ser. E *1973* (3) : 129 - 137, 1973.

17654 - **RUDOĬ, A.B.** : Matematicheskoe modelirovanie prevrashcheniya pigmentov na na-
chal'nom ètape zeleneniya ètiolirovannykh rasteniĭ. [Mathematical modelling
of pigment transformation at the beginning of greening of etiolated plants.]
- In : Upravlenie Skorost'yu i Napravlennost'yu Biosinteza u Rasteniĭ. Pp.
35 - 36. Krasnoyarsk 1973. [In R.]

17655 - **RUDOĬ, A.B.** : Sovremennye predstavleniya o vydelenii i svoĭstvakh khlorofillazy. [Recent ideas on the separation and properties of chlorophyllase.] - In :
Formirovanie Pigmentnogo Apparata Fotosinteza. Pp. 143 - 170. Nauka i Tekhnika, Minsk 1973. [In R.]

16656 - **RUETZ, W.F.** : The seasonal pattern of CO_2 exchange of *Festuca rubra* L. in a
montane meadow community in northern Germany. - Oecologia *13* : 247 - 269,
1973.

17657 - **RUNGE, M.** : IV. Standortslehre (Ökologische Geobotanik). - Fortschr. Bot. *35*:
347 - 357, 1973. [Ps.]

17658 - **RUNGE, M.** : A. Der biologische Energieumsatz in Land-Ökosystemen unter Einfluß des Menschen. - In : **ELLENBERG, H.** (ed.) : Ökosystemforschung. Pp. 123 -
- 141. Springer-Verlag, Berlin - Heidelberg - New York 1973.

17659 - **RUNGE, M.** : Energieumsätze in den Biozönosen terrestrischer Ökosysteme. (Untersuchungen im "Sollingprojekt".) - Scripta geobot. *4* : 1 - 77, 1973.

17660 - **RYAN, F.J., OMATA, S., HAN SAN KU, TOLBERT, N.E.** : Ratio of RUDP carboxylase/
/RUDP oxygenase activity. - Plant Physiol. *51* (Suppl.) : 40, 1973.

17661 - **RYBIN, I.A., MIKHEEVA, S.A.** : Deĭstvie strofantina na bioèlektricheskuyu reaktsiyu lista rasteniya pri vklyuchenii i vyklyuchenii sveta. [Effect of strop-
hanthin on bioelectrical response of plant leaf in the light and in the dark.]
- Fiziol. Rast. *20* : 86 - 89, 1973. [In R, ab : E.]

17662 - **RYBIN, I.A., MIKHEEVA, S.A.** : Ingibitory obmena kak vozmozhnyĭ instrument issledovaniya bioèlektricheskoĭ reaktsii list'ev vysshikh rasteniĭ pri vklyuchenii i vyklyuchenii sveta. [Metabolic inhibitors as a possible mean for studying bioelectric reaction of leaves of higher plants after switching on and
switching off light.] - In : Voprosy Regulyatsii Fotosinteza. Vol. 3. Pp. 58-
- 76, 161. Ural'. gos. Univ., Sverdlovsk 1973. [In R.]

17663 - **RYCZKOWSKI, M., SZEWCZYK, E.** : Photosynthesis in the developing embryo of *Hae-
manthus katharinae* BAK. (monocotyledonous plant). - Bull. Acad. pol. Sci.,
Sér. Sci. biol. *21* : 695 - 699, 1973.

17664 - **RYLE, G.J.A., BROCKINGTON, N.R., POWELL, C.E., CROSS, B.** : The measurement and
prediction of organ growth in a uniculm barley. - Ann. Bot. *37* : 233 - 246,
1973. [Ps.]

17665 - **RYZHOVA, E.F.** : Deĭstvie i posledeĭstvie sveta na izmenenie soderzhaniya ksan-
tofillov v list'yakh *Hydrangea hortensis* SMITH. [Effect and aftereffect of
light on the change of xanthophyll content in leaves of *Hydrangea hortensis*
SMITH.] - Bot. Zh. *58* : 540 - 545, 1973. [In R.]

17666 - **RZAEV, I.T., AMIRASLANOV, I.A., MAMEDOV, M.A., ABBASOV, G.S.** : Vliyanie udobreniĭ na nekotorye fiziologicheskie protsessy v khlopchatnike. [Effect of fertilizers on some physiological processes in cotton.] - Khim. sel'. Khoz. *11* (2) : 94 - 95, 1973. [Ch]; in R.]

17667 - **SAAKOV, V.S.** : Der Einfluss einiger Inhibitoren auf den Chlorophyllgehalt in grünen Zellen. - Biochem. Physiol.Pflanzen *164* : 199 - 212, 1973.

17668 - **SAAKOV, V.S.** : Die durch Hemmstoffe induzierten Umwandlungen der Karotinoidpigmente in Pflanzenzellen. - Biochem. Physiol. Pflanzen *164* : 213 - 227, 1973.

17669 - **SAAKOV, V.S., SHPOTAKOVSKIĬ, V.S.** : Metod proizvodnoĭ spektrofotometrii v izuchenii struktury apparata fotosinteza. [Method of derivative spectrophotometry in research of photosynthetic apparatus structure.] - In : Metody Kompleksnogo Izucheniya Fotosinteza. Vol. 2. Pp. 280 - 295. VASKHNIL, Leningrad 1973. [In R.]

17670 - **SABAD, Ĭ., SALAI, L.** : Perenos énergii élektronnogo vozbuzhdeniya v smeshannykh rastvorakh karotinoidov i khlorofillov. [Transfer of electronic excitation energy in mixed solutions of carotinoids and chlorophylls.] - Zh. prikl. Spektroskop. *19* : 423 - 427, 1973.

17671 - **SADLER, D.M., LEFORT-TRAN, M., POUPHILE, M.** : Structure of photosynthetic membranes of *Euglena* using X-ray diffraction. - Biochim. biophys. Acta *298* : 620 - 629, 1973.

17672 - **SAGROMSKY, H.** : Einfluß der Lichtintensität auf die Pigmentzusammensetzung in den Plastiden von *Antirrhinum majus*, Sippe 50, und zwei Mutanten davon. - Kulturpfkanze *21* : 111 - 118, 1973.

*17673 - **SAIFULLAH, S.M.** : The relation between standing crop and productivity of phytoplankton. - J. Sci. Phys. Sec. *1* (2) : 89 - 93, 1972.

17674 - **SAIJO, Y.** : The formation of the chlorophyll maximum in the Indian Ocean. - In : ZEITSCHEL, B. (ed.) : The Biology of the Indian Ocean. Pp. 171 - 173. Springer-Verlag, Berlin - Heidelberg - New York 1973.

17675 - **SAITO, K., NAKAYAMA, R., TAKEDA, K., KUWATA, H.** : [Studies on the breeding of the grass. IV. Differentiation of plants by anther culture in orchardgrass and smooth bromegrass.] - Bull. Fac. Agr. Hirosaki Univ. *21* : 1 - 8, 1973. [Stomata; in Jap., ab : E.]

17676 - **SAKALO, V.D., OKANENKO, A.S., VYVAL'KO, I.G.** : Osobennosti radial'nogo peredvizheniya assimilyatov v sakharnoĭ svekle. [Specific features of radial movement of assimilates in sugar beet plants.] - Tr. biol.-pochv. Inst., dal'nevost. nauch. Tsentr Akad. Nauk SSSR *20* (123) : 131 - 136, 1973. [In R, ab : E.]

17677 - **SAKANO, K., WILDMAN, S.G.** : Isolation of crystalline fraction I protein from etiolated tobacco leaves. - Plant Physiol. *51* (Suppl.) : 27, 1973.

*17678 - **SAKHAROVA, O.V., KORZH, B.V.** : Pigmenty inbrednykh i gibridnykh form kukuruzy i ikh svyaz' s protsessom rosta i urozhaĭnost'yu. [Pigments of inbred and hybrid forms of maize as related to growth and yields.] - In : Geneticheskie Aspekty Fotosinteza. Tezisy Dokladov. Pp. 113 - 114. Donish, Dushanbe 1972. [In R.]

17679 - **SAKSHAUG, E., MYKLESTAD, S.** : Studies on the phytoplankton ecology of the Trondheimsfjord. III. Dynamics of phytoplankton blooms in relation to environmental factors, bioassay experiments and parameters for the physiological state of the population. - J. exp. mar. Biol. Ecol. *11* : 157 - 188, 1973. [Ch].]

17680 - **SALAMON, Z.** : Energy transfer between carotene and chlorophyll. - Bull. Acad. pol. Sci., Sér. Sci. math., astron., phys. *21* : 1055 - 1060, 1973.

17681 - **SALAMON, Z., FRĄCKOWIAK, D.** : The protective action of carotene on chlorophyll fluorescence. - Bull. Acad. pol. Sci., Sér. Sci. math., astron., phys. *21* : 781 - 792, 1973.

17682 - **SALE, P.J.M.** : Productivity of vegetable crops in a region of high solar input. I. Growth and development of the potato (*Solanum tuberosum* L.). - Aust.
J. agr. Res. *24* : 733 - 749, 1973. [Growth analysis.]

17683 - **SALE, P.J.M.** : Productivity of vegetable crops in a region of high solar input. II. Yields and efficiencies of water use and energy. - Aust. J. agr. Res.
24 : 751 - 762, 1973.

17684 - **SALEMA, R., BRANDÃO, I.** : The use of PIPES buffer in the fixation of plant
cells for electron microscopy. - J. Submicroscop. Cytol. *5* : 79 - 96, 1973.
[Chl.]

17685 - **SALEMME, F.R., FREER, S.T., ALDEN, R.A., KRAUT, J.** : Atomic coordinates for
ferricytochrome c_2 of *Rhodospirillum rubrum*. - Biochem. biophys. Res. Commun.
54 : 47 - 52, 1973.

17686 - **SALIN, M.L.** : Changes of photorespiratory activity and glycolic acid metabolism with leaf age. - Diss. Abstr. Int. B *33* : 5140-B, 1973.

17687 - **SALIN, M.L., CAMPBELL, W.H., BLACK, C.C. Jr.** : Oxaloacetate as the Hill oxi-
dant in mesophyll cells of plants possessing the C_4-dicarboxylic acid cycle
of leaf photosynthesis. - Proc. nat. Acad. Sci. *70* : 3730 - 3734, 1973.

17688 - **SALIN, M.L., HOMANN, P.H.** : Glycolate metabolism in young and old tobacco
leaves, and effects of α-hydroxy-2-pyridinemethanesulfonic acid. - Can. J. Bot.
51 : 1857 - 1865, 1973.

17689 - **SALISBURY, F.B., KIMBALL, S.L., BENNETT, B., ROSEN, P., WEIDNER, M.** : Active
plant growth at freezing temperatures. - Space Life Sci. *4* : 124 - 138, 1973.
[Chl.]

17690 - **SALVADOR, G.F.** : Evolution du système plastidien d'*Euglena gracilis* etiolée
a l'obscurité après stimulation lumineuse. - Exp. Cell Res. *79* : 479 - 484,
1973.

17691 - **SAMISH, Y.B., PALLAS, J.E. Jr.** : A semiclosed compensating system for the control of CO_2 and water vapor concentrations and the calculation of their exchange rates. - Photosynthetica *7* : 345 - 350, 1973.

17692 - **SAMSEL, G.L. Jr.** : Effects of sedimentation on the algal flora of a small re-
creational impoundment. - Water Resources Bull. *9* : 1145 - 1152, 1973. [Chl.]

17693 - **SAMSONOVA, I.A., BĚTTKHER, F.** : Issledovanie mutabil'nosti plastoma. Soobshche-
nie III. Chastota spontannykh obratnykh mutatsiǐ u plastomnogo mutanta tomata
Pl-alb 1. [Mutability of the plastom. III. The frequency of spontaneous reversible mutations in the plastom mutant Pl-alb 1 of tomato.] - Genetika *9*
(5) : 43 - 51, 1973. [In R, ab : E.]

17694 - **SAMSONOVA, I.A., BÖTTCHER, F.** : Rückmutationen zu ergrünungsfähigen Plastiden
bei einer ǀPlastommutante von *Lycopersicon esculentum*. - Wiss. Z. Ernst-Mo-
ritz-Univ. Greifswald *22* (math.-naturwiss. Reihe 1/2) : 69 - 80, 1973.

17695 - **SAMUILOV, V.D., GRINYUS, L.L., SKULACHEV, V.P.** : Prevrashchenie énergii v
khromatoforakh fotosinteziruyushchikh bakteriǐ. [Energy conversion in chromatophores of photosynthetic bacteria.] - Tr. mosk. Obshch. Ispyt. Prirody, Otd.
biol. *49* : 198 - 203, 1973. [In R, ab : E.]

17696 - **SÁNCHEZ, R.A., BONNER, B.A.** : Spectral forms of protochlorophyll(ide) and
their phototransformations in cucumber seedlings at different stages of development. - Plant Physiol. *51* (Suppl.) : 28, 1973.

17697 - **SANGER, J.E., GORHAM, E.** : A comparison of the abundance and diversity of fos-
sil pigments in wetland peats and woodland humus layers. - Ecology *54* : 605 -
- 611, 1973.

17698 - **SANKHLA, N., ZIEGLER, H.** : Plastid development in response to a new herbicide
EMD-IT-5914. - Naturwissenschaften *60* : 157, 1973.

17699 - **SANTARIUS, K.A.** : Freezing. The effect of eutectic crystallization on biological membranes. - Biochim. biophys. Acta *291* : 38 - 50, 1973. [Ps.]

17700 - **SANTARIUS, K.A.** : The protective effect of sugars on chloroplast membranes
during temperature and water stress and its relationship to frost, desiccation
and heat resistance. - Planta *113* : 105 - 114, 1973.

*17701 - **SANTHANAM, R.** : Aspects of an experimental study concerning diatoms. - J. Annamalai Univ. Sci. *30* : 159 - 170, 1972. [Ps.]

17702 - **SAPOZHNIKOV, D.I.** : Investigation of the violaxanthin cycle. - Pure appl. Chem. *35* : 47 - 61, 1973.

17703 - **SAPOZHNIKOV, D.I., GABR, M.A., MASLOVA, T.G.** : O polozhenii svetovogo poroga reaktsii dezepoksidatsii violaksantina v list'yakh svetolyubivykh i tenevynoslivykh rastenii. [Position of light threshold in the reaction of violaxanthin de-epoxidation in leaves of photophilous and shade-tolerant plants.] - Bot. Zh. *58* : 1205 - 1209, 1973. [In R.]

17704 - **SARAIVA, M.C.** : Utilisation de la mesure polarographique de l'oxygène dissous dans la réalisation d'un test rapide de toxicité : action de quelques métaux sur *Dunaliella* et *Lebistes*. - Ann. Inst. océanogr. *49* : 145 - 150, 1973. [Ps.]

17705 - **SARIĆ, M.R., PETROVIĆ, M.J.** : The effect of quality and intensity of light on the weight of dry matter, content of pigment in chloroplasts and N, P, K and Ca in maize plants. - In : The Sun in the Service of Mankind. Pp. V.3-1 - V. 3-11. Unesco, Paris 1973.

177C - **SATOH, K., YAKUSHIJI, E., KATOH, S.** : Studies on cytochromes in photosynthetic electron transport system II. Participation of photosystem I in the light-induced oxidation-reduction of cytochrome b_{559} in spinach chloroplasts. - Plant Cell Physiol. *14* : 763 - 767, 1973.

17707 - **SAVCHENKO, G.E., AVERINA, N.G., KOSTYUK, N.N., CHAĬKA, M.T., PRUDNIKOVA, I.V., SHLYK, A.A.** : O regulirovanii biosinteza protokhlorofillida v rastenii. [Control of protochlorophyllide biosynthesis in a plant.] - In : Upravlenie Skorost'yu i Napravlennost'yu Biosinteza u Rastenii. Pp. 39 - 40. Krasnoyarsk 1973. [In R.]

17708 - **SAVEL'EVA, R.P.** : Maket perfokarty dlya obrabotki informatsii po spektral'nym koéffitsientam yarkosti rastitel'nogo pokrova. [The model of punched card for the treatment of information on spectral coefficients of brightness of the vegetation cover.] - Rast. Resursy *9* : 280 - 287, 1973. [In R.]

17709 - **SAWAMURA, M., HASHINAGA, F., OSAJIMA, Y.** : [^{14}C-studies on seasonal transitions of organic acids and sugars in Satsuma mandarin (Studies on the quality of citrus fruits. Part I).] - J. agr. chem. Soc. Jap. *47* : 571 - 576, 1973. [In Jap., ab : E.]

17710 - **SAWAMURA, M., OSAJIMA, Y.** : [Studies on translocation of ^{14}C-labeled compounds from leaves to fruit in Satsuma mandarin (Studies on the quality of citrus fruits. Part II).] - J. agr. chem. Soc. Jap. *47* : 733 - 735, 1973. [In Jap., ab : E.]

17711 - **SAWHNEY, R., CUMMING, B.G.** : The role of photosynthesis and HER in the flowering of a SD plant, *Chenopodium rubrum*. - Plant Physiol. *51* (Suppl.) : 29, 1973.

17712 - **SAXENA, J., SIKKA, H.C., SCHWELITZ, F., ZWEIG, G.** : Permeability changes in dichlone-treated *Chlorella pyrenoidosa* as influenced by pH and temperature. - Abstr. annu. Meet. amer. Soc. Microbiol. *73* : 170, 1973. [Ps.]

17713 - **SAXENA, J., SIKKA, H.C., ZWEIG, G.** : Effect of certain substituted naphthoquinones on growth and respiration of *Rhodospirillum rubrum*. - Pestic. Biochem. Physiol. *3* : 66 - 72, 1973.

17714 - **SCAIFE, M.A.** : The early relative growth rates of six lettuce cultivars as affected by temperature. - Ann. appl. Biol. *74* : 119 - 128, 1973.

17715 - **SCARSBROOK, C.E., DOSS, B.D.** : Leaf area index and radiation as related to corn yield. - Agron. J. *65* : 459 - 461, 1973.

*17716 - **SCHEGG, E.** : Produktion und Destruktion in der trophogenen Schicht. Untersuchungen ökologischer Parameter im polytrophen Rotsee und in der mesotrophen Horwer Bucht (Vierwaldstättersee). - Schweiz. Z. Hydrol. *33* : 425 - 532, 1971. [Primary production.]

*17717 - **SCHENK, H.E.A.** : Nachweis einer lysozymempfindlichen Stützmembran der Endocyanellen von *Cyanophora Paradoxa* KORSCHIKOFF. - Z. Naturforsch. *25b* : 656, 1970. [Biliproteins.]

17718 - SCHIFF, J.A. : The development, inheritance, and origin of the plastid in *Eug-lena*. - In : ABERCROMBIE, M., BRACHET, J., KING, T.J. (ed.) : Advances in Mor-phogenesis. Vol. 10. Pp. 265 - 312. Academic Press, New York - London 1973.

17719 - SCHILLING, N., FERGUSON, J.A., KANDLER, O. : Isotopenstudium zur Bildung und zum Umsatz von Tanninen in Blättern von *Rhus typhina*. - Ber. deut. bot. Ges. *86* : 393 - 401, 1973. [Ps.]

17720 - SCHINDLER, D.W., FEE, E.J. : Diurnal variation of dissolved inorganic carbon and its use in estimating primary production and CO_2 invasion in lake 227. - J. Fish. Res. Board Can. *30* : 1501 - 1510, 1973.

17721 - SCHINDLER, D.W., FROST, V.E., SCHMIDT, R.V. : Production of epilithiophyton in two lakes of the Experimental Lakes Area, northwestern Ontario. - J. Fish. Res. Board Can. *30* : 1511 - 1524, 1973. [Ps.]

17722 - SCHINDLER, D.W., KLING, H., SCHMIDT, R.V., PROKOPOWICH, J., FROST, V.E., REID, R.A., CAPEL, M. : Eutrophication of lake 227 by addition of phosphate and nit-rate : the second, third, and fourth years of enrichment 1970, 1971, and 1972. - J. Fish. Res. Board Can. *30* : 1415 - 1440, 1973. [Chl.]

17723 - SCHMIDT, A. : Sulfate reduction in a cell-free system of *Chlorella*. The fer-redoxin dependent reduction of a protein-bound intermediate by a thiosulfona-te reductase. - Arch. Mikrobiol. *93* : 29 - 52, 1973.

17724 - SCHMIDT, G.W. : Primary production of phytoplankton in the three types of Ama-zonian waters. I. Introduction. - Amazoniana (Kiel) *4* : 135 - 138, 1973.

17725 - SCHMIDT, G.W. : Primary production of phytoplankton in the three types of Ama-zonian waters. II. The limnology of a tropical flood-plain lake in central Amazonia (Lago do Castanho). - Amazoniana (Kiel) *4* : 139 - 203, 1973.

17726 - SCHMIDT, G.W. : Primary production of phytoplankton in the three types of Ama-zonian waters. III. Primary productivity of phytoplankton in a tropical flood-plain lake of central Amazonia, Lago do Castanho, Amazonas, Brazil. - Amazo-niana (Kiel) *4* : 379 - 404, 1973. [Chl.]

17727 - SCHMIDT, K., BAIER, D., LATZKO, E. : Kinetische Eigenschaften der photosyn-thetischen Fructose-Diphosphatase. - Hoppe-Seyler's Z. physiol. Chem. *354* : 1240, 1973.

17728 - SCHMIDT, K., LIAAEN-JENSEN, S. : Bacterial carotenoids. XLII. New keto-carote-noids from *Rhodopseudomonas globiformis (Rhodospirillaceae)*. - Acta chem. scand. *27* : 3040 - 3052, 1973.

17729 - SCHNARRENBERGER, C., OESER, A., TOLBERT, N.E. : Two isoenzymes each of gluco-se-6-phosphate dehydrogenase and 6-phosphogluconate dehydrogenase in spinach leaves. - Arch. Biochem. Biophys. *154* : 438 - 448, 1973.

17730 - SCHNEIDER, H.A.W. : δ-Aminolävulinatsynthetase, δ-Aminolävulinatanreicherung und Chlorophyllsynthese in Zellkulturen von Tabak. - Z. Pflanzenphysiol. *69* : 68 - 76, 1973.

17731 - SCHNEIDER, H.A.W. : Light mediated induction of δ-aminolevulinate synthetase and related enzymes and a tentative working model for control of chlorophyll biosynthesis. - Enzyme *16* : 108 - 114, 1973.

17732 - SCHNEIDER, H.A.W. : Limitierende Proteine in der Chlorophyllbiosynthese. Ex-perimente und Modellvorstellungen. - Ber. deut. bot. Ges. *86* : 431 - 436, 1973.

17733 - SCHNEIDER, H.A.W. : pH-dependent elution volume of some pyrroles chromatograp-hed on Sephadex G-15. - J. Chromatogr. *81* : 160 - 162, 1973.

17734 - SCHNEIDER, H.A.W. : Regulation der Chlorophyllbiosynthese. Licht- und ent-wicklungsbedingte Aktivitätsänderungen von vier aufeinander folgenden Enzymen der Porphyrin- und Chlorophyllbiosynthesekette. - Z. Naturforsch. *28c* : 45 - - 58, 1973.

17735 - SCHNEIDER, V. : *In vitro* inhibition of photophosphorylation by kaempferol. - Z. Pflanzenphysiol. *70* : 88 - 93, 1973.

17736 - SCHOCH, P.-G., CANDELARIO, L.S. : Croissance des feuilles de *Vigna sinensis*.

Bilan individuel de la productivité foliaire lors des phases diurnes et noc-
turnes. - Oecol. Plant. *8* : 301 - 308, 1973.

17737 - **SCHÖN, G., BIEDERMANN, M.** : Growth and adaptive hydrogen production of *Rhodospirillum rubrum* (F_1) in anaerobic dark cultures. - Biochim. biophys. Acta
304 : 65 - 75, 1973. [Chl.]

17738 - **SCHÖNBOHM, E.** : Kontraktile Fibrillen als aktive Elemente bei der Mechanik
der Chloroplastenverlagerung. - Ber. deut. bot. Ges. *86* : 407 - 422, 1973.

17739 - **SCHÖNBOHM, E.** : Die lichtinduzierte Verankerung der Plastiden im cytoplasma-
tischen Wandbelag : Eine phytochromgesteuerte Kurzzeitreaktion. - Ber. deut.
bot. Ges. *86* : 423 - 430, 1973.

17740 - **SCHÖNFELD, M., RAHAT, M., NEUMANN, J.** : Photosynthetic reactions in the marine
alga *Codium vermilara* I. CO_2 fixation and Hill reaction in isolated chloroplasts. - Plant Physiol. *52* : 283 - 287, 1973.

17741 - **SCHOR, S.L.** : Regular variations in chloroplast functional activity during the
division cycle of *Chlamydomonas reinhardi*. - Diss. Abstr. int. B *33* : 3487-B,
1973.

17742 - **SCHRAMM, W., OHNO, M.** : [Some observations on the influence of salinity on
biological activity of *Fucus vesiculosus*.] - Bull. jap. Soc. Phycol. *21* : 81-
- 85, 1973. [Ps; in Jap., ab : E.]

17743 - **SCHREIBER, U.** : Reversible inhibition of energy transfer between photosynthetic pigments by hydrostatic pressure. - Carnegie Inst. Year Book *72* : 327 -
- 330, 1973.

17744 - **SCHREIBER, U., VIDAVER, W.** : Hydrostatic pressure : a reversible inhibitor of
primary photosynthetic processes. - Z. Naturforsch. *28c* : 704 - 709, 1973.

17745 - **SCHREIBER, U., VIDAVER, W.** : Photosynthetic energy transfer reversibly inhibited by hydrostatic pressure. - Photochem. Photobiol. *18* : 205 - 208, 1973.

17746 - **SCHUEPP, P.H.** : Model experiments on free-convection heat and mass transfer
of leaves and plant elements. - Boundary-Layer Meteorol. *3* : 454 - 467, 1973.

17747 - **SCHULDINER, S., ROTTENBERG, H., AVRON, M.** : Stimulation of ATP synthesis by a
membrane potential in chloroplasts. - Europe. J. Biochem. *39* : 455 - 462, 1973.

17748 - **SCHULTZ, G.** : Reversible changes in the ratios of chlorophylls and carotenoids
in *Vaccinium myrtillus*. - Z. Pflanzenphysiol. *69* : 91 - 93, 1973.

17749 - **SCHULZE, E.-D., LANGE, O.L., KAPPEN, L., BUSCHBOM, U., EVENARI, M.** : Stomatal
responses to changes in temperature at increasing water stress. - Planta *110*:
29 - 42, 1973.

17750 - **SCHWARTZ, D.M., BAZZAZ, F.A.** : *In situ* measurements of carbon dioxide gradients in a soil-plant-atmosphere system. - Oecologia *12* : 161 - 167, 1973.

17751 - **SCHWELITZ, F.D., SIKKA, H.C., SAXENA, J., ZWEIG, G.** : Ultrastructural changes
in isolated spinach chloroplast and in *Chlorella pyrenoidosa* CHICK (Emerson
strain) treated with dichlone. - Plant Physiol. *51* (Suppl.) : 27, 1973.

17752 - **SCHWENKER, U., GINGRAS, G.** : A carotenoprotein from chromatophores of *Rhodospirillum rubrum*. - Biochem. biophys. Res. Commun. *51* : 94 - 99, 1973.

17753 - **SCHWENN, J.D., LILLEY, R.McC., WALKER, D.A.** : Inorganic pyrophosphatase and
photosynthesis by isolated chloroplasts. I. Characterisation of chloroplast
pyrophosphatase and its relation to the response of exogenous pyrophosphate.
- Biochim. biophys. Acta *325* : 586 - 595, 1973.

17754 - **SCHWIEBERT, H., LENZ, F.** : Einfluß von Simazin auf die Photosynthese fruchten-
der und nichtfruchtender Erdbeerpflanzen. - Erwerbsobstbau *15* : 84 - 86, 1973.

17755 - **SCOTT, R.K., OGUNREMI, E.A., IVINS, J.D., MENDHAM, N.J.** : The effect of sowing
date and season on growth and yield of oilseed rape (*Brassica napus*). - J. agr.
Sci. *81* : 277 - 285, 1973.

17756 - **SCOTT, R.K., OGUNREMI, E.A., IVINS, J.D., MENDHAM, N.J.** : The effect of fertilizers and harvest date on growth and yield of oilseed rape sown in autumn
and spring. - J. agr. Sci. *81* : 287 - 293, 1973.

17757 - **SEELY, G.R.** : Effects of spectral variety and molecular orientation on energy trapping in the photosynthetic unit : a model calculation. - J. theor. Biol. *40* : 173 - 187, 1973.

17758 - **SEELY, G.R.** : Energy transfer in a model of the photosynthetic unit of green plants. - J. theor. Biol. *40* : 189 - 199, 1973.

17759 - **SEIBERT, M., ALFANO, R.R., SHAPIRO, S.L.** : Picosecond fluorescent kinetics of *in vivo* chlorophyll. - Biochim. biophys. Acta *292* : 493 - 495, 1973.

17760 - **SELIM, A.K.A., MARWAN, M., EL-SAYED, S.I.** : Radiation induced mutations in tomato. - Ann. agr. Sci., Fac. Agr., Ain Shams Univ. Cairo *18* : 91 - 99, 1973. [Chl.]

17761 - **SELMAN, B.R., BANNISTER, T.T., DILLEY, R.A.** : Trypsin inhibition of electron transport. - Biochim. biophys. Acta *292* : 566 - 581, 1973. [Ps.]

17762 - **SELMAN, I.W., YAHAMPATH, A.C.I.** : Some physiological characteristics of two tomato cultivars, one tolerant and one susceptible to tobacco mosaic virus. - Ann. Bot. *37* : 853 - 865, 1973. [Growth analysis.]

17763 - **SEMKIN, B.I., BURTSEVA, R.A.** : Raspredelenie assimilyatov v slozhnom liste. [Photosynthate distribution in a compound leaf.] - Tr. biol.-pochv. Inst., dal'nevost. nauch. Tsentr Akad. Nauk SSSR *16* (119) - Fiziologiya i Ekologiya Drevesnykh Rasteniĭ Primor'ya : 34 - 37, 1973. [In R, ab : E.]

17764 - **SEMYCHAEVS'KYĬ, V.D., LOZOVA, G.I.** : Osoblyvosti kinetyky fotodestruktsiĭ feofitynu *a* u̐ systemakh z riznym stanom agregatsiĭ. [Peculiarities in kinetics of pheophytin *a* photodestruction in systems with different state of aggregation.] - Ukr. bot. Zh. *30* : 647 - 650, 676, 1973. [In Ukr., ab : E, R.]

17765 - **SENCHENKOVA, E.M.** : Razvitie predstavleniĭ ob avtonomii khloroplastov. [Development of ideas on chloroplast autonomy.] - In : Iz Istorii Biologii. Vol. 4. Pp. 41 - 62. Nauka, Moskva 1973. [In R, ab : E.]

17766 - **SENCHENKOVA, E.M., MEL'NIKOVA, L.V.** : M.S. Tsvet i sozdanie metoda khromatografii. [M.S. Tsvet and creation of the chromatographic method.] - Byull. mosk. Obshch. Ispyt. Prirody, Otd. Biol. *78* : 139 - 144, 1973. [Also Chl; in R.]

17767 - **SEN GUPTA, R., JANNASCH, H.W.** : Photosynthetic production and dark-assimilation of CO_2 in the Black Sea. - Int. Rev. ges. Hydrobiol. *58* : 625 - 632, 1973.

17768 - **SEREBRENIKOV, V.S., KIRYUKHIN, V.P., ANISIMOV, B.V.** : Izotopnyĭ metod v issledovaniyakh po kartofelyu. [The isotopic method in potato studies.] - Nauch. Tr. nauch.-issled. Inst. kartof. Khoz. *17* (Voprosy Fiziologo-Biokhimicheskikh Issledovaniĭ po Kul'ture Kartofelya) : 152 - 166, 1973. [Ps; in R.]

17769 - **SEROVA, Z.Ya., REUTSKAYA, L.N.** : Izmenenie dykhaniya i fotosinteza kartofelya pod vliyaniem porazheniya X-virusom. [Change in the respiration and photosynthesis of the potato affected by potato X virus.] - In : VECHER, A.S. (ed.) : Fotosintez i Ustoĭchivost' Rastenii. Pp. 129 - 146. Nauka i Tekhnika, Minsk 1973. [In R.]

17770 - **ŠESTÁK, Z., ČATSKÝ, J.** : Bibliography of reviews and methods of photosynthesis - 23, 24, 25, 26. - Photosynthetica *7* : 93 - 103, 195 - 203, 282 - 295, 414 - 433, 1973.

17771 - **ŠETLÍK, I., RIED, A., BERKOVÁ, E.** : The analysis of the Kok-effect in green and blue green algae. - Annu. Rep. algol. Lab. Třeboň *1970* : 119 - 142, 1973.

17772 - **ŠETLÍK, I., RIED, A., BERKOVÁ, E., BOSSERT, U.** : The effect of low irradiances on oxygen exchange in green and blue-green algae. 2. Variation in efficiency of steady-state oxygen exchange rates during the cell cycles of *Scenedesmus quadricauda*. - Photosynthetica *7* : 177 - 194, 1973.

17773 - **SHAGA, V.S., SHAGA, N.I., DANILOV, V.P.** : Dinamika produktivnosti i khimicheskiĭ sostav fitomassy poĭmennykh lugov Nizhnego Amura. [Dynamics of productivity and chemical composition of plant matter in the flood-plain meadows of the lower course of Amur.] - Rast. Resursy *9* : 440 - 443, 1973. [In R.]

17774 - **SHAH, N.M.** : Seasonal variation of phytoplankton pigments and some of the associated oceanographic parameters in the Laccadive Sea off Cochin. - In :

ZEITSCHEL, B. (ed.) : The Biology of the Indian Ocean. Pp. 175 - 185. Springer-Verlag, Berlin - Heidelberg - New York 1973.

17775 - SHAHAK, J., CHIPMAN, D., SHAVIT, N. : Studies of photophosphorylation and exchange reactions with ADP and ATP analogues in chloroplasts. - In : Ninth International Congress of Biochemistry. Stockholm, 1 - 7 July, 1973. Abstract
Book. P. 238. IUB, Stockholm 1973.

17776 - SHAHAK, Y., CHIPMAN, D.M., SHAVIT, N. : Photophosphorylation studies with
fluorescent adenine nucleotide analogs. - FEBS Lett. *33* : 293 - 296, 1973.

17777 - SHAKHOV, A.A. : Photoenergy of plants and the use of concentrated sunlight for
raising crop yields. - In : The Sun in the Service of Mankind. Pp. V.16-1 -
- V.16.10. Unesco, Paris 1973.

*17778 - SHAKHOV, A.A., BALAUR, N.S. : Formirovanie membrannoľ sistemy khloroplastov u
vysokoproduktivnykh fotomutantov. [Formation of the chloroplast membrane system in highly productive photomutants.] - In : Geneticheskie Aspekty Fotosinteza. Tezisy Dokladov. Pp. 78 - 79. Donish, Dushanbe 1972. [In R.]

1777' - SHAPIRO, J. : Blue-green algae : Why they become dominant. - Science *179* :
382 - 384, 1973. [Ps.]

17780 - SHAPOSHNIKOVA, M.G., KRASNOVSKIĬ, A.A. : Sravnitel'noe izuchenie fotookisleniya analogov khlorofilla v vodnykh rastvorakh detergentov. [Comparative investigation of photooxidation of chlorophyll analogs in aqueous detergent solutions.] - Biokhimiya *38* : 193 - 201, 1973. [In R, ab : E.]

17781 - SHARABASH, M.T.M., EL-BASTAWESY, F.I., EL-MASRY, R.R. : Effect of *gamma* rays
on wheat plants. - Egypt. J. Bot. *16* : 413 - 422, 1973. [Chl.]

17782 - SHĀRĀN, V.K. : On simulating the lower portion of the natural boundary layer.
- Atmos. Environm. *7* : 225 - 226, 1973.

17783 - SHARPE, P.J.H. : Adaxial and abaxial stomatal resistance of cotton in the field.
- Agron. J. *65* : 570 - 574, 1973.

*17784 - SHATILOV, I.S., KARPIN, V.I., PONOMAREV, A.V. : Ispol'zovanie fotosintetiches-
ki aktivnoľ radiatsii yachmenem. [Utilization of photosynthetically active radiation in barley.] - Dokl. mosk. sel'.-khoz. Akad. K.A. Timiryazeva *182* :
45 - 48, 1972. [Production; in R.]

*17785 - SHATILOV, I.S., ROZOV, N.F., SHURYGINA, T.D. : Osobennosti intensivnosti foto-
sinteza i dykhaniya rasteniľ, kul'tiviruemykh v iskusstvennykh usloviyakh
vneshneľ sredy. [Peculiarities of photosynthetic and respiration rates in
plants grown in artificial environmental conditions.] - Izv. TSKhA (Moskva)
1972 (5) : 13 - 19, 1972. [In R, ab : E.]

17786 - SHAWCROFT, R.W., LEMON, E.R., STEWART, D.W. : Estimation of internal crop water status from meteorological and plant parameters. - In : SLATYER, R.O.
(ed.) : Plant Response to Climatic Factors. Pp. 449 - 459. Unesco, Paris 1973.

17787 - SHCHERBATYUK, A.S., GOLSTOV, I.P. : Adaptatsiya seyantsev sosny k svetu raz-
lichnoľ intensivnosti. [Adaptation of pine seedlings to light of various intensity.] - In : Informatsionnye Materialy Akad. Nauk SSSR, Sibir. Otd., Sibir. Inst. Fiziol. Biokhim. Rast. Irkutsk *11* : 33 - 35, 1973. [In R.]

17788 - SHEEHY, J.E., COOPER, J.P. : Light interception, photosynthetic activity, and
crop growth rate in canopies of six temperate forage grasses. - J. appl. Ecol.
10 : 239 - 250, 1973.

17789 - SHEEN, S.J. : Correlation between chlorophyll and chlorogenic acid content in
tobacco leaves. - Plant Physiol. *52* : 422 - 426, 1973.

17790 - SHEPHARD, D.C., BIDWELL, R.G.S. : Photosynthesis and carbon metabolism in a
chloroplast preparation from *Acetabularia*. - Protoplasma *76* : 289 - 307, 1973.

17791 - SHERIDAN, R.P. : Hydrogen sulfide production by *Synechococcus lividus* Y52-s.
- J. Phycol. *9* : 437 - 445, 1973.

17792 - SHERIFF, D.W. : An infra-red psychrometer for detecting changes in the humidity of leaf boundary layers. - J. exp. Bot. *24* : 641 - 647, 1973.

17793 - SHERIFF, D.W., SINCLAIR, R. : Fluctuations in leaf water balance, with a period of 1 to 10 minutes. - Planta *113* : 215 - 228, 1973. [Stomatal resistance.]

*17794 - **SHERMAN, L.A., HASELKORN, R.** : Growth of the blue-green algae virus LPP-1 under conditions which impair photosynthesis. - Virology *45* : 739 - 746, 1971.

17795 - **SHERRATT, D., GIVAN, C.V.** : The apparent absence of a pathway for synthesis of acetyl coenzyme A in pea chloroplasts : lack of $^{14}CO_2$ incorporation into lipids by the isolated organelle. - Planta *113* : 47 - 52, 1973.

17796 - **SHERSTENIKINA, A.B., MARSHAKOVA, M.I.** : O sostoyanii fotosinteticheskogo apparata kartofelya v usloviyakh izbytka khlora. [State of the photosynthetic apparatus of potato in the presence of excess chlorine.] - In : VECHER, A.S. (ed.) : Fotosinteticheskaya Ustoĭchivost' Rasteniĭ. Pp. 33 - 38. Nauka i Tekhnika, Minsk 1973. [In R.]

*17797 - **SHESTAK, Z., CHATSKI, Zh., SOLAROVA, Zh., TICHA, I., ZIKMUNDOVA, Kh.** : Perenos dvuokisi ugleroda i fotokhimicheskaya aktivnost' kak faktory fotosinteza v ontogeneze list'ev fasoli. [Carbon dioxide transfer and photochemical activities as factors of photosynthesis during ontogenesis of bean leaves.] - In: Geneticheskie Aspekty Fotosinteza. Tezisy Dokladov. Pp. 50 - 51. Donish, Dushanbe 1972. [In R.]

17798 - **SHIBATA, H., OCHIAI, H.** : Effect of 4-thiouridine on photosystems of a higher plant. - Agr. biol. Chem. *37* : 471 - 476, 1973.

17799 - **SHIBATA, H., OCHIAI, H.** : [Studies on the chloroplast development in radish cotyledons (3). Effect of 4-thiouridine on the development of photosystem I and II in the early stage of greening.] - Bull. Fac. Agr., Shimane Univ. *7* : 123 - 128, 1973. [In Jap., ab : E.]

17800 - **SHIBATA, K.** : Dual wavelength scanning of leaves and tissues with opal glass. - Biochim. biophys. Acta *304* : 249 - 259, 1973.

*17801 - **SHIFF, Dzh. A.** : Vzaimodeĭstvie vnutrikletochnykh struktur v protsesse razvitiya chloroplastov u evgleny. [Interactions among cellular compartments in *Euglena* during chloroplast development.] - In : Geneticheskie Aspekty Fotosinteza. Tezisy Dokladov. Pp. 35 - 36. Donish, Dushanbe 1972. [In R.]

17802 - **SHIMURA, S., FUJITA, Y.** : Some properties of the chlorophyll fluorescence of the diatom *Phaeodactylum tricornutum*. - Plant Cell Physiol. *14* : 341 - 352, 1973.

17803 - **SHIMURA, S., ICHIMURA, S.** : Selective transmission of light in the ocean waters and its relation to phytoplankton photosynthesis. - J. oceanogr. Soc. Jap. *29* (6) : 257 - 266, 1973.

17804 - **SHINDY, W.W., KLIEWER, W.M., WEAVER, R.J.** : Benzyladenine-induced movement of ^{14}C-labeled photosynthate into roots of *Vitis vinifera*. - Plant Physiol. *51* : 345 - 349, 1973.

17805 - **SHIRYAEVA, G.A., GAMBURG, K.Z.** : Vliyanie auksina na soderzhanie karotinoidov v suspenzionnoĭ kul'ture tkani tabaka. [Effect of auxin on carotenoid content in suspension culture of tobacco tissue.] - Fiziol. Rast. *20* : 493 - 498, 1973. [In R, ab : E.]

17806 - **SHLYK, A.A.** : Biosintez khlorofillovogo apparata. [Biosynthesis of the chlorophyll apparatus.] - In : Sovremennye Problemy Fotosinteza. Pp. 85 - 108. Izd. mosk. Univ., Moskva 1973. [In R.]

17807 - **SHLYK, A.A., AVERINA, N.G.** : O kharaktere vliyaniya kinetina na protsess nakopleniya protokhlorofillida v ėtiolirovannykh i zelenykh list'yakh yachmenya. [Character of kinetin effect on the process of protochlorophyllide accumulation in etiolated and green leaves of barley.] - Dokl. Akad. Nauk SSSR *213* : 235 - 238, 1973. [In R.]

17808 - **SHLYK, A.A., AVERINA, N.G.** : Sovmestnoe deĭstvie kinetina i khloramfenikola na soderzhanie khlorofilla i obrazovanie aktivnogo protokhlorofillida v zelenykh list'yakh yachmenya. [Combined effect of kinetin and chloramphenicol on chlorophyll content and formation of active protochlorophyllide in green leaves of barley.] - Fiziol. Rast. *20* : 725 - 732, 1973. [In R, ab : E.]

17809 - **SHLYK, A.A., FRADKIN, L.I., KALININA, L.M.** : Issledovanie énergeticheskogo
vzaimodeĭstviya mezhdu pigmentami v svyazi s vzaimnoĭ lokalizatsieĭ ikh molekul. [Energetic interaction between pigments in connection with mutual localization of their molecules.] - Tr. mosk. Obshch. Ispyt. Prirody, Otd. biol.
49 (Problemy Biofotokhimii) : 122 - 131, 1973. [in R, ab : E.]

17810 - **SHLYK, A.A., GAPONENKO, V.I., BALEVA, E.F.** : O kharaktere razrusheniya raznykh
form khlorofillov *a* i *b* pri sovmestnom deĭstvii tritona X-100 i ul'trazvuka.
[Mode of decomposition of different forms of chlorophylls *a* and *b* during combined effect of Triton X-100 and ultrasound.] - Fiziol. Rast. *20* : 1013 -
- 1023, 1973. [in R, ab : E.]

17811 - **SHLYK, A.A., PRUDNIKOVA, I.V., KAMYSHENKO, L.K., LOSITSKAYA, T.V., MITSUK, Z.
I., GROZOVSKAYA, M.S.** : Vliyanie ingibitorov sinteza RNK na obrazovanie khlorofillov *a* i *b* v postetiolirovannykh prorostkakh. [Effect of RNA synthesis inhibitors on the formation of chlorophyll *a* and *b* in postetiolated sprouts.] -
- Dokl. Akad. Nauk SSSR *208* : 472 - 475, 1973. [in R.]

*17812 - **SHLYK, A.A., PRUDNIKOVA, I.V., SAVCHENKO, G.E., AVERINA, N.G., KOSTYUK, N.N.,
KAMYSHENKO, L.K., VLASENOK, L.I., GAPONENKO, V.I., BALEVA, E.F., PARAMONOVA,
T.K., LOSITSKAYA, T.V., VEZITSKIĬ, A.Yu.** : Svyaz' biosinteza khlorofilla s
sintezom belkov i RNK v zeleneyushchikh i zelenykh list'yakh. [Relationship
between chlorophyll biosynthesis and synthesis of proteins and RNA in greening and green leaves.] - In : Geneticheskie Aspekty Fotosinteza. Tezisy Dokladov. Pp. 37 - 39. Donish, Dushanbe 1972. [in R.]

17813 - **SHLYK, A.A., PRUDNIKOVA, I.V., SAVCHENKO, G.E., KAMYSHENKO, L.K., GROZOVSKA-
YA, M.S., MITSUK, Z.I., LOSITSKAYA, T.V.** : Vliyanie ingibitorov sinteza RNK
na biosintez protokhlorofillida v postetiolirovannykh i zelenykh prorostkakh
yachmenya. [Effect of RNA synthesis inhibitors on the biosynthesis of protochlorophyllide in postetiolated and green sprouts of barley.] - Dokl. Akad.
Nauk SSSR *211* : 744 - 747, 1973. [in R.]

17814 - **SHLYK, A.A., RUDOI, A.B., FRADKIN, L.I.** : Analysis of isotopic kinetics for
the study of biosynthesis and metabolic heterogeneity of chlorophylls. - Physiol. vég. *11* : 25 - 43, 1973.

17815 - **SHLYK, A.A., SEMENOVICH, N.D.** : Dokazatel'stvo sushchestvovaniya protsessa ob-
novleniya bakteriokhlorofilla. [Evidence of bacteriochlorophyll renovation
process.] - Dokl. Akad. Nauk SSSR *210* : 1240 - 1242, 1973. [in R.]

17816 - **SHNEYOUR, A.** : On the site of electron donation to the photosynthetic electron
transport chain by 1,5-diphenylcarbazide. - Biochem. biophys. Res. Commun. *51*:
391 - 398, 1973.

17817 - **SHNEYOUR, A., RAISON, J.K., SMILLIE, R.M.** : The effect of temperature on the
rate of photosynthetic electron transfer in chloroplasts of chilling-sensiti-
ve and chilling-resistant plants. - Biochim. biophys. Acta *292* : 152 - 161,
1973.

17818 - **SHOEMAKER, E.M., SRIVASTAVA, L.M.** : The mechanics of stomatal opening in corn
(*Zea mays* L.) leaves. - J. theor. Biol. *42* : 219 - 225, 1973.

17819 - **SHOMER-ILAN, A., WAISEL, Y.** : The effect of sodium chloride on the balance
between the C_3- and C_4-carbon fixation pathways. - Physiol. Plant. *29* : 190 -
- 193, 1973.

17820 - **SHOSHAN, V., SHAVIT, N.** : On the reconstitution of photophosphorylation in
chloroplast membranes. - Europe. J. Biochem. *37* : 355 - 360, 1973.

B17821 - **SHPOTA, L.A.** (ed.) : Polevye Metody i Pribory dlya Izucheniya Fiziologii Sel'-
skokhozyaĭstvennykh Rasteniĭ. (Temat. Sb. Vyp. 2.) [Field Methods and Appara-
tuses for Studying Physiology of Agricultural Plants.] - Kirg. gos. Univ.,
Frunze 1973. [Ps, Chl; in R.]

17822 - **SHRIVER, J.W., BINGHAM, S.W.** : Physiological effects of bromacil on Kentucky
bluegrass and orchardgrass. - Weed Sci. *21* : 212 - 217, 1973. [Ps.]

17823 - **SHRIVER, J.W., BINGHAM, S.W.** : Selectivity of bromacil on Kentucky bluegrass
and orchardgrass. - Crop Sci. *13* : 45 - 49, 1973. [Ps.]

17824 - **SHUL'GIN, I.A., LIPOVETSKAYA, O.N., VYGODSKAYA, N.N., NICHIPOROVICH, A.A.** :

Primenenie nomogramm dlya rascheta soderzhaniya pigmentov v liste. [Calcula-
tion of pigment content in the leaf by means of nomograms.] - Fiziol. Rast.
20 : 103 - 108, 1973. [In R, ab : E.]

17825 - SHUMOV, Yu.S., SEVASTYANOV, V.I., KOMISSAROV, G.G. : Fotovol'taicheskiĭ éffekt
v plenkakh pigmentov, kontaktiruyushchikh s elektrolitom II. Prigotovlenie ne-
poristykh plenok ftalotsianina i ikh vol'tampernye kharakteristiki. [Photo-
voltaic effect in pigment films contacting with electrolytes. II. Preparation
of non-porous phtalocyanin films and their voltage current.] - Biofizika 18 :
48 - 52, 1973. [In R, ab : E.]

17826 - SHUMWAY, L.K., KLEINHOFS, A. : Aspects of the biochemistry and ultrastructure
of a cytoplasmically inherited plastid defect (Dpl) of tobacco. - Biochem.
Genet. 8 : 271 - 280, 1973.

17827 - SHUVALOV, V.A., LITVIN, F.F. : Energetika tsentrov svecheniya fotosinteziruyushchikh organizmov i ikh svyaz' s reaktsionnymi tsentrami fotosinteza : sootnoshenie poslesvecheniya i izmeneniĭ kvantovogo vykhoda fluorestsentsii.
[Energetics of emission centers of photosynthesizing organisms and their relation to the reactive centers of photosynthesis : Correlation of afterglow
and changes of the quantum yield of fluorescence.] - Tr. mosk. Obshch. Ispyt.
Prirody 49 : 148 - 153, 1973. [In R.]

17828 - SHUVALOVA, N.P., KRUPENKO, A.N., VOLKOVA, T.V., BELL, L.N. : O sootnoshenii
mezhdu skorostyami fotosinteza khlorelly, izmeryaemymi po energo- i gazoobme-
nu pri nasyshchayushchikh intensivnostyakh sveta. [Relation between Chlorella
photosynthetic rates measured according to the energy and gas exchange at sa-
turating irradiances.] - Dokl. Akad. Nauk SSSR 210 : 478 - 480, 1973. [In R.]

17829 - SHVINKA, Yu.E., ADAMOVA, N.P., RUBIN, L.B. : Regulyatornaya i substratnaya
rol' sveta v razvitii Ectothiorhodospira shaposhnikovii. [Regulatory and substrate role of light in the growth of Ectothiorhodospira shaposhnikovii.] -
Mikrobiologiya 42 : 452 - 457, 1973. [Chl; in R, ab : E.]

*17830 - SIBEL'DINA, L.A. : Deactivation of the triplet state of porphyrins and chloro-
phyll in model and intact systems. - In : BRODA, E., LOCKER, A., SPRINGER-LE-
DERER, H. (ed.) : Proceedings of the First European Biophysics Congress. Vol.
4. Pp. 249 - 252. Verlag Wiener/Med. Akad., Wien 1971.

17831 - SIBERT, J., PARKER, R.R. : Effect of pulpmill effluent on dissolved oxygen in
a stratified estuary - II. Numerical model. - Water Res. 7 : 515 - 523, 1973.
[Ps, Chl.]

*17832 - SIDOROVA, K.K., KHVOSTOVA, V.V. : Metody indutsirovaniya khlorofil'nykh mutan-
tov gorokha i ikh geneticheskaya priroda. [Methods of induction of pea chlorophyll mutants and their genetic nature.] - In : Geneticheskie Aspekty Fotosinteza. Tezisy Dokladov. Pp. 67 - 68. Donish, Dushanbe 1972. [In R.]

17833 - SIEDOW, J., YOCUM, C.F., SAN PIETRO, A. : The reducing side of photosystem I.
- In : SANADI, D.R., PACKER, L. (ed.) : Current Topics in Bioenergetics. Vol.
5. Pp. 107 - 123. Academic Press, New York - London 1973.

17834 - SIEGEL, M.I., LANE, M.D. : Chemical and enzymatic evidence for the participation of a 2-carboxy-3-ketoribitol-1,5-diphosphate intermediate in the carboxylation of ribulose 1,5-diphosphate. - J. biol. Chem. 248 : 5486 - 5498, 1973.

17835 - SIEGENTHALER, P.-A. : Change in pH dependence and sequential inhibition of
photosynthetic activity in chloroplasts by unsaturated fatty acids. - Biochim.
biophys. Acta 305 : 153 - 162, 1973.

17836 - SIEGHARDT, H. : Strahlungsnutzung von Phragmites communis. - In : ELLENBERG, H.
(ed.) : Ökosystemforschung. Pp. 79 - 86. Springer-verlag, Berlin - Heidelberg
- New York 1973.

17837 - SIEGHARDT, H. : Utilization of solar energy and energy content of different
organs of Phragmites communis TRIN. - Pol. Arch. Hydrobiol. 20 : 151 - 156,
1973.

17838 - SIELICKI, M., BURNHAM, J.C. : The effect of selenium on the physiology and
structure of Phormidium luridum. - Abstr. annu. Meet. amer. Soc. Microbiol.
73 : 32, 1973. [Chloroplast.]

17839 - SIELICKI, M., BURNHAM, J.C. : The effect of selenite on the physiological and morphological properties of the blue-green alga *Phormidium luridum* var. *oliva-cea*. - J. Phycol. *9* : 509 - 514, 1973.

17840 - **SILAEVA, A.M.** : Sravnitel'noe issledovanie struktury dvukh tipov khloroplas-tov kukuruzy. [Comparative study of the structure of two chloroplast types in maize.] - Dokl. Akad. Nauk SSSR *211* : 1447 - 1449, 1973. [In R.]

17841 - **SIMMONS, G.M. Jr., NEFF, S.E.** : Observations on limnetic carbon assimilation rates in Mountain Lake, Virginia during its thermal stratification periods. - Virginia J. Sci. *24* : 206 - 211, 1973.

17842 - **SIMOLA, L.K.** : The origin and development of organelles in germinating embryos of *Bidens cernua*. Ultrastructural effects of cycloheximide, actinomycin *D* and chloramphenicol. - Ann. bot. fenn. *10* : 71 - 88, 1973. [Chloroplast.]

17843 - **SIMONIS, W., SEUBERLING, H.B.** : Sensitivity of chlorophyll formation and of increase of NADP-dependent GPD-activity in greening *Euglena gracilis* to X-rays and to other inhibitors of protein synthesis, - Rad. Bot. *13* : 297 - 300, 1973.

17844 - **SIMONIS, W., URBACH, W.** : Photophosphorylation *in vivo*. - Annu. Rev. Plant Physiol. *24* : 89 - 114, 1973.

17845 - **SIMONOV, I.N.** : Razlichnoe proiskhozhdenie vetok chernoĭ smorodiny i vliyanie étogo razlichiya na urozhaĭ yagod i nekotorye fiziologicheskie osobennosti. [Different origin of black currant branches and effect of this difference on berry yield and some physiological properties.] - Tr. VSKhIZO *68* (Voprosy Fiziologii Rasteniĭ) : 24 - 28, 1973. [Ps, Chl; in R.]

17846 - **SIMONOVA, E.I., OKUNTSOV, M.M., IL'INA, L.P.** : Issledovanie zaklyuchitel'nogo étapa biosinteza khlorofilla v étiolirovannykh rasteniyakh yachmenya na zele-nom svetu. [Study of the end phase of chlorophyll biosynthesis in etiolated barley plants under green light.] - In : Upravlenie Skorost'yu i Napravlen-nost'yu Biosinteza u Rasteniĭ. Pp. 41 - 42. Krasnoyarsk 1973. [In R.]

17847 - **SIMS, J.L., ATKINSON, W.O.** : Accumulation of dry matter and nitrogen content of burley tobacco growing in. fertilizer-induced acid soil. - Agron. J. *65* : 762 - 765, 1973.

17848 - **SINCLAIR, J., ARNASON, J.T.** : The rate limiting thermal reaction for photosyn-thetic oxygen evolution. - Plant Physiol. *51* (Suppl.) : 67, 1973.

17849 - **SINCLAIR, T.R., LEMON, E.R.** : The distribution of 660 and 730 nm radiation in corn canopies. - Solar Energy *15* : 89 - 97, 1973.

17850 - **SINCLAIR, T.R., SCHREIBER, M.M., HOFFER, R.M.** : Diffuse reflectance hypothesis for the pathway of solar radiation through leaves. - Agron. J. *65* : 276 - 283, 1973.

17851 - **SINESHCHEKOV, V.A., SHUBIN, V.V., LITVIN, F.F.** : Dokazatel'stvo sushchestvova-niya mnogikh izluchayushchikh i énergeticheski vzaimosvyazannykh nativnykh form khlorofilla metodom selektivnogo vozbuzhdeniya i izmereniya fluorestsentsii. [The proof of existence of many radiating and energetically interconnected na-tive forms of chlorophyll by the method of selective stimulation and fluores-cence measurement.] - Dokl. Akad. Nauk SSSR *211* : 1226 - 1229, 1973. [In R.]

17852 - **SINGH, H., JOHN, J., CAMA, H.R.** : Separation of β-apocarotenals and related compounds by reversed-phase paper and thin-layer chromatography. - J. Chroma-togr. *75* : 146 - 150, 1973.

17853 - **SINGH, J.S.** : A compartment model of herbage dynamics for Indian tropical grasslands. - Oikos *24* : 367 - 372, 1973.

17854 - **SINGH, J.S., COLEMAN, D.C.** : A technique for evaluating functional root bio-mass in grassland ecosystems. - Can. J. Bot. *51* : 1867 - 1870, 1973. [^{14}C.]

17855 - **SINGH, M., OGREN, W.L., WIDHOLM, J.M.** : Photosynthetic characteristics of se-veral plant species grown under different light intensities. - Plant Physiol. *51* (Suppl.) : 68, 1973.

17856 - **SINGH, O.S., SHARMA, V.K.** : Inhibition of chlorophyll synthesis by aflatoxin in cucumber (*Cucumis sativus* L.) cotyledons. - Indian J. exp. Biol. *11* : 471- - 473, 1973.

17857 - SINGH, O.S., SHARMA, V.K., MADAN, S.K. : Physiology of expansion & chlorophyll synthesis in isolated cotyledons of watermelon, *Citrullus vulgaris* SCHRAD. - Indian J. exp. Biol. *11* : 124 - 126, 1973.

17858 - SINGH, R.K., BEN-AZIZ, A., BRITTON, G., GOODWIN, T.W. : Biosynthesis of spheroidene and hydroxyspheroidene in *Rhodopseudomonas* species : Experiments with nicotine as inhibitor. - Biochem. J. *132* : 649 - 652, 1973.

17859 - SINGH, R.K., BRITTON, G., GOODWIN, T.W. : Carotenoid biosynthesis in *Rhodopseudomonas spheroides*. S-Adenosylmethionine as the methylating agent in the biosynthesis of spheroidene and spheroidenone. - Biochem. J. *136* : 413 - 419, 1973.

17860 - SINGH, T.N., PALEG, L.G., ASPINALL, D. : Stress metabolism. III. Variations in response to water deficit in the barley plant. - Aust. J. biol. Sci. *26* : 65 - 76, 1973. [Chl.]

17861 - SINGHAL, G.S., SZALAY, L., TOMBÁCZ, E. : Absorption and fluorescence spectra of monomers, dimers and polymers of chlorophyll-a in solution. - Acta phys. chem. (Szeged) *19* : 11 - 14, 1973.

17862 - SIREVÅG, R., LEVINE, R.P. : Transcription and translation for carotenoid synthesis in *Chlamydomonas reinhardtii*. - Planta *111* : 73 - 84, 1973.

*17863 - SIROKHI, G.S. : Optimizatsiya fotosinteticheskoĭ aktivnosti v svyazi s urozhaĭnost'yu. [Optimization of photosynthetic activity in relation to crop yields.] - In : Geneticheskie Aspekty Fotosinteza. Tezisy Dokladov. Pp. 47 - 48. Donish, Dushanbe 1972. [In R.]

17864 - SIRONVAL, C., BONOTTO, S., KIRCHMANN, R. : Sur l'hétérogénéité plastidiale chez l'Acétabulaire. Étude de la fluorescence émise à la température de l'azote liquide. - Plant Sci. Lett. *1* : 47 - 52, 1973.

17865 - SIRONVAL, C., STRASSER, R.J. : Pigment-protein-complexes as carrier of informations. - In : Ninth International Congress of Biochemistry. Stockholm, 1-7 July, 1973. Abstract Book. P. 220. IUB, Stockholm 1973.

17866 - SIROTKIN, Yu.D., ANUFRIEVA, V.G. : Osobennosti sezonnogo rosta sosny i eli v smeshannykh lesnykh kul'turakh. [Features of seasonal growth of pine and fir in mixed forest cultures.] - Lesoved. les. Khoz. *1973* (7) : 50 - 57, 1973. [Chl, Car.; in R.]

17867 - SIVTSEV, M.V. : Fotokhimicheskaya aktivnost' khloroplastov i prochnost' svyazi khlorofilla v komplekse u kul'turnykh rastenii pri deĭstvii gerbitsidov, zasoleniya i biologicheski aktivnykh veshchestv. [Photochemical activity of chloroplasts and bond strength of chlorophyll in complex in cultured plants during action of herbicides, salinization and biologically active compounds.] - Fiziol. Rast. *20* : 1176 - 1181, 1973. [In R, ab : E.]

17868 - SIVTSEV, M.V. : Vliyanie razlichnykh faktorov na soderzhanie pigmentov u rastenii. [Effect of different factors on the content of pigments in plants.] - Fiziol. Biokhim. kul't. Rast. *5* : 239 - 243, 1973. [In R, ab : E.]

17869 - SIVTSEV, M.V., KUZNETSOVA, E.A. : Vliyanie prometrina na sostoyanie plastidnykh pigmentov v list'yakh luka (*Alium cepa* L.). [Effect of prometrin on the state of plastid pigments in leaves of onion (*Alium cepa* L.).] - Fiziol. Biokhim. kul't. Rast. *5* : 163 - 166, 1973. [In R, ab : E.]

17870 - SIVTSEV, M.V., KUZNETSOVA, E.A. : Sostoyanie plastidnykh pigmentov v list'yakh morkovi v svyazi s primeneniem gerbitsida solana. [State of plastid pigments in carrot leaves in relation to the use of the herbicide solan.] - Agrokhimiya *1973* (2) : 134 - 137, 1973. [In R.]

17871 - SIVTSEV, M.V., PONOMAREVA, S.A. : Anatomicheskaya struktura, ovodnennost' list'ev tomatov i aktivnost' v nikh khlorofillazy pri khloridnom zasolenii. [Anatomical structure, leaf water content and chlorophyllase activity in tomato under chloride salinization.] - Nauch. Dokl. vyssh. Shkoly, biol. Nauki *16* (3) : 84 - 87, 1973. [In R.]

17872 - SIVTSEV, M.V., PONOMAREVA, S.A., KUZNETSOVA, E.A. : Aktivnost' khlorofillazy v list'yakh tomatov pod vliyaniem zasoleniya i gerbitsida. [Effect of salinization and herbicide on the activity of chlorophyllase in tomato leaves.] - Fiziol. Rast. *20* : 62 - 65, 1973. [In R, ab : E.]

17873 - SIVTSEV, M.V., PROTSENKO, D.F., PONOMAREVA, S.A. : Vliyanie zasoleniya khlo-
ristym natriem na pronitsaemost' protoplazmy, topografiyu i fotokhimicheskuyu
aktivnost' khloroplastov v list'yakh tomatov. [Effect of sodium chloride salinization on protoplasm permeability, topography and photochemical activity
of chloroplasts in tomato leaves.] - Fiziol. Biokhim. kul't. Rast. 5 : 401 -
- 406, 1973. [In R, ab : E.]

17874 - SIZOV, S.S., SIZOVA, S.L., SEMASH, D.P. : Vliyanie orosheniya na fiziologo-
-biokhimicheskie protsessy v pochkakh yabloni v zimniĭ period. [Effect of irrigation on physiological and biochemical processes in apple buds during winter period.] - Fiziol. Biokhim. kul't. Rast. 5 : 633 - 638, 1973. [In R, ab :
E.]

17875 - SKORKOVSKÁ, Z., VAVŘINEC, E. : Příprava velmi čistých vzorků chlorofylu a a
b. [Preparation of very pure samples of chlorophyll a and b.] - Chem. Listy
67 :307 - 311, 1973. [In Czech, ab : E.]

17876 - SKOŚKIEWICZ, K. : Stomatal movements in summer rape Bronowski IHAR (Brassica
napus L. ssp. oleifera (METZG.)SINSK. f. annua THEL.) in dependence on the
age of the leaf, water deficit, light intensity and CO_2 concentration. - Hodowla Rośl., Aklimat. Nasienn. 17 : 359 - 386, 1973.

17877 - SLÁMA, J. : Vliv molybdenu a kobaltu na syntézu bílkovin a obsah β-karotenu
v jetelovinách. [Effect of molybdenum and cobalt on the protein synthesis and
the β-carotene content in clovers.] - Agrochémia 13 : 179 - 184, 1973. [In
Czech, ab : E, G, R.]

17878 - SLATER, J.H., MORRIS, I. : Photosynthetic carbon dioxide assimilation by Rhodospirillum rubrum. - Arch. Mikrobiol. 88 : 213 - 223, 1973.

17879 - SLATER, J.H., MORRIS, I. : The pathway of carbon dioxide assimilation in Rhodospirillum rubrum grown in turbidostat continuous-flow culture. - Arch. Mikrobiol. 92 : 235 - 244, 1973.

17880 - SLATYER, R.O. : The effect of internal water status on plant growth, development and yield. - In : SLATYER, R.O. (ed.) : Plant Response to Climatic Factors. Pp. 177 - 191. Unesco, Paris 1973.

17881 - SLATYER, R.O. : Effects of short periods of water stress on leaf photosynthesis. - In : SLATYER, R.O. (ed.) : Plant Response to Climatic Factors. Pp. 271-
- 276. Unesco, Paris 1973.

17882 - SLAVÍK, B. : Transpiration resistance in leaves of maize grown in humid and
dry air. - In : SLATYER, R.O. (ed.) : Plant Response to Climatic Factors. Pp.
267 - 269. Unesco, Paris 1973. [Stomatal resistance.]

17883 - SŁOMKA, J., SŁOMKA, K. : Obliczanie energii promieniowania słonecznego aktyw-
nego w fotosyntezie (PhAR). [Calculation of energy of PhAR.] - Wiad. ekol.
19 : 186 - 193, 1973. [In Pol., ab : E.]

17884 - SLOOTEN, L. : Fluorescence excitation spectra and the relative numbers of pigment molecules in reaction centres from Rhodopseudomonas spheroides. - Biochim.
biophys. Acta 314 : 15 - 27, 1973.

17885 - SLOVACEK, R.E., BANNISTER, T.T. : The effects of carbon dioxide concentration
on oxygen evolution and fluorescence transients in synchronous cultures of
Chlorella pyrenoidosa. - Biochim. biophys. Acta 292 : 729 - 740, 1973.

17886 - SLOVACEK, R.E., BANNISTER, T.T. : NH_4Cl activation of the fluorescence yield
in CO_2-starved Chlorella pyrenoidosa. - Biochim. biophys. Acta 325 : 114 -
- 119, 1973.

17887 - SLUITERS-SCHOLTEN, C.M.T., van den BERG, F.M., STEGWEE, D. : Aminolaevulinate
dehydratase in greening leaves of Phaseolus vulgaris L. - Z. Pflanzenphysiol.
69 : 217 - 227, 1973.

*17888 - SMAĬLI, R.M. : Kontrol' ul'trastruktury khloroplastov, fiziologo-geneticheskie
i vneshnie faktory optimizatsii fotosinteticheskoĭ aktivnosti. [The implications of genetic, physiological and environmental control of chloroplast ultrastructure to the organization of photosynthetic activity.] - In : Geneticheskie Aspekty Fotosinteza. Tezisy Dokladov. P. 26. Donish, Dushanbe 1972.
[In R.]

*17889 - SMAÏLI, R.M. : Biokhimicheskie i geneticheskie osnovy temperaturnoĭ chuvstvi-
tel'nosti fotosinteza i razvitie teplolyubivykh rasteniĭ. [Biochemical and
genetical bases for the temperature sensitivity of photosynthesis and growth
in chilling sensitive plants.] - In : Geneticheskie Aspekty Fotosinteza. Tezisy Dokladov. Pp. 114 - 115. Donish, Dushanbe 1972. [In R.]

17890 - SMALL, E. : Photosynthetic ecology of normal and variegated *Aegopodium podag-
raria*. - Can. J. Bot. *51* : 1589 - 1592, 1973.

17891 - SMALLIDGE, R.L., QUACKENBUSH, F.W. : β,β-carotene-2,3,3'-triol : a new carotenoid in *Anacystis nidulans*. - Phytochemistry *12* : 2481 - 2482, 1973.

17892 - SMART, R.E., LANG, A.R.G. : A field method for estimating sun-leaf angles in
productivity studies. - Agr. Meteorol. *11* : 445 - 450, 1973.

17893 - SMILLIE, R.M., SCOTT, N.S., BISHOP, D.G. : Gene expression in chloroplasts
and regulation of chloroplast differentiation. - In : LEE, J.W., POLLAK, J.K.
(ed.) : The Biochemistry of Gene Expression in Higher Organisms. Pp. 479 -
- 503. Australia and New Zealand Book Co., Sydney 1973.

17894 - SMITH, A.E., LEINWEBER, C.L. : Incorporation of ^{14}C by little bluestem tillers
at two stages of phenological development. - Agron. J. *65* : 908 - 910, 1973.

17895 - SMITH, A.J. : Synthesis of metabolic intermediates. - In : CARR, N.G., WHIT-
TON, B.A. (ed.) : The Biology of Blue-Green Algae. Pp. 1 - 38. Blackwell sci.
Publications, Oxford - London - Edinburgh - Melbourne 1973. [Ps.]

17896 - SMITH, B.N., BROWN, W.V. : The Kranz syndrome in the *Gramineae* as indicated
by carbon isotopic ratios. - Amer. J. Bot. *60* : 505 - 513, 1973.

17897 - SMITH, B.N., BROWN, W.V. : C_4 photosynthesis in the *Gramineae* as indicated by
carbon isotopic ratios. - Plant Physiol. *51* (Suppl.) : 6, 1973.

17898 - SMITH, B.N., HERATH, H.M.W., CHASE, J.B. : Effect of growth temperature on
carbon isotopic ratios in barley, pea and rape. - Plant Cell Physiol. *14* :
177 - 182, 1973. [Ps.]

B17899 - SMITH, D.C. : The Lichen Symbiosis. - Oxford Univ. Press, London 1973. [Ps.]

17900 - SMITH, D.W., BROCK, T.D. : Water status and the distribution of *Cyanidium caldarium* in soil. - J. Phycol. *9* : 330 - 332, 1973. [Ps, Chl.]

17901 - SMITH, D.W., BROCK, T.D. : The water relations of the alga *Cyanidium caldarium* in soil. - J. gen. Microbiol. *79* : 219 - 231, 1973. [Ps.]

17902 - SMITH, R.I.L., WALTON, D.W.H. : Calorific values of South Georgian plants. -
Antarct. Surv. Bull. *36* : 123 - 127, 1973.

17903 - SMITH, S.V. : Carbon dioxide dynamics : A record of organic carbon production,
respiration, and calcification in the Eniwetok reef flat community. - Limnol.
Oceanogr. *18* : 106 - 120, 1973.

17904 - SMITH, W.R. Jr., SYBESMA, C., LITCHFIELD, W.J., DUS, K. : Photochemical systems of *Rhodospirillum rubrum*. Light-induced reactions and biological functions
of *c*-type cytochromes in relation to *P*-870. - Biochemistry *12* : 2665 - 2671,
1973.

17905 - SNAUWAERT, F., TOBBACK, P.P., VERHEES, J., MAES, E. : Influence of *gamma* rays
on the chlorophyll content in peas (*Pisum sativum*). - In : Radiation Preservation of Food. Pp. 61 - 72. Int. at. Energy Agency, Vienna 1973.

17906 - SNYTKO, A.I. : Fotosinteticheskaya deyatel'nost' sakharnoĭ svekly v usloviyakh
zapadnoĭ Sibiri. [Photosynthetic activity of the sugar beet in Western Sibi-
ria.] - Sel'.-khoz. Biol. *8* : 777 - 778, 1973. [In R.]

17907 - SOFROVÁ, D. : Fosforylační reakce v rostlinné buňce a jejich inhibitory.
[Phosphorylation reactions in the plant cell and their inhibitors.] - Stud.
Inform. ÚVTI, zákl. Vědy Zeměd. (Praha) *73* (2) : 1 - 87, 1973. [In Czech, ab:
E, R.]

17908 - SOFROVÁ, D., LEBLOVÁ, S. : Photochemical reactions of the blue-green algae
Plectonema boryanum. - In : Ninth International Congress of Biochemistry.
Stockholm, 1 - 7 July, 1973. Abstract Book. P. 241. IUB, Stockholm 1973.

17909 - **SOKOLOVA, Z.N.** : Vliyanie tsinka na pigmentnuyu sistemu vinogradnoĭ lozy.
[Effect of zinc on pigment system of grape-vine.] - Tr. kishinev. sel'skókhoz.
Inst. M.V. Frunze *118* (Vinogradarstvo) : 84 - 90, 106, 1973. [In R.]

17910 - **SOKOL'SKIĬ, A.F.** : Pervichnaya i bakterial'naya produktsiya nekotorykh rybo-
vodnykh prudov del'ty r. Volgi. [Primary and bacterial production of some
fishery ponds in the Volga estuary.] - Gidrobiol. Zh. *9* (6) : 77 - 82, 1973.
[In R.]

17911 - **SOLÁROVÁ, J.** : Changes in minimal diffusive resistances of leaf epidermes during ageing of primary leaves of *Phaseolus vulgaris* L. - Biol. Plant. *15* :
237 - 240, 1973.

17912 - **SOLOV'EVA, A.A.** : Pervichnaya produktsiya fitoplanktona zalivov vostochnogo
Murmana. [Primary phytoplankton productivity in the Eastern Murman bays.] -
Gidrobiol. Zh. *9* (4) : 14 - 19, 1973. [Chl; in R, ab : E.]

17913 - **SORAN, V., SMITH, B., VINTILA, R.** : Absorption spectra of *Zamioculcas Boivi-
nii* chloroplasts recorded *in vivo* and isolation by micro-spectrophotometric
method. - Z. Pflanzenphysiol. *69* : 100 - 108, 1973.

17914 - **SORENSEN, F.C., FERRELL, W.K.** : Photosynthesis and growth of Douglas-fir seedlings when grown in different environments. - Can. J. Bot. *51* : 1689 - 1698,
1973.

17915 - **SOROKIN, E.M.** : Kvantovyĭ vykhod pervichnogo akta preobrazovaniya énergii pri
fotosinteze. [Quantum yield of the primary act of energy transformation in
the photosynthesis.] - Dokl. Akad. Nauk SSSR *209* : 1227 - 1229, 1973. [In R.]

17916 - **SOROKIN, E.M.** : Netsiklicheskiĭ transport élektronov i svyazannye s nim voprosy. [Non-cyclic electron transport and related problems.] - Fiziol. Rast. *20* :
733 - 741, 1973. [In R, ab : E.]

17917 - **SOROKIN, E.M.** : Netsiklicheskiĭ transport élektronov i fluorestsentsiya - zavisimost' ot intensivnosti sveta. [Non-cyclic electron transport and fluorescence : Effect of irradiance.] - Fiziol. Rast. *20* : 978 - 987, 1973. [In R,
ab : E.]

17918 - **SOROKIN, Yu.I.** : O produktivnosti priberezhnykh tropicheskikh vod zapadnoĭ
chasti Tikhogo okeana. [Productivity of shallow tropical waters of the western
Pacific.] - Okeanologiya *13* : 669 - 676, 1973. [In R, ab : E.]

17919 - **SOURNIA, A.** : Comments on the diel periodicity of phytoplankton photosynthesis, with an example from the Indian ocean. - Spec. Publ. mar. biol. Ass. India *1973* : 52 - 59, 1973.

17920 - **SOURNIA, A.** : État actuel des recherches océanographiques dans le domaine de
la respiration cellulaire. - Bull. Union Océanogr. Fr. *5* (4) : 35, 1973. [Ps.]

17921 - **SOURNIA, A.** : La production primaire planctonique en Méditerranée. Essai de
mise à jour. - Bull. Étude Commun. Méditerranée *5* (Num. spéc.) : 1 - 128, 1973.

B17922 - Sovremennye Problemy Fotosinteza. [Contemporary Problems of Photosynthesis.]
- Izd. mosk. Univ., Moskva 1973. [In R.]

17923 - **SPENCE, D.H.N., CAMPBELL, R.M., CHRYSTAL, J.** : Specific leaf areas and zona-
tion of freshwater macrophytes. - J. Ecol. *61* : 317 - 328, 1973. [Ps.]

17924 - **SPENCER, P.W.** : Incorporation of ^{14}C-leucine into apple leaf protein and its
inhibition by protein synthesis inhibitors during growth and senescence. -
Plant Physiol. *52* : 151 - 155, 1973. [Chl.]

17925 - **SPENCER, P.W., TITUS, J.S.** : Apple leaf senescence : leaf disc compared to
attached leaf. - Plant Physiol. *51* : 89 - 92, 1973. [Chl.]

17926 - **SPERLING, J.A., HALE, G.M.** : Patterns of radiocarbon uptake by a thermophilic
blue-green alga under varying conditions of incubation. - Limnol. Oceanogr.
18 : 658 - 662, 1973.

17927 - **SPERLING, P.G., WALNE, P.L., SCHWARZ, O.J., TRIPLETT, L.L.** : Studies on characterization of pigments from isolated eyespots of euglenoid flagellates. -
J. Phycol. *9* (Suppl.) : 20, 1973.

17928 - **SPIERTZ, J.H.J.** : Effects of successive applications of maneb and benomyl on growth and yield of five wheat varieties of different heights. - Neth. J. agr. Sci. *21* : 282 - 296, 1973. [Growth analysis.]

17929 - **SPODNIEWSKA, I., HILLBRICHT-ILKOWSKA, A.** : Experimentally increased fish stock in the pond type lake Warniak. VI. Biomass and production of phytoplankton. - Ekol. pol. *21* : 519 - 532, 1973.

17930 - **SPRENT, J.I.** : Growth and nitrogen fixation in *Lupinus arboreus* as affected by shading and water supply. - New Phytol. *72* : 1005 - 1022, 1973. [Growth analysis.]

17931 - **SPREY, B.** : Lichtinduzierte Entwicklung von Etioplasten zu Chloroplasten : Induktion und Regulation der Membranbildung. - Ber. Kernforschungsanlage Jülich *1019* : 1 - 244, 1973.

*17932 - **SREENIVASAN, A.** : Energy transformations through primary productivity and fish production in sóme tropical freshwater impoundments and ponds. - In : KAJAK, Z., HILLBRICHT-ILKOWSKA, A. (ed.) : Productivity Problems of Freshwaters. Pp. 505 - 514. Pol. Sci. Publ., Warszawa 1972. [Ps.]

17933 - **STADELMAN, P., MUNAWAR, M.** : Biomass parameters and primary production at a nearshore and a midlake station of Lake Ontario during IFYGL (IFYGL). - In : Proceedings of the 17th Conference on Great Lakes Research. Pp. 109 - 119. Int. Ass. Great Lakes Res., 1974.

*17934 - **STADELMANN, P.** : Stickstoffkreislauf und Primärproduktion im mesotrophen Vierwaldstättersee (Horwer Bucht) und im eutrophen Rotsee, mit besonderer Berücksichtigung des Nitrats als limitierenden Faktors. - Schweiz. Z. Hydrol. *33* : 1 - 65, 1971.

17935 - **STAFFORD, H.A., BLISS, M.** : The effect of greening of sorghum leaves on the molecular weight of a complex containing 4-hydroxycinnamic acid hydroxylase activity. - Plant Physiol. *52* : 453 - 458, 1973.

17936 - **STAHL, C.L., SOJKA, G.A.** : Growth of *Rhodopseudomonas capsulata* on L- and D- -malic acid. - Biochim. biophys. Acta *297* : 241 - 245, 1973. [Ps.]

17937 - **STANEV, V., VASSILEV, G.** : Effect of the allylthioureidosalicilic acid and potassium deficiency on chlorophyll state and protein and nucleic acid content in bean leaves. - Dokl. SKhA G. Dimitrova (Sofia) *8* : 109 - 113, 1973.

17938 - **STANHILL, G., FUCHS, M., BAKKER, J., MORESHET, S.** : The radiation balance of a glasshouse rose crop. - Agr. Meteorol. *11* : 385 - 404, 1973. [Chl.]

17939 - **STANHILL, G., ISRAELI, M., ROSENZWEIG, D.** : The solar radiation balance in scrub forest and pasture on the Carmel mountain, Israel : a comparative study. - Ecology *54* : 819 - 828, 1973.

17940 - **STANIER, R.Y.** : Autotrophy and heterotrophy in unicellular blue-green algae. - In : CARR, N.G., WHITTON, B.A. (ed.) : The Biology of Blue-Green Algae. Bot. Monogr. 9. Pp. 501 - 518. Blackwell sci. Publ., Oxford - London - Edinburgh - Melbourne 1973. [Ps.]

17941 - **STANTON, M.G.** : Digital light integrator for ecology. - New Phytol. *72* : 1375 - 1379, 1973.

17942 - **STARCK, Z.** : Effect of kinetin and shading on the subsequent distribution of ^{14}C-assimilates in *Raphanus sativus*. - Proc. Res. Inst. Pomol., Skierniewice, Ser. E *1973* (3) : 171 - 179, 1973.

17943 - **STARCK, Z.** : Preliminary investigation on the interaction between some physiological processes in donors and acceptors of assimilates. - Acta Soc. Bot.Pol. *42* : 143 - 162, 1973.

17944 - **STARCK, Z.** : The effect of shading during growth on the subsequent distribution of ^{14}C-assimilates in *Raphanus sativus*. - Bull. Acad. pol. Sci., Sér. Sci. biol. *21* : 309 - 314, 1973.

17945 - **STARCK, Z.** : Effect of kinetin and shading on the subsequent distribution of ^{14}C-assimilates in *Raphanus sativus*. - Bull. Acad. pol. Sci., Sér. Sci. biol. *21* : 573 - 580, 1973.

*17946 - **STARTSEV, G.A., YULDASHEV, O.Kh.** : Khlorofil'nye mutanty *Arabidopsis thaliana*, indutsirovannye rubinovym lazerom. [Chlorophyll mutants of *Arabidopsis thalia-na*, induced by a ruby lazer.] - In : Geneticheskie Aspekty Fotosinteza. Tezisy Dokladov. Pp. 68 - 69. Donish, Dushanbe 1972. [In R.]

17947 - **STAVIS, R.L., HIRSCHBERG, R.** : Phototaxis in *Chlamydomonas reinhardtii*. - J. Cell Biol. *59* : 367 - 377, 1973. [Ps, Chl.]

17948 - **STEER, B.T.** : Control of ribulose-1,5-diphosphate carboxylase activity during expansion of leaves of *Capsicum frutescens* L. - Ann. Bot. *37* : 823 - 829, 1973.

17949 - **STEER, B.T.** : Diurnal variations in photosynthetic products and nitrogen metabolism in expanding leaves. - Plant Physiol. *51* : 744 - 748, 1973.

*17950 - **STEFANOVIĆ, K.** : Uporedno proučavanje produkcije CO_2 u zajednici *Querco-Carpinetum serbicum* RUDSKI i na otvorenom polju na Fruškoj Gori. [A comparative study of CO_2 production in the community *Querco-Carpinetum serbicum* RUDSKI and in an open field on the mountain Fruška Gora.] - Zemljište Biljka *21* : 105 - - 112, 1972. [CO_2 profiles.]

17951 - **STEINBIβ, H.-H., SCHMITZ, K.** : CO_2-Fixierung und Stofftransport in benthischen marinen Algen. V. Zur autoradiographischen Lokalisation der Assimilattransportbahnen im Thallus von *Laminaria hyperborea*. - Planta *112* : 253 - 263, 1973.

17952 - **STEMLER, A., GOVINDJEE** : Effect of bicarbonate ion on Hill reaction by maize chloroplasts. - Plant Physiol. *51* (Suppl.) : 67, 1973.

17953 - **STEMLER, A., GOVINDJEE** : Bicarbonate ion as a critical factor in photosynthetic oxygen evolution. - Plant Physiol. *52* : 119 - 123, 1973.

17954 - **ŠTĚRBA, S.** : Fotosyntetická aktivita listových pletiv na sazenici topolu. [Photosynthetic activity of leaf tissues of poplar seedling.] - Práce VÚLHM *43* : 37 - 47, 1973. [In Czech, ab : E, R.]

17955 - **STETLER, D.** : Non-photoconvertible protochlorophyllide in etiolated tissue lacking prolamellar bodies. - Amer. J. Bot. *60* (4 Suppl.) : 14, 1973.

17956 - **STETLER, D.A.** : Nonphotoconvertible protochlorophyllide in etiolated tissue lacking prolamellar bodies. - Bot. Gaz. *134* : 290 - 295, 1973.

17957 - **STEVENS, S.E. Jr., van BAALEN, C.** : Characteristics of nitrate reduction in a mutant of the blue-green alga *Agmenellum quadruplicatum*. - Plant Physiol. *51*: 350 - 356, 1973. [Ps.]

17958 - **STEVENS, S.E. Jr., PATTERSON, C.O.P., MYERS, J.** : The production of hydrogen peroxide by blue-green algae : a survey. - J. Phycol. *9* : 427 - 430, 1973.

17959 - **STEWART, W.D.P.** : Nitrogen fixation by photosynthetic microorganisms. - Annu. Rev. Microbiol. *27* : 283 - 316, 1973.

17960 - **STIGTER, C.J., BIRNIE, J., LAMMERS, B.** : Leaf diffusion resistance to water vapour and its direct measurement. II. Design, calibration and pertinent theory of an improved leaf diffusion resistance meter. - Med. Landbouwhogesch. Wageningen *73* (15) : 1 - 55, 1973.

17961 - **ŞTIRBAN, M., ŢARA, G.** : Variaţii ale conţinutului de pigmenţi asimilatori la diferite categorii de lăstari ai viţei de vie. [Variations in the assimilatory pigment content in different types of vine sprouts in vineyards.] - Stud. Cercet. Biol., Ser. bot. *25* : 159 - 166, 1973. [In Roum., ab : E.]

17962 - **ŞTIRBAN, M., ŢARA, G.** : Dinamica diurnă si sezonieră a fotosintezei la cîteva soiuri de viţă de vie. [Daily and seasonal dynamics of photosynthesis in various cultivars of grape-vine.] - Stud. Cercet. Biol., Ser. bot. *25* : 227 - - 235, 1973. [In Roum., ab : E.]

17963 - **STOJANOV, Zh.V.** : Determining the heat transfer coefficients of leaves. - In : **SLATYER, R.O.** (ed.) : Plant Response to Climatic Factors. Pp. 61 - 62. Unesco, Paris 1973.

17964 - **STOJANOV, Zh.V., FLOROV, R.J.** : Influence global de la lumiére, de la température et de l'humidité sur la photosynthèse et la respiration. - In : **SLATYER, R.O.** (ed.) : Plant Response to Climatic Factors. Pp. 137 - 139. Unesco, Paris 1973.

17965 - **STOJANOV, Zh.V., FLOROV, R.J.** : Recherches sur la diminution de l'énergie de

dissipation et l'augmentation de l'utilisation de la radiation. - In : The
Sun in the Service of Mankind. Pp. V.13-1 - V.13-9. Unesco, Paris 1973.

*17966 - STOLBOVA, A.V. : Geneticheskiĭ analiz svetochuvstvitel'nykh mutantov khlami-
domonady. [Genetic analysis of *Chlamydomonas* mutants sensitive to light.] -
In : Geneticheskie Aspekty Fotosinteza. Tezisy Dokladov. Pp. 71 - 72. Donish,
Dushanbe 1972. [In R.]

17967 - STOLOVITSKY, Yu.M., SHKUROPATOV, A.Ya., KADOSHNIKOV, S.I., EVSTIGNEEV, V.B. :
On the photoelectrochemical effect in solid chlorophyll and chlorophyll-protein films. - FEBS Lett. *34* : 147 - 149, 1973.

17968 - STRALEY, S.C., CLAYTON, R.K. : Extraction of oxidized bacteriochlorophyll
from illuminated photosynthetic reaction center particles. - Biochim. biophys.
Acta *292* : 685 - 691, 1973.

17969 - STRALEY, S.C., PARSON, W.W., MAUZERALL, D.C., CLAYTON, R.K. : Pigment content
and molar extinction coefficients of photochemical reaction centers from *Rhodopseudomonas spheroides*. - Biochim. biophys. Acta *305* : 597 - 609, 1973.

17970 - STRAŠKRABA, M. : Limnological basis for modeling reservoir ecosystems. - In :
ACKERMANN, W.C., WHITE, G.F., WORTHINGTON, E.B. (ed.) : Man-Made Lakes : Their
Problems and Environmental Effects. Geophys. Monogr. Ser. Vol. 17. Pp. 517 -
- 535. Amer. Geophys. Union, Washington, D.C. 1973. [Productivity.]

17971 - STRAŠKRABA, M., JAVORNICKÝ, P. : Limnology of two re-regulation reservoirs in
Czechoslovakia. - In : HRBÁČEK, J., STRAŠKRABA, M. (ed.) : Hydrobiological
Studies. Vol. 2. Pp. 249 - 316. Academia, Praha 1973. [Ps, Chl.]

17972 - STRASSER, R.J. : Induction phenomena in green plants when the photosynthetic
apparatus starts to work. - Arch. int. Physiol. Biochim. *81* : 935 - 955, 1973.

17973 - STRASSER, R.J. : Studies on the induction of the photosynthetic electron transport. - In : Ninth International Congress of Biochemistry. Stockholm, 1 - 7
July, 1973. Abstract Book. P. 241. IUB, Stockholm 1973.

*17974 - STRASSER, R.J., ERISMANN, K.H., METZNER, H. : Die Photooxidation von Sulfid
durch Grana. - Verhandl. schweiz. naturforsch. Ges. *150* : 235 - 238, 1970.
$~$1970.

*17975 - STRASSER, R.J., RÜGGEBERG, B., METZNER, H. : Lichtinduzierte Grössenverände-
rungen bei *Chlorella*. - Verhandl. schweiz. naturforsch. Ges. *150* : 156 - 158,
1970. [Ps.]

17976 - STRASSER, R.J., SIRONVAL, C. : Induction of PS II activity and induction of
a variable part of the fluorescence emission by weak green light in flashed
bean leaves. - FEBS Lett. *29* : 286 - 288, 1973.

17977 - STRAUB, V., LICHTENTHALER, H.K. : Die Wirkung von β-Indolessigsäure auf die
Bildung der Chloroplastenpigmente, Plastidenchinone und Anthocyane in *Raphanus*-
-Keimlingen. - Z. Pflanzenphysiol. *70* : 34 - 45, 1973.

17978 - STRAUB, V., LICHTENTHALER, H.K. : Die Wirkung von Gibberellinsäure A_3 und Kinetin auf die Bildung der Photosynthesepigmente, Lipochinone und Anthocyane
in *Raphanus*-Keimlingen. - Z. Pflanzenphysiol. *70* : 308 - 321, 1973.

17979 - STRAUSS, G., TIEN, H.T.: Energy transfer from carotenoids to chlorophyll *a*
in black lipid membranes. - Photochem. Photobiol. *17* : 425 - 431, 1973.

17980 - STREBEYKO, P. : Theoretical principles of gas exchange in plants. - Hodowla
Rośl., Aklimat. Nasienn. *17* : 287 - 295, 1973. [Ps.]

17981 - STREBEYKO, P., BACŁAWSKA-KRZEMIŃSKA, Z., JARECKA, M., WRÓBLEWSKA, H. : Influence of water deficit on photosynthetic activity of some crop plants. - Hodowla Rośl., Aklimat. Nasienn. *17* : 413 - 416, 1973.

17982 - STREICHER, S.L., VALENTINE, R.C. : Comparative biochemistry of nitrogen fixation. - Annu. Rev. Biochem. *42* : 279 - 302, 1973. [Also Ps.]

*17983 - STRICKLAND, J.D.H., HOLM-HANSEN, O., EPPLEY, R.W., LINN, R.J. : The use of a
deep tank in plankton ecology. I. Studies of the growth and composition of
phytoplankton crops at low nutrient levels. - Limnol. Oceanogr. *14* : 23 - 34,
1969. [Chl, Car.]

17984 - STRINGAM, G.R. : Inheritance and allelic relationships of seven chlorophyll-
-deficient mutants in *Brassica campestris* L. - Can. J. Genet. Cytol. *15* : 335-
- 339, 1973.

17985 - STROSS, R.G., CHISHOLM, S.W., DOWNING, T.A. : Causes of daily rhythms in photosynthetic rates of phytoplankton. - Biol. Bull. *145* : 200 - 209, 1973.

17986 - STROTMANN, H., HESSE, H., EDELMANN, K. : Quantitative determination of coupling factor CF_1 of chloroplasts. - Biochim. biophys. Acta *314* : 202 - 210,
1973.

17987 - STROTMANN, H., THIEL, A. : Zum Mechanismus der Entkopplung der Photophosphorylierung durch Anionen schwacher Säuren. - Ber. deut. bot. Ges. *86* : 209 - 212,
1973.

17988 - STUART, A.L., WASSERMAN, A.R. : Purification of cytochrome b_6. A tightly bound
protein in chloroplast membranes. - Biochim. biophys. Acta *314* : 284 - 297,
1973.

17989 - SUBBA RAO, D.V. : Effects of environmental perturbations on short-term phytoplankton production off Lawson's Bay, a tropical coastal embayment. - Hydrobiologia *43* : 77 - 91, 1973.

17990 - SUBBOTOVICH, A.S., PERSTNEV, N.D., FOKSHA, M.G. : Aktivnost' fiziologo-biokhimicheskikh protsessov okulirovok v zavisimosti ot sposobov ikh podgotovki
k posadke v shkolku. [Activity of physiological and biochemical processes in
dependence on their preparat'on for planting into nursery.] - Tr. kishinev.
sel'skokhoz. Inst. M.V. Frunze *118* (Vinogradarstvo) : 8 - 13, 102, 1973. [Ps,
Chl; in R.]

17991 - SUD'INA, E.G., LOZOVAYA, G.I., DOVBYSH, E.F., BABENKO, E.I., FOMISHINA, R.N.:
K voprosu o lokalizatsii khlorofillazy v khloroplaste. [Chlorophyllase localization in chloroplasts.] - Fiziol. Biokhim. kul't. Rast. *5* : 154 - 158,
1973. [In R, ab : E.]

17992 - SUD'INA, E.G., SEMICHAEVSKIĬ, V.D., BABENKO, E.I. : Izuchenie kinetiki khlo-
rofillaznoĭ reaktsii metodom differentsial'noĭ spektroskopii. [Chlorophyllase
reaction kinetics studied by differential spectroscopy.] - Fiziol. Biokhim.
kul't. Rast. *5* : 489 - 494, 1973. [In R, ab : E.]

17993 - SUD'ĬNA, O.G., DOVBYSH, K.P., GOLOD, M.G., FOMISHYNA, R.M.; BABENKO, E.G. :
Do vyvchennya energiĭ aktyvatsiĭ khlorofilaznoĭ reaktsiĭ u filogenetychno riznykh grup roslyn. [Chlorophyllase reaction activation energy in phylogenetically different groups of plants.] - Ukr. bot. Zh. *30* : 763 - 770, 811, 1973.
[In Ukr., ab : E, R.]

17994 - SUD'ĬNA, O.G., LOZOVA, G.I., DOVBYSH, K.P., FOMISHYNA, R.M., BABENKO, E.G. :
Doslidzhennya roli strukturnoĭ organizatsiĭ dlya khlorofilaznoĭ reaktsiĭ na
prykladi shtuchnykh kompleksiv. [Study of the role of structural organization
for the reaction of chlorophyllase on an example of artificial complexes.] -
Ukr. bot. Zh. *30* : 155 - 162, 265, 1973. [In Ukr., ab : E, R.]

*17995 - SUGANO, N., MIYA, S., NISHI, A. : Carotenoid synthesis in a suspension culture
of carrot cells. - Plant Cell Physiol. *12* : 525 - 531, 1971.

17996 - SUGIMOTO, K. : [Studies on transpiration and water requirement of indica and
japonica rice plants. I. Relationship of transpiration to leaf area and to meteorological factors.] - Jap. J. trop. Agr. *16* : 260 - 264, 1973. [Growth analysis; in Jap., ab : E.]

17997 - SUGIMOTO, K. : [Studies on transpiration and water requirement of indica and
japonica rice plants. II. Relationship of dry matter production to transpiration and to water requirement.] - Jap. J. trop. Agr. *16* : 265 - 269, 1973.
[Growth analysis; in Jap., ab : E.]

17998 - SUGIYAMA, T. : Purification, molecular, and catalytic properties of pyruvate
phosphate dikinase from the maize leaf. - Biochemistry *12* : 2862 - 2868, 1973.

17999 - SUNDARARAJ, V., KRISHNAMURTHY, K. : Photosynthetic pigments and primary production. - Curr. Sci. *42* : 185 - 189, 1973.

18000 - SUNDQVIST, C. : The relationship between chlorophyllide accumulation, the a-
mount of protochlorophyllide$_{636}$ and protochlorophyllide$_{650}$ in dark grown

wheat leaves treated with δ-aminolevulinic acid. - Physiol. Plant. *28* : 464 - - 470, 1973.

18001 - SUNDQVIST, C. : The influence of varying light intensities on the phototrans-formation of protochlorophyllide$_{636}$ in dark grown wheat leaves treated with δ-aminolevulinic acid. - Physiol. Plant. *29* : 434 - 439, 1973.

18002 - SUNDSTRÖM, K.-R., HÄLLGREN, J.-E. : Using lichens as physiological indicators of sulfurous pollutants. - Ambio *2* : 13 - 21, 1973. [Ps, Chl.]

B18003 - SUPPAN, P. : Principles of Photochemistry. - Chem. Soc., London 1973.

18004 - SUZUKI, M., MORTIMER, D.C. : Sugar concentration gradients of the sugar beet plant in relation to translocation. - Can. J. Bot. *51* : 1733 - 1739, 1973.

18005 - ŠVIHRA, J., SZABO, D., GÁL, D. : Produkčný proces jarného jačmeňa v závislosti od hustoty sejby. [Production process of spring barley as depending on sowing rate.] - Rostlinná Výroba (Praha) *19* : 505 - 516, 1973. [In Slovak, ab : E, R.]

18006 - SVOBODA, J. : Primary production of plant communities of the Truelove Lowland, Devon Island, Canada - beach ridges. - In : BLISS, L.C., WIELGOLASKI, F.E. (ed.) : Primary Production and Production Processes, Tundra Biome. Pp. 15 - - 26. Tundra Biome Steering Comm., Edmonton 1973.

18007 - SWAN, A.G., RAWSON, H.M. : A low-cost, portable system for the measurement of $^{14}CO_2$ in gas streams. - Photosynthetica *7* : 325 - 329, 1973.

18008 - SWANK, W.T., SCHREUDER, H.T. : Temporal changes in biomass, surface area, and net production for a *Pinus strobus* L. forest. - In : IUFRO Biomass Studies. Pp. 171 - 182. Univ. Maine, Orono 1973.

18009 - SWANSON, E.S., THOMSON, W.W., MUDD, J.B. : Effect of ozone on leaf cell mem-branes. - Can. J. Bot. *51* : 1213 - 1219, 1973. [Chloroplast.]

18010 - SWIFT, L.W., KNOERR, K.R. : Estimating solar radiation on mountain slopes. - Agr. Meteorol. *12* : 329 - 336, 1973.

18011 - SYRETT, P.J. : Regulation of growth and metabolism in eukaryotic algae. - J. gen. Microbiol. *75* (2) : III - IV, 1973. [Ps.]

18012 - SZAJNOWSKI, F. : The relation between the leaf area and production of the abo-veground parts of common reed (*Phragmites communis* TRIN.). - Pol. Arch. Hydro-biol. *20* : 157 - 158, 1973.

18013 - SZAJNOWSKI, F. : Relationship between leaf area index and shoot production of *Phragmites communis* TRIN. - Pol. Arch. Hydrobiol. *20* : 257 - 268, 1973.

18014 - SZALAY, L., HEVESI, J., LEHOCZKI, E. : Molecular fluorescence and photosynthe-sis. - Acta phys. chem. (Szeged) *19* : 403 - 416, 1973.

18015 - SZALAY, L., SINGHAL, G.S., TOMBÁCZ, E., KOZMA, L. : Light absorption and fluo-rescence of highly diluted chlorophyll solutions. - Acta phys. Acad. Sci. hung. *34* : 341 - 350, 1973.

18016 - SZAREK, S.R., JOHNSON, H.B., TING, I.P. : Drought adaptation in *Opuntia basi-laris*. Significance of recycling carbon through crassulacean acid metabolism. - Plant Physiol. *52* : 539 - 541, 1973.

18017 - SZCZEPAŃSKA, W., SZCZEPAŃSKI, A. : Emergent macrophytes and their role in wet-land ecosystems. - Pol. Arch. Hydrobiol. *20* : 41 - 50, 1973. [Growth analysis.]

18018 - SZCZEPAŃSKI, A. : Chlorophyll in the assimilation parts of helophytes. - Pol. Arch. Hydrobiol. *20* : 67 - 71, 1973.

18019 - SZCZOTKA, Z., BORYS, M.W., WOJCIECHOWSKI, J. : Relation between K$^+$-nutrition of potatoes and their leaflets' resistance to *Phytophthora infestans* de BY. - Phytopathol. Z. *76* : 57 - 66, 1973. [Chl.]

18020 - SZEICZ, G., VAN BAVEL, C.H.M., TAKAMI, S. : Stomatal factor in the water use and dry matter production by sorghum. - Agr. Meteorol. *12* : 361 - 389, 1973.

18021 - **TACHIKI, K.H., PON, N.G.** : Isolation and characterization of chlorophyll-protein complex I from spinach chloroplast lamellae. - In : Ninth International Congress of Biochemistry. Stockholm, 1 - 7 July, 1973. Abstract Book. P. 232. IUB, Stockholm 1973.

*18022 - **TAGEEVA, S.V., SEMENOVA, G.A., LADYGIN, V.G.** : Geneticheskaya regulyatsiya membrannoĭ organizatsii khloroplastov na primere pigmentnykh mutantov. [Genetic control of the membrane organization of chloroplast in pigment mutants taken as a model.] - In : Geneticheskie Aspekty Fotosinteza. Tezisy Dokladov. Pp. 27 - 28. Donish, Dushanbe 1972. [In R.]

*18023 - **TAGUCHI, S., NAKAJIMA, K.** : Plankton and seston in the sea surface of three inlets of Japan. - Bull. Plankton Soc. Jap. *18* : 20 - 36, 1971. [Chl.]

18024 - **TAKABE, T., AKAZAWA, T.** : Oxidative formation of phosphoglycolate from ribulose-1,5-diphosphate catalysed by *Chromatium* ribulose-1,5-diphosphate carboxylase. - Biochem. biophys. Res. Commun. *53* : 1173 - 1179, 1973.

18025 - **TAKAHASHI, K., SATO, K., HABASHITA, J.** : The effects of gibberellic acid on the growth of rice plant under different temperatures. - Tohoku J. agr. Res. *23* : 175 - 183, 1973. [Chl.]

1802? - **TAKAHASHI, M., FUJII, K., PARSONS, T.R.** : Simulation study of phytoplankton photosynthesis and growth in the Fraser River estuary. - Mar. Biol. *19* : 102- - 116, 1973.

18027 - **TAKAHASHI, M., NASH, F.** : The effect of nutrient enrichment on algal photosynthesis in Great Central Lake, British Columbia, Canada. - Arch. Hydrobiol. *71* : 166 - 182, 1973.

18028 - **TAKAMIYA, A.** : Distribution of photoconvertible, water-soluble chlorophyll protein complex CP668 in plants related to *Chenopodium album*. — Carnegie Inst. Year Book *72* : 330 - 336, 1973.

18029 - **TAKEBE, I., OTSUKI, Y., HONDA, Y., NISHIO, T., MATSUI, C.** : Fine structure of isolated mesophyll protoplasts of tobacco. - Planta *113* : 21 - 27, 1973. [Chloroplast.]

18030 - **TAKEDA, T., AKIYAMA, T.** : [Studies on dry matter production in corn plant. II. Dry matter production of seedlings as affected by dense planting.] - Proc. Crop Sci. Soc. Jap. *42* : 302 - 306, 1973. [In Jap., ab : E.]

18031 - **TAKEMOTO, J., LASCELLES, J.** : Bacteriochlorophyll and membrane protein synthesis in *Rhodopseudomonas spheroides*. - Abstr. annu. Meet. amer. Soc. Microbiol. *73* : 156, 1973.

18032 - **TAKIMOTO, S., CHIN, K., OKUKADO, N., YAMAGUCHI, M.** : Dehydration of zeaxanthin and xanthophyll. - Mem. Fac. Sci., Kyushu Univ., Ser. C *8* : 197 - 202, 1973.

18033 - **TALLING, J.F.** : The application of some electrochemical methods to the measurement of photosynthesis and respiration in fresh waters. - Freshwater Biol. *3* : 335 - 362, 1973.

18034 - **TANAKA, M., KARASAWA, I.** : [The existing state of pigments in food. (Part 1.) Spectrophotometric study of the physical state of carotenes in orange carrots.] - J. Home Economics [Kaseigaku Zasshi] *24* : 27 - 34, 1973. [In Jap., ab : E.]

18035 - **TANAKA, M., KARASAWA, I.** : [The existing state of pigments in food. (Part 2.) Spectrophotometric study of the physical state of carotenes in red carrots.] - J. Home Economics [Kaseigaku Zasshi] *24* : 353 - 358, 1973. [In Jap., ab : E.]

18036 - **TAO, K.-L. J., JAGENDORF, A.T.** : The ratio of free to membrane-bound chloroplast ribosomes. - Biochim. biophys. Acta *324* : 518 - 532, 1973.

B18037 - **TARASOV, V.M., KOVALENKO, V.F.** : Usykhanie Pobegov Yabloni (Nedostatochnost' Medi). [Wilting of Apple Tree Shoots (Copper Deficiency).] - Rossel'khozizdat, Moskva 1973. [Ps, Chl; in R.]

18038 - **TARASOVA, T.N.** : Pervichnaya produktsiya, produktsiya bakterioplanktona i destruktsiya organicheskogo veshchestva v Gor'kovskom vodokhranilishche. [Primary and bacterioplankton production and destruction of organic substance in Gor'kovskoe reservoir.] - Gidrobiol. Zh. *9* (3) : 5 - 11, 1973. [In R.]

*18039 - TARCHEVSKIĬ, I.A., CHIKOV, V.I., IVANOVA, A.P., SULEĬMANOVA, A.M. : Fotosintez yarovoĭ pshenitsy s geneticheski obuslovlennoĭ razlichnoĭ dlinoĭ steblya. [Photosynthesis in spring wheat with genetically caused difference of stem length.] - In : Geneticheskie Aspekty Fotosinteza. Tezisy Dokladov. Pp. 115 - 116. Donish, Dushanbe 1972. [In R.]

18040 - TARCHEVSKIĬ, I.A., IVANOVA, A.P., BIKTEMIROV, U.A. : K voprosu o peredvizhenii assimilyatov u pshenitsy i vliyanii mineral'nogo pitaniya na étot protsess. [Transport of photosynthates in wheat and the effect of mineral nutrition on this process.] - Tr. biol.-pochv. Inst., dal'nevost. nauch. Tsentr Akad. Nauk SSSR 20 (123) : 174 - 178, 1973. [In R, ab : E.]

18041 - TARCHEVSKY, I.A., BEZUGLOV, V.K., ZABOTIN, A.I., PETROV, V.E., CHERNOV, I.A.: The contribution of various processes to the energetic balance of photosynthesizing cells and isolated chloroplasts. - In : Ninth International Congress of Biochemistry. Stockholm, 1 - 7 July, 1973. Abstract Book. P. 239. IUB, Stockholm 1973.

18042 - TATENO, K., OJIMA, M. : [Growth analysis of grain sorghum as affected by planting density and amount of nitrogen.] - Proc. Crop Sci. Soc. Jap. 42 : 555 - 559, 1973. [In Jap., ab : E.]

18043 - TATSUMI, M. : [Studies on the photosynthesis of vegetable crops. III Diurnal changes in the photosynthesis of tomato and cucumber seedlings.] - Bull. hort. Res. Sta., Ser. A (Hiratsuka) 12 : 101 - 112, 1973. [In Jap., ab : E.]

18044 - TEARE, I.D., KANEMASU, E.T., POWERS, W.L., JACOBS, H.S. : Water-use efficiency and its relation to crop canopy area, stomatal regulation, and root distribution. - Agron. J. 65 : 207 - 211, 1973.

18045 - TEARE, I.D., RAO, M.R.M., KANEMASU, E.T. : Correlation of transpiration rates by cobalt chloride method and stomatal-diffusion porometer. - Indian J. agr. Sci. 43 : 639 - 642, 1973.

18046 - TEERI, J.A. : Polar desert adaptations of a high arctic plant species. - Science 179 : 496 - 497, 1973. [Ps.]

18047 - TEL-OR, E., FUCHS, S., AVRON, M. : Antibodies to plant ferredoxin. - FEBS Lett. 29 : 156 - 158, 1973.

*18048 - TEMPER, É.E. : Raznoobrazie pigmentnykh mutantov. [The variety of pigment mutants.] - In : Geneticheskie Aspekty Fotosinteza. Tezisy Dokladov. Pp. 69 - 70. Donish, Dushanbe 1972. [In R.]

18049 - TERENT'EV, V.M., FEDYUN'KIN, D.V., GOLOVIN, V.N., GOLOVNEVA, N.B., KOSHELEVA, L.L., PRIKUPETS, L.B. : O fotobiologicheskoĭ reaktsii molodykh rasteniĭ na obluchenie uzkopolosnym svetom. [Photobiological reaction of young plants to the irradiation with narrow-band light.] - In : Upravlenie Skorost'yu i Napravlennost'yu Biosinteza u Rasteniĭ. Pp. 172 - 173. Krasnoyarsk 1973. [Pigments; In R.]

18050 - TERRY, N., ULRICH, A. : Effects of phosphorus deficiency on the photosynthesis and respiration of leaves of sugar beet. - Plant Physiol. 51 : 43 - 47, 1973.

18051 - TERRY, N., ULRICH, A. : Effects of potassium deficiency on the photosynthesis and respiration of leaves of sugar beet. - Plant Physiol. 51 : 783 - 786, 1973.

18052 - TERRY, N., ULRICH, A. : Effects of potassium deficiency on the photosynthesis and respiration of leaves of sugar beet under conditions of low sodium supply. - Plant Physiol. 51 : 1099 - 1101, 1973.

18053 - TERSKOV, I.A., SID'KO, F.Ya., BELYANIN, V.N., BERESNEV, G.F., ALYPOV, V.F. : Fotosinteticheskaya aktivnost' kletok khlorelly pri stupenchatykh izmeneniyakh svetovogo rezhima. [Photosynthetic activity of Chlorella cells during stepped changes of light regime.] - Fiziol. Rast. 20 : 999 - 1006, 1973. [In R, ab : E.]

18054 - THEIβ-SEUBERLING, H.-B. : Einfluβ von Röntgenstrahlen und Hemmstoffen der Proteinsynthese auf die Synthese von Chlorophyll und NADP-abhängiger Glycerinaldehyd-3-Phosphat-Dehydrogenase in ergrünender Euglena gracilis. - Arch. Mikrobiol. 92 : 331 - 344, 1973.

18055 - **THIBAULT, P.** : Photojet de CO_2 et réserves internes en gaz carbonique chez *Zea mays*. - Planta *114* : 109 - 118, 1973.

18056 - **THIESS, D.E., LICHTENTHALER, H.K.** : Hemmung der Samenkeimung durch niedere einwertige Alkohole. - Naturwissenschaften *60* : 302, 1973. [Chl.]

18057 - **THINH, L. VAN, GRIFFITHS, D.J.** : Nucleic acid synthesis accompanying the recovery of cell division and chloroplast development in "giant" cells of the Emerson strain of *Chlorella*. - Plant Cell Physiol. *14* : 497 - 504, 1973. [Chl.]

18058 - **THINH, L. VAN, GRIFFITHS, D.J.** : The role of photosynthesis and respiration in the light-induced recovery of "giant" cells of the Emerson strain of *Chlorella*. - Plant Cell Physiol. *14* : 1177 - 1185, 1973.

18059 - **THOMAS, J.** : Continuous culture of filamentous blue-green algae. - In : CARR, N.G., WHITTON, B.A. (ed.) : The Biology of Blue-Green Algae. Pp. 531 - 535. Blackwell sci. Publ., Oxford - London - Edinburgh - Melbourne 1973.

18060 - **THOMAS, J., NAGARAJA, R.** : Absorption spectra of algal & chloroplast suspensions : A simple method using conventional spectrophotometers. - Indian J. Biochem. Biophys. *10* : 266 - 268, 1973. [Chl.]

18061 - **THOMAS, J.B.** : Physico-chemistry of photopigments. - In : CHECCUCCI, A., WEALE, R.A. (ed.) : Primary Molecular Events in Photobiology. Pp. 79 - 97. Elsevier, Amsterdam - London - New York 1973.

18062 - **THOMAS, J.B.** : Light absorption, energy transfer, and photosynthetic units. - In : CHECCUCCI, A., WEALE, R.A. (ed.) : Primary Molecular Events in Photobiology. Pp. 99 - 124. Elsevier, Amsterdam - London - New York 1973.

18063 - **THOMAS, J.P., O'KELLEY, J.C.** : The photoreversible nature of a pigment system in the green alga *Protosiphon botryoides* KLEBS. - Photochem. Photobiol. *17* : 469 - 472, 1973.

18064 - **THOMPSON, P.J.** : Retardation of chloroplast senescence by light in the apical portion of the wheat leaf. - Diss. Abstr. Int. B *33* : 3515-B, 1973.

18065 - **THOMSON, W.W., PLATT, K.** : Plastid ultrastructure in the barrel cactus, *Echinocactus acanthodes*. - New Phytol. *72* : 791 - 797, 1973.

18066 - **THÓRDARDÓTTIR, T.** : Successive measurements of primary production and composition of phytoplankton at two stations west of Iceland. - Norw. J. Bot. *20* : 257 - 270, 1973.

18067 - **THORHAUG, A.L., BACH, S.D.** : Productivity of red and green macro-algae in a South Florida estuary before and after the opening of a thermal effluent canal. - J. Phycol. *9* (Suppl.) : 10, 1973.

18068 - **THORNBER, J.P., HIGHKIN, H.R.** : The relationship between isolated chlorophyll-protein complexes and plant photosystems. - Plant Physiol. *51* (Suppl.) : 66, 1973.

18069 - **THROM, G.** : Untersuchungen zur Beziehung zwischen der lichtabhängigen und der redoxabhängigen Änderung des Membranpotentials bei *Griffithsia setacea*. - Planta *112* : 273 - 284, 1973. [Ps.]

*18070 - **TIEN, H.T.** : Photoelectric effects in thin and bilayer lipid membranes in aqueous media. - J. phys. Chem. *72* : 4512 - 4519, 1968. [Chloroplast pigments.]

18071 - **TIEN, H.T., HUEBNER, J.S.** : An analysis of flash-induced electrical transients of a BLM containing chloroplast lamella extracts. - J. Membrane Biol. *11* : 57 - 74, 1973.

18072 - **TIESZEN, L.L.** : Photosynthesis and respiration in arctic tundra grasses : Field light intensity and temperature responses. - Arct. alp. Res. *5* : 239 - 251, 1973.

18073 - **TIESZEN, L.L., SIGURDSON, D.C.** : Effect of temperature on carboxylase activity and stability in some Calvin cycle grasses from the arctic. - Arct. alp. Res. *5* : 59 - 66, 1973.

18074 - **TILLY, L.J.** : Comparative productivity of four Carolina lakes. - Amer. Midland Naturalist *90* : 356 - 365, 1973. [Ps, Chl.]

*18075 - **TILNEY-BASSETT, R.A.E.** : Genetics and plastid physiology in *Pelargonium*. III. Effect of cultivar and plastids on fertilisation and embryo survival. - Heredity 25 : 89 - 103, 1970.

18076 - **TILZER, M.** : B. Dynamik der planktischen Urproduktion unter den Extrembedingungen des Hochgebirgssees. - In : **ELLENBERG, H.** (ed.) : Ökosystemforschung. Pp. 51 - 59. Springer-Verlag, Berlin - Heidelberg - New York 1973.

18077 - **TILZER, M.M.** : Diurnal periodicity in the phytoplankton assemblage of a high mountain lake. - Limnol. Oceanogr. 18 : 15 - 30, 1973.

18078 - **TING, I.P., OSMOND, C.B.** : Activation of plant P-enolpyruvate carboxylases by glucose-6-phosphate : a particular role in Crassulacean acid metabolism. - Plant Sci. Lett. 1 : 123 - 128, 1973.

18079 - **TING, I.P., OSMOND, C.B.** : Photosynthetic phosphoenolpyruvate carboxylases. Characteristics of alloenzymes from leaves of C_3 and C_4 plants. - Plant Physiol. 51 : 439 - 447, 1973.

18080 - **TING, I.P., OSMOND, C.B.** : Multiple forms of plant phosphoenolpyruvate carboxylase associated with different metabolic pathways. - Plant Physiol. 51 : 448 - 453, 1973. [Chl.]

18081 - **TISHCHENKO, N.N., MIROSLAVOVA, S.A.** : Osobennosti azotnogo metabolizma v list'-yakh C-3 i C-4 rastenii pri razlichnom rezhime azotnogo pitaniya. [Peculiarities of nitrogen metabolism in leaves of C3 and C4 plants under different nitrogen supply.] - In : Upravlenie Skorost'yu i Napravlennost'yu Biosinteza u Rastenii. Pp. 63 - 64. Krasnoyarsk 1973. [Ps; in R.]

18082 - **TISHCHENKO, N.N., MIROSLAVOVA, S.A., VASIL'EV, B.R.** : Sintez azotistykh soedinenii na svetu v list'yakh rastenii s razlichnym tipom fiksatsii CO_2. [Synthesis of nitrogenous compounds in light in plant leaves with different type of carbon dioxide fixation.] - Vestn. leningrad. Univ., Biol. 1973 [9 (2)] : 103 - 111, 1973. [Ps, Chl; in R, ab : E.]

18083 - **TITLYANOV, E.A., LI, B.D.** : Sravnenie dvukh vidov morskikh zelenykh vodoroslei po ikh sposobnosti adaptirovat'sya k intensivnosti osveshcheniya. [Comparison of two species of marine green algae according to their adaptability to illuminance.] - In : Informatsionnye Materialy Akad. Nauk SSSR, Sibir. Otd., Sibir. Inst. Fiziol. Biokhim. Rast., Irkutsk 11 : 40 - 42, 1973. [In R.]

18084 - **TITLYANOV, E.A., MAGOMEDOV, I.M.** : Lokalizatsiya obmena organicheskikh kislot v zelenoi rastitel'noi kletke. [Localization of organic acid metabolism in a green plant cell.] - Tr. biol.-pochv. Inst., nov. Ser. 13 (Biokhimicheskie Issledovaniya na Sovetskom Dal'nem Vostoke) : 68 - 96, 1973. [Ps; in R, ab : E.]

18085 - **TITLYANOV, E.A., PESHEKHOD'KO, V.M.** : O transporte assimilyatov v tallomakh morskikh prikreplennykh vodoroslei. [Transport of photosynthates in thalli of fringing seaweeds.] - Tr. biol.-pochv. Inst., dal'nevost. nauch. Tsentr Akad. Nauk SSSR 20 (123) : 137 - 141, 1973. [In R, ab : E.]

*18086 - **TIŢU, H., BREZEANU, A., HURGHIŞIU, I.** : The influence of ^{60}Co *gamma* rays on the ultrastructure of cellular organelles and on auxin content in *Zerna inermis* (LEYS) LINDM. plants. - Rev. roum. Biol., Sér. Bot. 16 : 397 - 404, 1971. [Chloroplast.]

18087 - **TIŢU, H., HURGHIŞIU, I., BREZEANU, A.** : Influenţa razelor X asupra ultrastructurii cloroplastelor şi conţinutului lor în aminoacizi liberi la plantele de spanac (*Spinacia oleracea* L.). [Effect of X-rays on chloroplast ultrastructure and content of free amino acids in spinach plants.] - Stud. Cercet. Biol., Ser. bot. 25 : 107 - 112, 1973. [In Roum., ab : E.]

18088 - **TKACHEVA, Z.G., GOGOTOV, I.N.** : Gidrogenaznaya aktivnost' *Chloropseudomonas ethylica* shtamm 51. [Hydrogenase activity of *Chloropseudomonas ethylica*, strain 51.] - Izv. Akad. Nauk SSSR, Ser. biol. 1973 : 763 - 766, 1973. [Ps; in R, ab : E.]

18089 - **TODOROVA-TRIPHONOVA, A.D., KOLEVA-MARUDOVA, A.P.** : On the influence of pH of the medium in *Chlorella vulgaris* BEYER. - Natura (Plovdiv) 6 (1.) : 165 - 170, 1973. [Chl, Car.]

18090 - **TOLBERT, N.E.** : Activation of polyphenol oxidase of chloroplasts. - Plant Physiol. *51* : 234 - 244, 1973.

*18091 - **TOMBESI, L., ROMANO, E., LAUCIANI, E.** : Ricerche sulla evapotraspirazione e sulla fotosintezi compiute negli anni 1967 e 1968. Research on evapotraspiration and photosynthesis carried out in the years 1967 and 1968. - Ist. sper. Nutriz. Piante (Roma) *1969* : 2 - 24, 1969. [In Ital., E.]

18092 - **TOOMING, Kh.G., KALLIS, A.G.** : Znachenie i nekotorye rezul'taty issledovaniya KPD rastenii i rastitel'nogo pokrova. [Significance and some results of the study of efficiency coefficient of plants and stand.] - In : Problemy Biogeotsenologii. Pp. 203 - 213. Nauka, Moskva 1973. [In R.]

18093 - **TOTTERDELL, C.J., RAINS, A.B.** : Plant reflectance and colour infrared photography. - J. appl. Ecol. *10* : 401 - 407, 1973.

18094 - **TOVEY, D.A., GLASZIOU, K.T., FARQUHAR, R.H., BULL, T.A.** : Variability in radiation received by small plots of sugarcane due to differences in canopy heights. - Crop Sci. *13* : 240 - 242, 1973.

18095 **TOWLE, D.W., PEARSE, J.S.** : Production of the giant kelp, *Macrocystis*, estimated by *in situ* incorporation of ^{14}C in polyethylene bags. - Limnol. Oceanogr. *18* : 155 - 159, 1973.

18096 - **TRACZYK, T., TRACZYK, H., MOSZYŃSKA, B.** : Herb layer production of two pinewood communities in the Kampinos National Park. - Ekol. pol. *21* (3) : 37 - - 55, 1973.

18097 - **TRANTER, D.J.** : Seasonal studies of a pelagic ecosystem (meridian 110° E). - In : ZEITSCHEL, B. (ed.) : The Biology of the Indian Ocean. Pp. 487 - 520. Springer-Verlag, Berlin - Heidelberg - New York 1973. [Primary production, Chl.]

18098 - **TRAVERS, A., TRAVERS, M.** : Données sur quelques facteurs de l'écologie du plancton dans la région de Marseille. 3 - La lumière. - Tethys *5* : 7 - 30, 1973.

18099 - **TREBST, A., REIMER, S.** : Energy conservation in photoreductions by photosystem II. Reversal of dibromothymoquinone inhibition of Hill-reactions by phenylenediamines. - Z. Naturforsch. *28* C : 710 - 716, 1973.

18100 - **TREBST, A., REIMER, S.** : Properties of photoreductions by photosystem II in isolated chloroplasts. An energy-conserving step in the photoreduction of benzoquinones by photosystem II in the presence of dibromothymoquinone. - Biochim. biophys. Acta *305* : 129 - 139, 1973.

18101 - **TREBST, A., REIMER, S.** : Properties of photoreductions by Photosystem II in isolated chloroplasts. III. The effect of uncouplers of phenylenediamine shuttles across the membrane in the presence of dibromothymoquinone. - Biochim. biophys. Acta *325* : 546 - 557, 1973.

18102 - **TREFFRY, T.** : Chloroplast development in etiolated peas : Reformation of prolamellar bodies in red light without accumulation of protochlorophyllide. - J. exp. Bot. *24* : 185 - 195, 1973.

18103 - **TRÉMOLIÈRES, A., JACQUES, R., MAZLIAK, P.** : Régulation par la lumière de l'accumulation de l'acide linolénique dans la jeune feuille de Pois. Spectre d'action, influence de l'intensité de l'éclairement et rôle du phytochrome. - Physiol. vég. *11* : 239 - 251, 1973. [Chl.]

18104 - **TRENCH, R.K.** : Further studies on the mucopolysaccharide secreted by the pedal gland of the marine slug *Tridachia crispata (Opisthobranchia, Sacoglossa)*. - Bull. mar. Sci. *23* : 299 - 312, 1973. [Symbiotic chloroplasts.]

18105 - **TRENCH, R.K., BOYLE, J.E., SMITH, D.C.** : The association between chloroplasts of *Codium fragile* and the mollusc *Elysia viridis*. I. Characteristics of isolated *Codium* chloroplasts. - Proc. roy. Soc. London *184* : 51 - 61, 1973.

18106 - **TRENCH, R.K., BOYLE, J.E., SMITH, D.C.** : The association between chloroplasts of *Codium fragile* and the mollusc *Elysia viridis*. II. Chloroplast ultrastructure and photosynthetic carbon fixation in *E. viridis*. - Proc. roy. Soc. London *184* : 63 - 81, 1973.

18107 - **TRIFONOVA, I.S.** : Sootnoshenie intensivnosti fotosinteza i biomassy fitoplanktona v ozere Krasnom. [The ratio of photosynthetic rate and biomass of phytoplankton in the Red lake.] - In : Limnologiya Severo-Zapada SSSR. Pp. 137 -
- 140. Tallin 1973. [In R.]

18108 - **TRLICA, M.J., DYE, A.J., MOIR, W.H., BROWN, L.F., JAMESON, D.A., RICE, W.A.** :
A field laboratory for gas exchange measurements of grassland swards. - Photosynthetica 7 : 257 - 261, 1973.

18109 - **TROSPER, T., ALLEN, C.F.** : Carotenoid composition of spinach chloroplast grana and stroma lamellae. - Plant Physiol. 51 : 584 - 585, 1973.

18110 - **TROXLER, R.F., DOKOS, J.M.** : Formation of carbon monoxide and bile pigment in
red and blue-green algae. - Plant Physiol. 51 : 72 - 75, 1973.

18111 - **TROXLER, R.F., FARIS, B., FRANZBLAU, C.** : Effects of azetidine-2-carboxylic
acid on chlorophyll and phycocyanin synthesis in the alga, *Cyanidium caldarium*.
- Plant Physiol. 51 (Suppl.) : 28, 1973.

18112 - **TRUSCOTT, T.G., LAND, E.J., SYKES, A.** : The *in vitro* photochemistry of biological molecules - III. Absorption spectra, lifetimes and rates of oxygen quenching of the triplet states of β-carotene, retinal and related polyenes. - Photochem. Photobiol. 17 : 43 - 51, 1973.

18113 - **TSAREGORODTSEVA, S.O., NOVITSKAYA, Yu.E.** : O sostoyanii pigmentov v pochkakh
khvoĭnykh rasteniĭ v zimne-vesenniĭ period. [State of pigments in the buds of
coniferous trees during the winter-spring period.] - Fiziol. Rast. 20 : 1052-
- 1056, 1973. [In R, ab : E.]

18114 - **TSAREGORODTSEVA, S.O., NOVITSKAYA, Yu.E.** : Vliyanie usloviĭ osveshcheniya na
soderzhanie i sostoyanie pigmentov v pochkakh eli. [Effect of illumination
conditions on the content and state of pigments in spruce buds.] - Izv. vyssh.
ucheb. Zav., lesnoĭ Zh. 1973 (4) : 31 - 34, 1973. [In R.]

18115 - **TSCHAKALOVA, E.S., HOFFMANN, P.E.** : Das Interzellularvolumen in den oberirdis-
chen Organen der Weizenkeimpflanze in Verlauf der Entwiclung. - Dokl. bolg.
Akad. Nauk 27 : 533 - 535, 1973.

18116 - **TSEĬTLIN, V.B.** : Ob opredelenii velichiny pervichnoĭ produktsii po vertikal'-
nomu vynosu biogennykh elementov. [Determination of primary production values
from the vertical transport of nutrient salts.] - Okeanologiya 13 : 867 - 871,
1973. [In R, ab : E.]

18117 - **TSEL'NIKER, Yu.L.** : Ritmy rosta tkaneĭ, khloroplastov i determinatsiya priznakov svetovoĭ i tenevoĭ struktury lista u klena ostrolistnogo. [Growth rhythms
of tissues and chloroplasts and determination of characteristics of light and
dark leaf structure in Norway maple.] - Fiziol. Rast. 20 : 1182 - 1190, 1973.
[In R, ab : E.]

18118 - **TSIMILLI-MICHAEL, M., ISAAKIDOU, J., PAPAGEORGIOU, G.** : A study of the heterogeneity of chlorophyll *a in vivo* by means of exogenous quenchers of its e-
lectronic excitation. - Biochem. Biophys. News Lett. 1973 (4) : 2 - 3, 1973.

18119 - **TSOGLIN, L.N., SEMENENKO, V.E., KAYUSHIN, L.P., KUTYSHENKO, V.P., LAZAREVA, A.
V., OKON, M.S., SIBEL'DINA, L.A., CHEKULAEVA, L.N.** : Osobennosti rosta *Chlo-
rella* sp. K v D_2O. [Peculiarities of growth of *Chlorella* sp. in D_2O.] - Fiziol.
Rast. 20 : 1204 - 1208, 1973. [Ps; in R, ab : E.]

18120 - **TSUJIMOTO, H.Y., CHAIN, R.K., ARNON, D.I.** : Photoreduction of ferredoxin-NADP
in the presence and absence of ferredoxin-reducing substance (FRS). - Biochem.
biophys. Res. Commun. 51 : 917 - 923, 1973.

18121 - **TUBOI, S., HAYASAKA, S.** : Interconversion between the active and inactive forms
of δ-aminolevulinate synthetase in *Rhodopseudomonas spheroides*. - Enzyme 16 :
86 - 93, 1973.

*18122 - **TUGARINOV, V.V.** : Izuchenie kharaktera mutabil'nosti svetochuvstvitel'nykh
shtammov *Chlamydomonas reinhardi*. [Mutability in light-sensitive strains of
Chlamydomonas reinhardi.] - In : Geneticheskie Aspekty Fotosinteza. Tezisy Dokladov. Pp. 70 - 71. Donish, Dushanbe 1972. [In R.]

18123 - **TUKHLIEV, D.T.** : Vliyanie vodoobespechennosti rasteniĭ na soderzhanie khloro-

filla i soedinenii azota v list'yakh khlopchatnika. [Effect of water supply
on the content of chlorophyll and nitrogen compounds in cotton leaves,] -
Nauch. Tr. tashkent. sel',-khoz. Inst. *37* (Voprosy Fiziologii, Biokhimii Khlopchatnika i Drugikh Sel'skokhozyaistvennykh Kul'tur) : 50 - 55, 1973. [In R.]

18124 - TUMIDAJOWICZ, D. : The dynamics of biomass and primary production of herb layer plants in the deciduous *Tilio-Carpinetum* association of the Niepolomice
forest. - Bull. Acad. pol. Sci., Sér. Sci. biol. *21* : 101 - 107, 1973.

18125 - TUNG, H.F., BROUGHTON, W.J., LENZ, F. : Effects of fruit on ribulosediphosphate carboxylase activity in *Citrus madurensis* leaves. - Experientia *29* : 271 -
- 272, 1973.

18126 - TUPYK, N.D. : Vyvchennya vmistu vitaminiv grupy *B* kul'turi *Microcystis aeruginosa* KUETZ. emend. ELENK. zalezhno vid umov zhyvlennya, [Content of group *B*
vitamins in the *Microcystis aeruginosa* KUETZ. emend. ELENK. culture in dependence of nutrition.] - Ukr. bot. Zh. *30* : 324 - 328, 402, 1973. [Relation to
Ps; in Ukr., ab : E, R.]

18127 - TURGEON, R., WEBB, J.A. : Leaf development and phloem transport in *Cucurbita
pepo* : transition from import to export, - Planta *113* : 179 - 191, 1973,

18128 - TURNER, N.C. : Action of fusicoccin on the potassium balance of guard cells
of *Phaseolus vulgaris*, - Amer. J. Bot. *60* : 717 - 725, 1973.

18129 - TURNER, N.C. : Illumination and stomatal resistance to transpiration in three
field crops. - In : SLATYER, R.O. (ed.) : Plant Response to Climatic Factors.
Pp. 63 - 68. Unesco, Paris 1973.

18130 - TURNER, N.C., BEGG, J.E. : Stomatal behavior and water status of maize, sorghum, and tobacco under field conditions. I. At high soil water potential. -
Plant Physiol. *51* : 31 - 36, 1973. [Stomatal resistance.]

18131 - TYLER, G., GULLSTRAND, C., HOLMQUIST, K.-A., KJELLSTRAND, A.-M. : Primary production and distribution of organic matter and metal elements in two heath
ecosystems. - J. Ecol. *61* : 251 - 268, 1973.

18132 - TYLER, J.E. : Lux *vs.* quanta. - Limnol. Oceanogr. *18* : 810, 1973. [Chl,]

18133 - UDOVENKO, G.V. : Obrazovanie, ottok i mobilizatsiya assimilyatov v rasteniyakh
pri zasolenii. [Formation, utilization and outflow of photosynthates in plants
under salinization.] - Tr. biol.-pochv. Inst., dal'nevost. nauch. Tsentr Akad.
Nauk SSSR *20* (123) : 195 - 199, 1973. [In R, ab : E.]

18134 - UFFEN, R.L. : Growth properties of *Rhodospirillum rubrum* mutants and fermenta-
tion of pyruvate in anaerobic, dark conditions. - J. Bacteriol. *116* : 874 -
- 884, 1973. [Ps.]

18135 - UFFEN, R.L. : Effect of low-intensity light on growth response and bacteriochlorophyll concentration in *Rhodospirillum rubrum* mutant C. - J. Bacteriol.
116 : 1086 - 1088, 1973.

*18136 - UKAI, Y. : [Studies on varietal differences in radiosensitivity in rice. II.
The radiosensitivities with respect to the reduction in seedling height, co-
leoptile length, pollen fertility and seed fertility, and to the frequency of
chlorophyll mutation.] - Jap. J. Breed. *18* : 127 - 138, 1968. [In Jap., ab :
E.]

18137 - ULLRICH-EBERIUS, C.I. : Beziehungen der Aufnahme von Nitrat, Nitrit und Phos-
phat zur photosynthetischen Reduktion von Nitrat und Nitrit und zum ATP-Spie-
gel bei *Ankistrodesmus braunii*. - Planta *115* : 25 - 36, 1973.

18138 - ULTSCH, G.R. : The effects of water hyacinths (*Eichhornia crassipes*) on the
microenvironment of aquatic communities. - Arch. Hydrobiol. *72* : 460 - 473,
1973.

18139 - ULTSCH, G.R., ANTHONY, D.S. : The role of the aquatic exchange of carbon dioxide in the ecology of the water hyacinth (*Eichhornia crassipes*). - Florida
Scientist *36* : 16 - 22, 1973.

18140 - UMAROV, Kh., ZAKIROV, A., ASADOV, N., GAFUROV, B. : Vliyanie ponizhennoi temperatury pochvy na soderzhanie khlorofilla v list'yakh khlopchatnika. [Effect

of low soil temperature on chlorophyll content in cotton leaves.] - Uz. biol.
Zh. *1973* (6) : 23 - 24, 1973. [In R.]

18141 - **UMOESSIEN, S.N.** : Effect of gibberellic acid on the distribution of products
of photosynthesis in sunflower. - Diss. Abstr. Int. B *33* : 4695-B, 1973.

18142 - **UMRIKHINA, A.V., BUBLICHENKO, N.V., KRASNOVSKIĬ, A.A.** : Svetoindutsirovannye
signaly EPR pri fotovosstanovlenii i fotookislenii khlorofilla i ego analogov.
[Light-induced EPR signals in photochemical oxidoreduction of chlorophyll and
its analogues.] - Biofizika *18* : 565 - 568, 1973. [In R, ab : E.]

18143 - **UNSER, G., GRUBER, P., TOLBERT, N.E.** : Polyphenol oxidase in subchloroplast
fractions. - Plant Physiol. *51* (Suppl.) : 41, 1973.

18144 - **UPMEYER, D.J., KOLLER, H.R.** : Diurnal trends in net photosynthetic rate and
carbohydrate levels of soybean leaves. - Plant Physiol. *51* : 871 - 874, 1973.

18145 - **URIBE, E.G.** : ATP synthesis driven by a K^+-valinomycin-induced charge imbalan-
ce across chloroplast grana membranes. - FEBS Lett. *36* : 143 - 147, 1973.

18146 - **URIBE, E.G., LI, B.C.Y.** : Stimulation and inhibition of membrane-dependent ATP
syntnesis in chloroplasts by artificially induced K gradients. - J. Bioenerg.
4 : 435 - 444,

18147 - **URSINO, D.J.** : Effects of chronic internal β-radiation from photoassimilated
$^{14}CO_2$ on the retention and distribution of ^{14}C in young white pine plants. -
Plant Physiol. *51* : 954 - 959, 1973.

18148 - **URSINO, D.J.** : The translocation of ^{14}C-photoassimilate in single tree progeny of white spruce (*Picea glauca* (MOENCH) VOSS). - Can. J. Forest Res. *3* :
315 - 318, 1973.

18149 - **URSINO, D.J., MOSS, A., STIMAC, J.** : γ-radiation effects on the rates of photosynthesis in 1% and 21% oxygen and on dark respiration. - Plant Physiol. *51*
(Suppl.) : 42, 1973.

18150 - **URSINO, D.J., PAUL, J.** : The long-term fate and distribution of ^{14}C photoassimilated by young white pines in late summer. - Can. J. Bot. *51* : 683 - 687,
1973.

18151 - **USACHEVA, M.N., DAIN, B.Ya.** : O vzaimodeĭstvii vozbuzhdennykh molekul rastitel'nykh pigmentov s rastvoritelem. [Interaction of excited molecules of plant
pigments with a solvent.] - Biofizika *18* : 379 - 382, 1973. [In R, ab : E.]

18152 - **USIK, G.E.** : Formirovanie tsennykh khozyaĭstvenno-biologicheskikh priznakov u
tomatov pri vyrashchivanii rassady pod sinteticheskoĭ plenkoĭ. [Formation of
valuable commercial properties in tomatoes as a result of the cultivation of
seedlings under synthetic sheets.] - Sel'.-khoz. Biol. *8* : 783 - 785, 1973.
[Ps, Ch]; in R.]

*18153 - **USMANOV, P.D., BATALOV, R.B.** : Kolichestvennaya kharakteristika pigmentnogo
sostava mutantnykh form *Arabidopsis thaliana*. [Quantitative estimation of pigment composition in mutants of *Arabidopsis thaliana*.] - In : Geneticheskie
Aspekty Fotosinteza. Tezisy Dokladov. Pp. 74 - 75. Donish, Dushanbe 1972. [In
R.]

*18154 - **USMANOV, P.D., BATALOV, R.B., ABDULLAEV, Kh.A., SAKHIBNAZAROV, Sh.** : Ėksperi-
mental'nyĭ mutagenez i morfo-funktsional'nye modeli pigmentnykh mutatsiĭ *Arabidopsis thaliana*. [Experimental mutagenesis and morpho-functional models of
pigment mutations in *Arabidopsis thaliana*.] - In : Geneticheskie Aspekty Fotosinteza. Tezisy Dokladov. Pp. 73 - 74. Donish, Dushanbe 1972. [In R.]

18155 - **USMANOV, P.D., USMANOVA, O.V.** : On the genetic control of the chloroplast size
and number in mesophyll cells of species of *Arabidopsis*. - *Arabidopsis* Inform.
Serv. *10* : 20 - 21, 1973.

*18156 - **USMANOVA, O.V.** : Kharakter izmenchivosti assimilyatsionnoĭ poverkhnosti khloroplastov u mutantov, ėkologicheskikh ras i vidov roda *Arabidopsis*. [The nature of variations in assimilatory surface of chloroplasts in mutants, ecological races and species of *Arabidopsis*.] : In : Geneticheskie Aspekty Fotosinteza.
Tezisy Dokladov. Pp. 72 - 73. Donish, Dushanbe 1972. [In R.]

18157 - **USUDA, H., SAMEJIMA, M., MIYACHI, S.** : Distribution of radioactivity in carbon

atoms of malic acid formed during light-enhanced dark $^{14}CO_2$-fixation in maize
leaves. - Plant Cell Physiol. *14* : 423 - 426, 1973.

18158 - **UTURGAURI, A.I., DANELIYA, B.K., MKERVALI, V.G., KONTRIDZE, A.N.** : Vliyanie
uvedennykh veshchestv na protsessy fotosinteza limonnogo rasteniya. [Effect
of administered substances on photosynthesis in citrus plants.] - Subtrop.
Kul't. *1973* (1) : 65 - 68, 1973. [In R.]

18159 - **VAALA, A.R., MADJID, A.H., TORRADO, M.T.** : On the growing of large single
crystals of the biological carotenoid pigment, β-carotene. - J. Crystal Growth
18 : 39 - 44, 1973.

18160 - **VAARAMA, A., VALANNE, T.** : On the taxonomy, biology and origin of *Betula tor-
tuosa* LEDEB. - Ann. Univ. turkuensis, Ser. A II *53* : 70 - 84, 1973. Rep. Kevo
subarctic Res. Sta. *10* : 70 - 84, 1973. [Chl, Car.]

18161 - **VACCHI, C., PICCARI, G.G., PUPILLO, P.** : Characterization of NADP-linked gly-
ceraldehyde-3-phosphate dehydrogenase of *Euglena Gracilis*. - Z. Pflanzenphysiol. *69* : 351 - 358, 1973.

18162 - **VACEK, K., VAVŘINEC, E., KALOUSEK, I.** : Fluorescence of chlorophyll *a* excited
by a He-Ne laser. - Photochem. Photobiol. *17* : 63 - 64, 1973.

18163 - **VÁCLAVÍK, J.** : Effect of different leaf age on the relationship between the
CO_2 uptake and water vapour efflux in tobacco plants. - Biol. Plant. *15* :
233 - 236, 1973.

18164 - **VAĬNSHTEĬN, E.A.** : Nekotorye voprosy fiziologii lishaĭnikov. II. Fotos.intez.
[Some problems of lichen physiology. II. Photosynthesis.] - Bot. Zh. *58* :
454 - 464, 1973. [In R.]

18165 - **VAĬNSHTEĬN, M.B., DEVYATKIN, V.G., MITROPOL'SKAYA, I.V.** : Fotosinteticheskaya
aktivnost' fitoplanktona Ivan'kovskogo vodokhranilishcha v zone vliyaniya po-
dogretykh vod Konakovskoĭ GRĖS. [Photosynthetic phytoplankton activity in the
Ivanovskoe reservoir, in the zone of heated water from the Konakovskaya state
regional electric power station influence.] - Gidrobiol. Zh. *9* (6) : 22 - 29,
1973. [In R, ab : E.]

18166 - **VAKLINOVA, S.G., CHOBANOVA, Y., MOSKOVA, D.D.** : Influence of certain inhibitors on the activity of the glycolate oxidase in leaves from maize and pea
plants. - Dokl. bolg. Akad. Nauk *26* : 1533 - 1535, 1973.

18167 - **VAKLINOVA, S.G., FEDINA, I.S.** : Oxygen uptake by chromatophores and cells of
mutants of *Scenedesmus obliquus*. - Dokl. bolg. Akad. Nauk *26* : 281 - 284, 1973.

18168 - **VAKLINOVA, S.G., MOSKOVA, D.D.** : Influence of the nitrate and ammonium nitrogen on the activity of the glycolate oxidase in certain higher plants. - Dokl.
bolg. Akad. Nauk *26* : 1529 - 1532, 1973.

18169 - **VALLEJOS, R.H.** : Uncoupling of photosynthetic phosphorylation by benzophenan-
thridine alkaloids. - Biochim. biophys. Acta *292* : 193 - 196, 1973.

18170 - **VANDEN DRIESSCHE, T.** : The chloroplasts of *Acetabularia*. The control of their
multiplication and activities. - Sub-cell. Biochem. *2* : 33 - 67, 1973.

18171 - **VANDEN DRIESSCHE, T., HARS, R.** : Ultrastructure of the chloroplasts of *Acetabularia mediterranea* and rate of physiological activities. - Arch. Biol. *84* :
539 - 551, 1973.

18172 - **VANDEN DRIESSCHE, T., HELLIN, J., HARS, R.** : Limitations in chloroplast multiplication in *Acetabularia mediterranea*. - Protoplasma *76* : 465 - 472, 1973.

18173 - **VANDER MEULEN, D.L., GOVINDJEE** : Is there a triplet state in photosynthesis ?
- J. sci. ind. Res. *32* : 62 - 69, 1973.

18174 - **VAN HEUVELEN, A.** : Electron transfer mechanisms in electron transfer chains.
- J. biol. Phys. *1* : 215 - 243, 1973.

18175 - **VANSÉVEREN, J.P.** : Variations saisonnières de la teneur en chlorophylles des
feuilles de la strate arborescente. - Bull. Soc. roy. Bot. Belg. *106* : 279 -
- 288, 1973.

18176 - **VANSÉVEREN, J.P.** : Evolution saisonnière de la masse foliaire, de la quantité de chlorophylles et de l'index foliaire. - Bull. Soc. roy. Bot. Belg. *106* : 289 - 303, 1973.

18177 - **VANSÉVEREN-VAN ESPEN, N.** : Effets du saccharose sur le contenu en chlorophylles de protocormes de *Cymbidium* SW. *(Orchidaceae)* cultivés *in vitro*. - Bull. Soc. roy. bot. Belg. *106* : 107 - 115, 1973.

18178 - **VARTAPETIAN, B.B.** : Role of endogenous metabolic water in plants under conditions of water deficit. - In : SLATYER, R.O. (ed.) : Plant Response to Climatic Factors. Pp. 227 - 231. Unesco, Paris 1973. [Ps.]

18179 - **VASIL'EVA, V.E., PINEVICH, V.V., LEVITIN, M.G.** : K voprosu o nalichii epoksiksantofillov u sinezelenykh vodorosley. [Presence of epoxy-xanthophylls in blue-green algae.] - Vestn. leningrad. Univ. *1973* (21) : 145 - 146, 1973. [In R, ab : E.]

18180 - **VATER, J.** : The action of indophenols and nitrophenols on the deactivation reactions in the water-splitting system of photosynthesis. - Biochim. biophys. Acta *292* : 786 - 795, 1973.

18181 - **VATER, J.** : On the nature of the deactivator of the water-splitting system in photosynthesis. - Biochim. biophys. Acta *325* : 149 - 156, 1973.

18182 - **VÉBER, K., ONDOK, P.** : Gross assimilation and respiration rate of cucumbers cultivated under outdoor and greenhouse conditions. - Annu. Rep. algol. Lab. Třeboň *1970* : 183 - 186, 1973.

B18183 - **VECHER, A.S.** (ed.) : Fotosintez i Ustoichivost' Rastenii. [Photosynthesis and Plant Resistance.] - Nauka i Tekhnika, Minsk 1973. [In R.]

18184 - **VECHER, A.S., KOVAL'CHUK, R.A., MAS'KO, A.A., RESHETNIKOV, V.N.** : Izmeneniya sostava khloroplastov sakharnoy svekly pri povyshenii ploidnosti. [Changes in chloroplast composition in sugar beet with increased ploidy.] - Vestsi Akad. Navuk belarus. SSR, Ser. biyal. Navuk *1973* (5) : 109 - 110, 1973. [Chl; in R.]

*18185 - **VECHER, A.S., RESHETNIKOV, V.N., BULKO, O.P., PREDKEL', K.I., KOVAL'CHUK, R. A., MAS'KO, A.A.** : Razlichiya kletochnykh yader i khloroplastov iz list'ev di- i tetraploidnoy rzhi. [Differences between cellular nuclei and chloroplasts from the leaves of di- and tetraploid rye.] - In : Geneticheskie Aspekty Fotosinteza. Tezisy Dok.adov. Pp. 56 - 57. Donish, Dushanbe 1972. [In R.]

18186 - **VECHER, A.S., RESHETNIKOV, V.N., KOVAL'CHUK, R.A., MAS'KO, A.A., BULKO, O.P., PREDKEL', K.I., MASNYI, M.N., KRYLOVA, M.Ya.** : O razlichiyakh kletochnykh yader i khloroplastov di- i tetraploidnoy rzhi. [Differences in cellular nuclei and chloroplasts of di- and tetraploid rye.] - Vestsi Akad. Navuk belarus. SSR, Ser. biyal. Navuk *1973* (2) : 105 - 108, 1973. [Chl; in R.]

18187 - **VEDERNIKOV, V.I., KONOVALOV, B.V., KOBLENTS-MISHKE, O.I.** : Rezul'taty primeneniya spektrofotometricheskogo metoda opredeleniya feofitina "A" v probakh morskoy vody. [Spectrophotometric determinations of phaeophytin "a" in the sea water samples.] - Tr. Inst. Okeanol. *95* (Formirovanie Biologicheskoy Produktivnosti i Donnykh Osadkov v Svyazi s Osobennostyami Tsirkulyatsii Vod v Yugo--vostochnoy Chasti Atlanticheskogo Okeana) : 138 - 146, 1973. [In R, ab : E.]

18188 - **VELDE, H.H. van der** : The natural occurrence in red algae of two phycoerythrins with different molecular weights and spectral properties. - Biochim. biophys. Acta *303* : 246 - 257, 1973.

18189 - **VELDE, H.H. van der** : The use of phycoerythrin absorption spectra in the classification of red algae. - Acta bot. neer. *22* : 92 - 99, 1973.

18190 - **VELTHUYS, B.R., AMESZ, J.** : The effect of dithionite on fluorescence and luminescence of chloroplasts. - Biochim. biophys. Acta *325* : 126 - 137, 1973.

18191 - **VENRICK, E.L., McGOWAN, J.A., MANTYLA, A.W.** : Deep maxima of photosynthetic chlorophyll in the Pacific Ocean. - Fish. Bull. *71* : 41 - 52, 1973.

18192 - **VERBEEK, L., LICHTENTHALER, H.K.** : Der Einfluß von Stickstoffmangel auf die Lipochinon- und Isoprenoid-Synthese der Chloroplasten von *Hordeum vulgare* L. - Z. Pflanzenphysiol. *70* : 245 - 258, 1973.

18193 - **VERKHOTUROV, V.N., TULBU, G.V.** : Indutsirovannoe svetom izmenenie rasseivayushchikh svoystv khloroplastov list'ev gorokha. [Photo-induced changes of light-

-scattering in isolated pea chloroplasts.] - Biofizika *18* : 1052 - 1057, 1973.
[In R, ab : E.]

18194 - **VERMEGLIO, A., MATHIS, P.** : Photooxidation of cytochrome b_{559} and the electron
donors in chloroplast Photosystem II. - Biochim. biophys. Acta *292* : 763 -
- 771, 1973.

18195 - **VERMEGLIO, A., MATHIS, P.** : Photoreduction of C-550 and oxidation of cytochrome b_{559} in chloroplasts : Dependence on the state of Photosystem II. - Biochim.
biophys. Acta *314* : 57 - 65, 1973.

18196 - **VERNON, L.P.** : Important events in photosynthesis. - Photochem. Photobiol. *18*:
529 - 531, 1973.

18197 - **VERNOTTE, C., BRIANTAIS, J.-M., BENNOUN, P.** : Effets d'un cation divalent
(Mg^{++}) sur les spectres d'action des deux photosystèmes dans les chloroplastes
isolés. - Compt. rend. Acad. Sci. Paris, Sér. D *277* :1695 - 1698, 1973.

18198 - **VERNOTTE, C., MOYA, I.** : Action de la température sur la durée de vie de fluorescence et le rendement de fluorescence de la *C*-Phycocyanine en solution. -
Photochem. Photobiol. *17* : 245 - 254, 1973.

18199 - **VERVELDE, G.J.** : Biologische produktie. [Biological production.] - Vakblad
Biol. *53* (4) : 60 - 64, 1973. [In Holl.]

18200 - **VESELOVA, T.V., BYUTNER, E.G., GRINENKO, V.V., VESELOVSKIĬ, V.V., FISENKO, V.
Yu.** : Fiziologicheskaya nesovmestimost' u yabloni pri privivke i sposoby ee
obnaruzheniya. [Physiological incompatibility in apple tree grafts and methods
of its detection.] - Sel'skokhoz. Biol. *8* : 459 - 461, 1973. [Chl; in R.]

18201 - **VESELOVA, T.V., VESELOVSKIĬ, V.A., GRINENKO, V.V., MARENKOV, V.S., POSPELOVA,
Yu.S., FISENKO, V.Yu.** : Dlitel'noe poslesvechenie list'ev zemlyaniki pri raznykh urovnyakh ovodnennosti. [Prolonged afterglow of strawberry leaves during
dehydration.] - Fiziol. Rast. *20* : 229 - 232, 1973. [In R, ab : E.]

18202 - **VESELOVA, T.V., VESELOVSKIĬ, V.A., GRINENKO, V.V., TARUSOV, B.N., POSPELOVA,
Yu.S., STETSENKO, I.I.** : Vliyanie obezvozhivaniya na dlitel'noe poslesvechenie list'ev vinograda. [Effect of dehydration on delayed light emission of
grape leaves.] - Fiziol. Rast. *20* : 47 - 53, 1973. [In R, ab : E.]

18203 - **VETTERMANN, W.** : Mechanism of the light-dependent accumulation of starch in
chloroplasts of *Acetabularia*, and its regulation. - Protoplasma *78* : 261 -
- 278, 1973.

*18204 - **VETTSHTEĬN, D.** : Strukturnye i regulyatornye geny biosinteza khlorofillov.
[Structural and regulatory genes for chlorophyll biosynthesis.] - In : Geneticheskie Aspekty Fotosinteza. Tezisy Dokladov. Pp. 21 - 22. Donish, Dushanbe
1972. [In R.]

18205 - **VIDAVER, W., SCHREIBER, U.** : Reversible changes of chlorophyll fluorescence
induction with desiccation in *Porphyra perforata*. - Plant Physiol. *51* (Suppl.):
67, 1973.

18206 - **VIDOVIČ, J., POKORNÝ, V.** : The effect of different sowing densities and nutrient levels on leaf area index, production and distribution of dry matter in
maize (*Zea mays* L.). - Biol. Plant. *15* : 374 - 382, 1973.

18207 - **VIEIRA DA SILVA, J.B.** : Influence de la sécheresse sur la photosynthèse et la
croissance du Cotonnier. - In : SLATYER, R.O. (ed.) : Plant Response to Climatic Factors. Pp. 213 - 220. Unesco, Paris 1973.

18208 - **VIELGOLASKI, F.E.** : Tipy rastitel'nosti i biomassa rasteniĭ tundry. [Tundra
vegetation types and plant biomass.] - Ėkologiya *4* (2) : 19 - 36, 1973. [In R.]

18209 - **VIEWEG, G.H., DE FEKETE, M.A.R.** : Regulation des Stoffwechsels der Stärke in
Blättern von *Zea mays*. - Ber. deut. bot. Ges. *86* : 233 - 239, 1973.

18210 - **VIGNES, D., CARLES, J.** : Essais de bilans énergétique et glucidique journali-
ers d'une feuille de vigne vierge (*Parthenocissus tricuspidata*). - Oecol. Plant
8 : 71 - 93, 1973.

18211 - **VIGNES, D., GAY, M., CARLES, J.** : Étude de la photosynthèse des disques de
feuilles de vigne vierge, au cours des premières heures après leur section. -
Bull. Soc. hist. nat. Toulouse *109* : 351 - 359, 1973.

18212 - **VIGNES, D., GAY, M., LASCOMBES, G., CARLES, J.** : Effets du traumatîsme sur la photosynthèse des disques de feuilles. - Compt. rend. Acad. Sci. Paris, Sér. D *277* : 569 - 572, 1973.

*18213 - **VIÏL, Yu.A., PYARNIK, T.R.** : Del'stvie kisloroda na assimilyatsiyu CO_2 u rastenii s razlichnoi potentsial'noi skorost'yu fotosinteza. [Effect of oxygen on CO_2 assimilation in plants with different photosynthetic capacity.] - In : Genetîcheskie Aspekty Fotosinteza. Tezisy Dokladov. Pp. 43 - 44. Donish, Dushanbe 1972. [In R.]

18214 - **VINCE, S., VALIELA, I.** : The effects of ammonium and phosphate enrichments on clorophyll *a*, pigment ratio and species composition of phytoplankton of Vineyard Sound. - Mar. Biol. *19* : 69 - 73, 1973.

18215 - **VINER, A.B.** : Responses of a mixed phytoplankton population to nutrient enrichments of ammonia and phosphate, and some associated ecological implications. - Proc. roy. Soc. London, Ser. B - biol. Sci. *183* : 351 - 370, 1973. [Ps.]

18216 - **VINOGRADOV, A.P.** : Fotosintez i biosfera. [Photosynthesis and biosphere.] - In : Sovremennye Problemy Fotosinteza. Pp. 8 - 16. Izd. mosk. Univ., Moskva 1973. [In R.]

18217 - **VINTÉJOUX, C.** : Variations saisonnières des constituants ultrastructuraux, dans les plastes foliares, chez l'*Utricularia neglecta* L. - Compt. rend. Acad. Sci. Paris, Sér. D *276* : 1693 - 1696, 1973. [Chloroplast.]

18218 - **VIRGIN, H.I., FRENCH, C.S.** : The light induced protochlorophyll-chlorophyll *a*-transformation and the succeeding interconversions of the different forms of chlorophyll. - Physiol. Plant. *28* : 350 - 357, 1973.

18219 - **VĪTOLA, Ā., KRISTKALNE, S.** : Fotosintēzes apstākļu ietekme uz asimilātu uzkrāšanos lapās. [Influence of conditions of photosynthesis on photosynthate accumulation in leaves.] - In : Tautsaimniecība Dergo Augu Agrotehnika un Selekcija. Pp. 105 - 113. Zinatne, Rīgā 1973. [In Latv., ab : R.]

18220 - **VLASOVA, M.P., OSIPOVA, O.P.** : Vliyanie intensivnosti sveta na tonkuyu strukturu khloroplastov rasteni*i Vicia faba.* [Effect of irradiance on chloroplast ultrastructure of *Vicia faba.*] - Fiziol. Rast. 20 : 742 - 746, 1973. [In R, ab : E.]

*18221 - **VLASOVA, M.P., VOSKRESENSKAYA, N.P.** : Sravnitel'noe issledovanie tonkoi struktury khloroplastov normal'nykh i mutantnykh rastenii gorokha (*Pisum sativum*), vyrashchennykh na svetu razlichnogo spektral'nogo sostava. [Comparison of chloroplast fine structure in normal and mutant *Pisum sativum* plants grown under different illuminance.] - In : Genetîcheskie Aspekty Fotosinteza. Tezisy Dokladov. Pp. 84 - 85. Donish, Dushanbe 1972. [In R.]

18222 - **VLASOVA, M.P., VOSKRESENSKAYA, N.P.** : Tonkaya struktura khloroplastov normal'nykh i mutantnykh rastenii gorokha, vyrashchennykh na svetu razlichnogo spektral'nogo sostava. [Ultrastructure of chloroplasts of normal and mutant pea plants grown in the light of different spectral composition.] - Fiziol. Rast. 20 : 96 - 102, 1973. [In R, ab : E.]

18223 - **VLASYUK, P.A., LAVRENTOVICH, D.I., CHERNOVA, L.M.** : Vliyanie margantsa na fotosintez, uglevodnyi obmen i urozhai sakharnoi svekly. [Effect of manganese on photosynthesis, carbohydrate metabolism and yield of sugar beet.] - Nauch. Tr. USKhA *71* (Khimizatsiya v Rastenievodstve) : 77 - 80, 1973. [In R.]

*18224 - **VOICA, C.** : L'influence du B sur les plantes de Tabac variété Bărăgan 226. - Rev. roum. Biol., Sér. Bot. *16* : 63 - 71, 1971. [Ps, Chl.]

*18225 - **VOICA, C.** : Influence de différentes concentrations du $ZnSO_4$ sur certains processus physiologiques chez les plantes de tabac. - Rev. roum. Biol., Sér. Bot. *16* : 281 - 288, 1971. [Ps, Chl.]

18226 - **VOLDENG, H.D., BLACKMAN, G.E.** : An analysis of the components of growth which determine the course of development under field conditions of selected inbreds and their hybrids of *Zea mays*. - Ann. Bot. *37* : 539 - 552, 1973. [Growth analysis.]

18227 - VOLDENG, H.D., BLACKMAN, G.E. : The influence of seasonal changes in solar radiation and air temperature on the growth in the early vegetative phase of *Zea
mays*. - Ann. Bot. *37* : 553 - 563, 1973.

18228 - VOLKOV, G.A. : Bioelectrical response of the *Nitella flexilis* cell to illumination : A new possible state of plasmalemma in a plant cell. - Biochim. biophys. Acta *314* : 83 - 92, 1973. [Ps inhibitors.]

18229 - VOLODARSKIĬ, N.I., BYSTRYKH, E.E. : Aktivnost' reaktsii Khilla v ontogeneze
podsolnechnika pri zasukhe. [Hill reaction activity during sunflower ontogenesis under drought.] - Sel'skokhoz. Biol. *8* : 652 - 657, 1973. [In R, ab :
E.]

18230 - VOLOVIK, O.I., KANIVETS, N.P., VASILENOK, L.I., VOLKOVA, N.V., BERSHTEĬN, B.
I., ZAĬTSEVA, N.A., MUSHKETIK, L.S., OKANENKO, A.S., OSTROVSKAYA, L.K., REIN-
GARD, T.A., YASNIKOV, A.A. : Ob uchastii atsetolfosfata-P^{32} v fotofosforiliro-
vanii v khloroplastakh gorokha. [Participation of acetolphosphate-^{32}P in photophosphorylation in pea chloroplasts.] - Fiziol. Biokhim. kul't. Rast. *5* :
582 - 586, 1973. [In R, ab : E.]

*18231 - VOROB'EV, V.N. : Sostoyanie pigmentnogo apparata pri razlichnom urovne smolo-
obrazovatel'nykh i generativnykh protsessov. [State of pigment apparatus at
various level of resin-forming and generative processes.] - In : Materialy
Konferentsii po Fiziologii i Biokhimii Rastenii, Posvyashchennoĭ 50-Letiyu
Obrazovaniya SSSR. Pp. 19 - 21. Krasnoyar. gos. Univ., vses. bot. Obshch.,
Krasnoyarsk 1972. [Chl; in R.]

*18232 - VOROB'EVA, L.M., LANG, F. : Monomernye i agregirovannye formy khlorofilla v
list'yakh i ikh izmeneniya pod vliyaniem osveshcheniya. [Monomer and aggregat-
ed chlorophyll forms in leaves and their transformation upon illumination.] -
- In : Geneticheskie Aspekty Fotosinteza. Tezisy Dokladov. P. 85. Donish, Dushanbe 1972. [In R.]

18233 - VORONKOVA, N.M. : Iskusstvennoe izmenenie kharaktera raspredeleniya assimilyatov u mnogoyarusnogo luka (*Allium proliferum* SCHRAD.). [Artificial change in
patterns of photosynthate distribution in *Allium proliferum* SCHRAD.] - Tr.
biol.-pochv. inst., dal'nevost. nauch. Tsentr Akad. Nauk SSSR *20* (123) : 214-
- 220, 1973. [In R, ab : E.]

18234 - VORONKOVA, N.M., SEMKIN, B.I., BELIKOV, I.F. : Peredvizhenie assimilyatov u
nekotorykh predstavitelei korneplodnykh i lukovichnykh rastenii. [Transport
of photosynthates in some representatives of root and bulbous plants.] - Tr.
biol.-pochv. inst., dal'nevost. nauch. Tsentr Akad. Nauk SSSR *20* (123) : 161-
- 167, 1973. [In R, ab : E.]

18235 - VOSKRESENSKAYA, N.P., KHODZHIEV, A.Kh. : Posledeĭstvie krasnogo i sinego sveta na aktivnost' glikolatoksidazy i glioksilataminotransferaz u rastenii bobov
i kukuruzy. [After-effect of red and blue light on the activity of glycolate
oxidase and glyoxylate aminotransferases in broad-bean and maize plants.] -
Fiziol. Rast. *20* : 309 - 316, 1973. [In R, ab : E.]

*18236 - VOSKRESENSKAYA, N.P., POYARKOVA, N.M., KHODZHIEV, A., DROZDOVA, I.S. : Osoben-
nosti fotosinteticheskogo metabolizma ugleroda nekotorykh vysshikh rastenii i
regulyatornoe deĭstvie sinego sveta na aktivnost' karboksiliruyushchikh fermentov i fermentov glikolatnogo puti. [The peculiarities of the photosynthetic carbon metabolism in some higher plants and blue light effects on carboxylating and glycolate pathway enzymes.] - In : Geneticheskie Aspekty Fotosinteza. Tezisy Dokladov. Pp. 44 - 45. Donish, Dushanbe 1972. [In R.]

18237 - VOZILOVA, L.D. : Vliyanie sveta na soderzhanie pigmentov khvoi prorostkov sosny, vyrashchennoĭ v temnote. [Effect of light on pigment content in needles of
dark grown pine seedlings.] - In : Voprosy Botaniki, Zoologii i Pochvovedeniya. Sbornik Rabot Molodykh Uchenykh, No. 1. Pp. 3 - 9. Izd. tomsk. Univ.,
Tomsk 1973. [In R.]

18238 - VOZILOVA, L.D., SIMONOVA, E.I., IL'INA, L.P., STVOLOVA, A.P., NOVIKOVA, N.S.:
Prevrashchenie predshestvennikov khlorofilla v zavisimosti ot uslovii osveshcheniya. [Transformation of chlorophyll precursors in dependence on illumination conditions.] - In : Sbornik Materialov Pyatoĭ Nauchnoĭ Konferentsii Fizio-
logov, Biokhimikov i Farmakologov Zap.-Sib. Ob"edin. Fiziol. Obshchestva Akad.
Nauk SSSR. P. 231. Tomsk 1973. [In R.]

18239 - **VOZNEṢENSKIĬ, V.L.** : Nekotorye voprosy interpretatsii rezul'tatov issledova-niĭ, provodimykh s pomoshch'yu radioaktivnykh izotopov. [Interpretation of results of investigations carried out with radioactive tracers.] - Tr. biol.--pochv. inst., dal'nevost. nauch. Tsentr Akad. Nauk SSSR, N.S. *20* (123) - Transport Assimilyatov i Otlozhenie Veshchestv v Zapas u Rastenii : 282 - 288, 1973. [In R, ab : E.]

18240 - **VREDENBERG, W.J., HOMANN, P.H., TONK, W.J.M.** : Light-induced potential changes across the chloroplast enclosing membranes as expressions of primary events at the thylakoid membrane. - Biochim. biophys. Acta *314* : 261 - 265, 1973.

18241 - **VREDENBERG, W.J., TONK, W.J.M.** : Photosynthetic energy control of an electro-genic ion pump at the plasmalemma of *Nitella translucens*. - Biochim. biophys. Acta *298* : 354 - 368, 1973.

18242 - **VRUBLEVSKAYA, K.G., ZAĬTSEVA, T.A.** : Vliyanie sveta na biosintez adenozinfos-fatov v rasteniyakh pshenitsy. [Effect of light on the biosynthesis of adeno-sinephosphates in wheat plants.] - In : Upravlenie Skorost'yu i Napravlennost'-yu Biosinteza u Rastenii. Pp. 7 - 8. Krasnoyarsk 1973. [In R.]

18243 - **VSEVOLODOV, N.N., KOSTIKOV, L.P., KAYUSHIN, L.P., GORBATENKOV, V.I.** : Dvukh-fotonnoe pogloshchenie lazernogo izlucheniya khlorofiliom *a* i nekotorymi or-ganicheskimi krasitelyami. [Two-photon absorption of laser radiation by chlo-rophyll *a* and some organic dyes.] - Biofizika *18* : 755 - 757, 1973. [In R, ab : E.]

18244 - **VYAS, L.N., GARG, R.K., RANAWAT, M.P.S.** : Plant biomass and net production of *Anogeissus latifolia* WALL. in forests of semiarid zone of Rajasthan (India). - Biol. Plant. *15* : 280 - 285, 1973.

18245 - **VYAS, L.N., GARG, R.K., RANAWAT, M.P.S., SHRIMAL, R.L.** : Studies on the pro-duction relations of deciduous forests of semiarid zone of Rajasthan (India). Plant biomass and net production of *Soymida febrifuga* JUSS. - Biológia (Bra-tislava) *28* : 499 - 506, 1973.

18246 - **VYSKOT, M.** : Root biomass of silver fir (*Abies alba* MILL.). - Acta Univ. Agr. (Brno), Ser. C *42* : 215 - 261, 1973.

18247 - **WADA, Y.** : [Physiological studies on the chlorophyll-deficient tobacco plants. I. Changes in the chlorophyll content and the photosynthetic activity during leaf growth.] - Bull. Hatano Tobacco exp. Sta. *73* : 1 - 6, 1973. [In Jap., ab : E.]

18248 - **WADA, Y.** : [Physiological studies on the chlorophyll-deficient tobacco plants. II. Effect of shading on the pigment and free amino acid content.] - Bull. Ha-tano Tobacco exp. Sta. *73* : 6 - 11, 1973. [In Jap., ab : E.]

18249 - **WADA, Y.** : [Physiological studies on the chlorophyll-deficient tobacco plants. III. Effect of shading on the pigment content and the fine structure of chlo-roplast.] - Bull. Hatano Tobacco exp. Sta. *73* : 11 - 15, 1973. [In Jap., ab : E.]

18250 - **WAGNER, J.F., PARKER, M.** : Primary production and limiting nutrients in a small, subalpine Wyoming Lake. - Trans. amer. Fish. Soc. *102* : 698 - 706, 1973.

18251 - **WAITE, D.T., DUTHIE, H.C., MATTHEWS, J.R.** : A note on two liquid scintillation fluors useful for primary production work. - Hydrobiologia *43* : 231 - 234, 1973.

18252 - **WALI, M.K., DEWALD, G.W., JALAL, S.M.** : Ecological aspects of some bluestem communities in the Red River Valley. - Bull. Torrey bot. Club *100* : 339 - 348, 1973. [Productivity.]

18253 - **WALKER, D.A.** : Photosynthetic induction phenomena and the light activation of ribulose diphosphate carboxylase. - New Phytol. *72* : 209 - 235, 1973.

18254 - **WALKER, D.A., KOSCIUKIEWICZ, K., CASE, C.** : Photosynthesis by isolated chloro-plasts : some factors affecting induction in CO_2-dependent oxygen evolution. - New Phytol. *72* : 237 - 247, 1973.

18255 - **WALLENTINUS, H.-G.** : Above-ground primary production of a *Juncetum gerardi* on a Baltic sea-shore meadow. - Oikos *24* : 200 - 219, 1973.

18256 - WALLENTINUS, H.-G., GUSTAFSSON, K., SÖDERSTRÖM, B. : Bladvassen, *Phragmites communis* TRIN., Í Brunnsviken, Stockholm 1971. [The reed, *Phragmites communis* TRIN., in the Brunnsviken, Stockholm 1971.] - Svensk bot. Tidskr. *67* : 81 - - 96, 1973. [Growth analysis; in Swed., ab : E.]

18257 - WALLIHAN, E.F. : Portable reflectance meter for estimating chlorophyll concentrations in leaves. - Agron. J. *65* : 659 - 662, 1973.

*18258 - WALSH, J.J., KELLEY, J.C., DUGDALE, R.C., FROST, B.W. : Gross features of the Peruvian upwelling system with special reference to possible diel variation. - Inv. Pesq. *35* : 25 - 42, 1971. [Chl.]

18259 - WALSTAD, J.D., NIELSEN, D.G., JOHNSON, N.E. : Effect of the pine needle scale on photosynthesis of Scots pine. - Forest Sci. *19* : 109 - 111, 1973.

18260 - WALTHER, W.G., EDMUNDS, L.N. Jr. : Studies on the control of the rhythm of photosynthetic capacity in·synchronized cultures of *Euglena gracilis* (Z). - Plant Physiol. *51* : 250 - 258, 1973.

18261 - WALTON, D.W.H. : Changes in standing crop and dry matter production in an *Acaena* community on South Georgia. - In : BLISS, L.C., WIELGOLASKI, F.E. (ed.) : Primary Production and Production Processes, Tundra Biome. Pp. 185 - 190. Tundra Biome Steering Comm., Edmonton 1973.

18262 - WANG, A.Y.-I., PACKER, L. : Mobility of membrane particles in chloroplasts. - Biochim. biophys. Acta *305* : 488 - 492, 1973.

18263 - WANG, R.T., CLAYTON, R.K. : Isolation of photochemical reaction centers from a carotenoidless mutant of *Rhodospirillum rubrum*. - Photochem. Photobiol. *17*: 57 - 61, 1973.

18264 - WANG, R.T., MYERS, J. : Energy transfer between photosynthetic units analyzed by flash oxygen yield *vs.* flash intensity. - Photochem. Photobiol. *17* : 321 - - 332, 1973.

18265 - WARD, F.J., NAKANISHI, M. : A comparison of liquid scintillation and Geiger- -Müller estimates of primary productivity in an *in situ* experiment. - J. Fish. Res. Board Can. *30* : 708 - 711, 1973.

18266 - WARDEN, J.T. Jr. : An electron-spin resonance study of the primary processes in plant and algal photosynthesis using flash photolysis. - Diss. Abstr. int. B *33* : 5167-B - 5168-B, 1973.

18267 - WARDEN, J.T. Jr., BOLTON, J.R. : Simultaneous quantitative comparison of the optical changes at 700 nm (*P700*) and electron spin resonance signals in system I of green plant photosynthesis. - J. amer. chem. Soc. *95* : 6435 - 6436, 1973.

18268 - WAREMBOURG, F.R., PAUL, E.A. : The use of $C^{14}O_2$ canopy techniques for measuring carbon transfer through the plant-soil system. - Plant Soil *38* : 331 - - 345, 1973.

18269 - WASSMAN, E.R., RAMUS, J. : Primary-production measurements for the green seaweed *Codium fragile* in Long Island Sound. - Mar. Biol. *21* : 289 - 297, 1973. [Ps, Chl.]

18270 - WATANABE, I. : Mechanism of varietal differences in photosynthetic rate of soybean leaves. I. Correlations between photosynthetic rates and some chloroplast characters. - Proc. Crop Sci. Soc. Jap. *42* : 377 - 386, 1973.

18271 - WATANABE, I. : Mechanism of varietal differences in photosynthetic rate of soybean leaves. II. Varietal differences in the balance between photochemical activities and dark reaction activities. - Proc. Crop Sci. Soc. Jap. *42* : 428- - 436, 1973.

17272 - WATANABE, I., TABUCHI, K. : Mechanism of varietal differences in photosynthetic rate of soybean leaves. III. Relationship between photosynthetic rate and some leaf-characters such as fresh weight,' dry weight or mesophyll volume per unit leaf area. - Proc. Crop Sci. Soc. Jap. *42* : 437 - 441, 1973.

18273 - WATSCHKE, T.L., SCHMIDT, R.E., CARSON, E.W., BLASER, R.E. : Temperature influence on the physiology of selected cool season turfgrasses and bermudagrass. - Agron. J. *65* : 591 - 594, 1973. [Ps.]

18274 - **WATSON, M.W.** : Optical anisotropy and the location of the carotenoid pigment
in chlorophycean eyespots. - J. Phycol. *9* (Suppl.) : 20, 1973.

18275 - **WEARE, N.M., BENEMANN, J.R.** : Nitrogen fixation by *Anabaena cylindrica*. II.
Nitrogenase activity during induction and aging of batch cultures. - Arch.
Mikrobiol. *93* : 101 - 112, 1973. [Ps inhibitors.]

18276 - **WEBER, A.** : Über die Chlorophylle und Carotinoide einiger *Ulotrichales*. -
Mitt. Staatsinst. allgem. Bot. Hamburg *14* : 25 - 29, 1973.

18277 - **WEBER, F., LASKAWY, G., GROSCH, W.** : Enzymatischer Carotinabbau in Erbsen, So-
jabohnen, Weizen und Leinsamen. - Z. Lebensmittel-Untersuch. -Forsch. *152* :
324 - 331, 1973.

18278 - **WEERDHOF, T. van de, WIERSUM, M.L., REISSENWEBER, H.** : Application of liquid
chromatography in food analysis. - J. Chromatogr. *83* : 455 - 460, 1973. [Car.]

18279 - **WEGMANN, K., MÜHLBACH, H.-P.** : Photosynthetic CO_2 incorporation by isolated
leaf cell protoplasts. - Biochim. biophys. Acta *314* : 79 - 82, 1973.

18280 - **WEIDNER, M., KÜPPERS, U.** : Phosphoenolpyruvat-Carboxykinase und Ribulose-1,5-
-Diphosphat-Carboxylase von *Laminaria hyperborea* (GUNN.) FOSL. : Das Verteil-
ungsmuster der Enzymaktivitäten im Thallus. - Planta *114* : 365 - 372, 1973.

18281 - **WEISSENBÖCK, G.** : Untersuchungen zur Lokalisation von Flavonoiden in Plastiden. II. Vergleich des Flavonoidmusters isolierter Etioplasten und Chloroplasten von *Avena sativa* L. - Ber. deut. bot. Ges. *86* : 351 - 364, 1973.

18282 - **WELLBURN, A.R., WELLBURN, F.A.M.** : Developmental changes of etioplasts in isolated suspensions and *in situ*. - Ann. Bot. *37* : 11 - 19, 1973.

18283 - **WELLBURN, F.A.M., WELLBURN, A.R.** : Response of etioplasts *in situ* and in isolated suspensions to pre-illumination with various combinations of red, far-
-red and blue light. - New Phytol. *72* : 55 - 60, 1973.

18284 - **WELLBURN, F.A.M., WELLBURN, A.R., STODDART, J.L., TREHARNE, K.J.** : Influence
of gibberellic and abscisic acids and the growth retardant, CCC, upon plastid
development. - Planta *111* : 337 - 346, 1973.

18285 - **WELLER, H.G. Jr., TIEN, H.T.** : Determination of chlorophyll porphyrin ring ori
entation in black lipid membranes by photovoltage spectroscopy. - Biochim.
biophys. Acta *325* : 433 - 440, 1973.

18286 - **WESSELIUS, J.C.** : Influence of light intensity on growth and energy conversion
in mass cultures of *Scenedesmus* sp. - In : SLATYER, R.O. (ed.) : Plant Response to Climatic Factors. Pp. 51 - 56. Unesco, Paris 1973.

18287 - **WESSELIUS, J.C.** : Influences of external factors on the energy conversion and
productivity of *Scenedesmus* sp. in mass culture. - Med. Landbouwhogesch. (Wageningen) *73-6* : 1 - 97, 1973. [Ps, Chl.]

18288 - **WESSELS, J.S.C., van ALPHEN-van WAVEREN, O., VOORN, G.** : Isolation and properties of particles containing the reactioncenter complex of photosystem II from
spinach chloroplasts. - Biochim. biophys. Acta *292* : 741 - 752, 1973.

18289 - **WEST, J., HALL, D.O.** : Anomalous effects of phlorrhizin (phloridzin) on photosynthetic oxygen production. - Biochem. Soc. Trans. *1* : 888 - 890, 1973.

*18290 - **WEST, J., MANGAN, J.L.** : The digestion of chloroplasts in the rumen of sheep
and the effect of disruption and glutaraldehyde treatment. - Proc. Nutr. Soc.
31 : 108 A, 1972.

18291 - **WEST, J., MANGAN, J.L.** : A comparison of glutaraldehyde and formaldehyde fixation of isolated pea chloroplasts and its implications for the treatment of
herbage for nutritional studies. - J. agr. Sci. *80* : 399 - 406, 1973. [Ps.]

18292 - **WEST, K.R., WISKICH, J.T.** : Evidence for two phosphorylation sites associated
with the electron transport chain of chloroplasts. - Biochim. biophys. Acta
292 : 197 - 205, 1973.

18293 - **WEST, S.H.** : Carbohydrate metabolism and photosynthesis of tropical grasses
subjected to low temperatures. - In : SLATYER, R.O. (ed.) : Plant Response to
Climatic Factors. Pp. 165 - 168. Unesco, Paris 1973.

18294 - **WETZEL, R.G.** : Productivity investigations of interconnected marl lakes. (I)
The eight lakes of the Oliver and Walters chains, Northeastern Indiana. -
Hydrobiol. Stud. *3* : 91 - 143, 1973.

18295 - **WETZEL, R.G., HOUGH, R.A.** : Productivity and role of aquatic macrophytes in
lakes. An assessment. - Pol. Arch. Hydrobiol. *20* : 9 - 19, 1973.

18296 - **WETZEL, R.G., RICH, P.H.** : Carbon in freshwater systems. - In : **WOODWELL, G.M.,
PECAN, E.V.** (ed.) : Carbon and the Biosphere. Pp. 241 - 263. Nat. tech. Inform.
Serv., Springfield, Virginia 1973. [Ps.]

18297 - **WHEELER, L., STROTHER, A.** : Comparison of NCS solubilization and a wet combustion technique for ^{14}C measurement in dried biological tissues. - Anal.
Biochem. *53* : 42 - 48, 1973.

18298 - **WHELAN, T., SACKETT, W.M., BENEDICT, C.R.** : Enzymatic fractionation of carbon
isotopes by phosphoenolpyruvate carboxylase from C_4 plants. - Plant Physiol.
51 : 1051 - 1054, 1973.

18299 - **WHITTINGHAM, C.P.** : Botany Department. - Rep. Rothamsted exp. Sta. *1972* :
90 - 109, 1973. [Ps.]

18300 - **WICKRAMASINGHE, R.H.** : Iron-sulphur proteins : Their possible place in the
origin of life and the development of early metabolic systems. - Space Life
Sci. *4* : 341 - 352, 1973. [Ps.]

18301 - **WIEDNER, G., SIMON, B., THOMAS, L.** : Carbonic anhydrase : a new method of detection on polyacrylamide gels using conductivity measurements. - Anal. Biochem. *55* : 93 - 97, 1973.

18302 - **WIELGOLASKI, F.E.** : Influence of climate on primary production in a northern
mountain area. - Int. J. Biometeorol. *17* : 355 - 357, 1973.

18303 - **WIELGOLASKI, F.E., KJELVIK, S.** : Production of plants (vascular plants and
cryptogams) in alpine tundra, Hardangervidda. - In : **BLISS, L.C., WIELGOLAS-
KI, F.E.** (ed.) : Primary Production and Production Processes, Tundra Biome.
Pp. 75 - 86. Tundra Biome Steering Comm., Edmonton 1973.

18304 - **WIELGOLASKI, F.E., KJELVIK, S.** : Mineral elements and energy of plants at Hardangervidda, Norway. - In : **BLISS, L.C., WIELGOLASKI, F.E.** (ed.) : Primary
Production and Production Processes, Tundra Biome. Pp. 231 - 238. Tundra Biome Steering Comm., Edmonton 1973. [Production.]

18305 - **WIEN, H.C., WALLACE, D.H.** : Light-induced leaflet orientation in *Phaseolus
vulgaris* L. - Crop Sci. *13* : 721 - 724, 1973.

18306 - **WILD, A., KE, B., SHAW, E.R.** : The effect of light intensity during growth of
Sinapis alba on the electron-transport components. - Z. Pflanzenphysiol. *69* :
344 - 350, 1973.

18307 - **WILD, A., MÜLLENBECK, E.** : Untersuchung zur Photosyntheseleistung von *Zea mays*
nach Anzucht unter verschiedenen Lichtintensitäten. - Z. Pflanzenphysiol. *70* :
235 - 244, 1973.

18308 - **WILDMAN, S.G., LU-LIAO, C., WONG-STAAL, F.** : Maternal inheritance, cytology
and macromolecular composition of defective chloroplasts in a variegated mutant of *Nicotiana tabacum*. - Planta *113* : 293 - 312, 1973. [Chl.]

18309 - **WILDNER, G.F.** : Ribulosediphosphate-carboxylase and its role in photorespiration. - In : Ninth International Congress of Biochemistry. Stockholm, 1 - 7
July, 1973. Abstract Book. P. 241. IUB, Stockholm 1973.

18310 - **WILKINS, M.B.** : An endogenous circadian rhythm in the rate of carbon dioxide
output of *Bryophyllum*. VI. Action spectrum for the induction of phase shifts
by visible radiation. - J. exp. Bot. *24* : 488 - 496, 1973. [Ps.]

18311 - **WILKINSON, T.G., BARNES, R.L.** : Effects of ozone on $^{14}CO_2$ fixation patterns
in pine. - Can. J. Bot. *51* : 1573 - 1578, 1973.

18312 - **WILLENBRINK, J., KREMER, B.P.** : Lokalisation der Mannitbiosynthese in der marinen Braunalge *Fucus serratus*. - Planta *113* : 173 - 178, 1973. [Ps.]

18313 - **WILLENBRINK, J., LÜNING, K., SCHMITZ, K.** : On translocation of ^{14}C-labelled
assimilates in the *Phaeophyceae Laminaria hyperborea* and *L. saccharina*. -
Proc. Res. Inst. Pomol., Skierniewice, Ser. E *1973* (3)' : 159 - 160, 1973.

*18314 - **WILLIAMS, C.N.** : Growth and productivity of tapioca (*Manihot utilissima*). II.
Stomatal functioning and yield. - Exp. Agr. *7* : 49 - 62, 1971.

18315 - **WILLIAMS, G.J. III, MARKLEY., J.L.** : The photosynthetic pathway type of North
American shortgrass prairie species and some ecological implications. - Photosynthetica *7* : 262 - 270, 1973.

18316 - **WILLIAMS, J.A.** : A considerably improved method for preparing plastic epidermal imprints. - Bot. Gaz. *134* : 87 - 91, 1973. [Stomata.]

18317 - **WILLMER, C., KANAI, R., PALLAS, J.E. Jr., BLACK, C.C. Jr.** : Detection of high
levels of phosphoenolpyruvate carboxylase in leaf epidermal tissue and its
significance in stomatal movements. - Life Sci., Part II *12* (4) : 151 - 155,
1973.

18318 - **WILLMER, C.M., PALLAS, J.E. Jr., BLACK, C.C. Jr.** : CO_2 metabolism in leaf epidermal tissue. - Plant Physiol. *51* (Suppl.) : 42, 1973.

18319 - **WILLMER, C.M., PALLAS, J.E. Jr., BLACK, C.C. Jr.** : Carbon dioxide metabolism
in leaf epidermal tissue. - Plant Physiol. *52* : 448 - 452, 1973.

18320 - **WILLMOT, A., MOORE, P.D.** : Adaptation to light intensity in *Silene alba* and
S. dioica. - Oikos *24* : 458 - 464, 1973. [Ps.]

18321 - **WILSON, D.** : Physiology of light utilization by swards. - In : BUTLER, G.W.,
BAILEY, R.W. (ed.) : Chemistry and Biochemistry of Herbage. Vol. 2. Pp. 57 -
- 101. Academic Press, London - New York 1973.

18322 - **WILSON, J.H., CLOWES, M.St.J., ALLISON, J.C.S.** : Growth and yield of maize at
different altitudes in Rhodesia. - Ann. appl. Biol. *73* : 77 - 84, 1973.
[Growth analysis.]

18323 - **WINTER, K.** : CO_2-Gaswechsel von an hohe Salinität adaptiertem *Mesembryanthemum
crystallinum* bei Rückführung in glykisches Anzuchtmedium. - Ber. deut. bot.
Ges. *86* : 467 - 476, 1973.

18324 - **WINTER, K.** : CO_2-Fixierungsreaktionen bei der Salzpflanze *Mesembryanthemum
crystallinum* unter variierten Außenbedingungen. - Planta *114* : 75 - 85, 1973.

18325 - **WINTER, K.** : NaCl-induzierter Crassulaceensäurestoffwechsel bei einer weite-
ren Aizoacee : *Carpobrotus edulis*. - Planta *115* : 187 - 188, 1973.

18326 - **WINTER, S.R., OHLROGGE, A.J.** : Leaf angle, leaf area, and corn (*Zea mays* L.)
yield. - Agron. J. *65* : 395 - 397, 1973.

18327 - **WIRTH, V., TÜRK, R.** : Über Standort, Verbreitung und Soziologie der borealen
Flechten *Cetraria sepincola* (EHRH.)ACH. und *Parmelia olivacea* s. ampl. in Mit-
teleuropa. - Veröff. Landesst. N. L. Bd.-Wttb. *41* : 88 - 117, 1973. [Ps.]

18328 - **WITT, H.T., BOECK, M.** : Further studies on the correlation between electrical
field generation, field decay and phosphorylation on the functional membrane
of photosynthesis. - In : Ninth International Congress of Biochemistry. Stockholm, 1 - 7 July, 1973. Abstract Book. P. 232. IUB, Stockholm 1973.

18329 - **WITT, K.** : Some photoreactions of spinach chloroplasts in sucrose syrup at
high and low temperature. - FEBS Lett. *38* : 112 - 115, 1973.

18330 - **WITT, K.** : Further evidence of X-320 as a primary acceptor of photosystem II
in photosynthesis. - FEBS Lett. *38* : 116 - 118, 1973.

18331 - **WOHLER, J.R., HARTMANN, R.T.** : Some characteristics of an *Oscillatoria*-domi-
nated metalimnetic phytoplankton community. - Ohio J. Sci. *73* : 297 - 307,
1973. [Ps, Chl.]

18332 - **WOJCIESKA, U.** : Dynamika wzrostu i produktywność fotosyntezy żyta ozimego (*Secale cereale* L.). Część I. Przyrost masy, aktywność fotosyntetyczna i dystry-
bucja asymilatów. [Dynamics of growth and productivity of photosynthesis in
winter rye (*Secale cereale* L.). Part I. Increase of matter, photosynthetic
activity and distribution of photosynthates at different stages of growth and
development of plants.] - Pamięt. puław. - Prace IUNG *56* : 7 - 30, 1973. [In
Pol., ab : E, R.]

18333 - **WOJCIESKA, U.** : Dynamika wzrostu i produktywność fotosyntezy żyta ozimego (*Secale cereale* L.). Część II. Wpływ chlorku chlorocholiny na wzrost i rozwój

roślin oraz na aktywność fotosyntetyczną i akumulację asymilatów. [Dynamics of growth and productivity of photosynthesis of winter rye (*Secale cereale* L.). Part II. Influence of chlorocholine chloride on the growth and development of plants as well as on the photosynthetic activity and accumulation of assimilates.] - Pamięt. puław. - Prace IUNG *56* : 31 - 50, 1973. [In Pol., ab : E, R.]

18334 - **WOJCIESKA, U.** : Dynamika wzrostu i produktywność fotosyntezy żyta ozimego (*Secale cereale* L.). Część III. Wpływ żywienia azotem na długość życia liści oraz na wzrost roślin i ich aktywność fotosyntetyczną. [Dynamics of growth and productivity of photosynthesis of winter rye (*Secale cereale* L.). Part III. Influence of nitrogen nutrition on the duration of leaf life as well as on the growth of plants and their photosynthetic activity.] - Pamięt. puław. - Prace IUNG *56* : 51 - 74, 1973. [In Pol., ab : E, R.]

18335 - **WOJCIESKA, U.** : Współzależność pomiędzy plonem, intensywnością fotosyntezy i aktywnością enzymów biorących udział w wiązaniu węgla. [Correlation between the yield, photosynthetic rate and the activity of enzyme involved in carbon dioxide fixation.] - Postępy Nauk rol. *20* : 45 - 54, 1973. [In Pol.]

18336 **WOJCIESKA, U., SZCZYPA-WOLSKA, E., ŚLUSARCZYK, M.** : Photosynthesis, translocation and accumulation of assimilates in winter rye and winter wheat plants. - Proc. Res. Inst. Pomol., Skierniewice, Ser. E *1973* (3) : 139 - 150, 1973.

18337 - **WØLDIKE, K.** : Phytoplankton in the oligohaline lake, Selsø. Primary production and standing crop. - Ophelia *12* : 27 - 44, 1973. [Chl.]

18338 - **WOLEDGE, J.** : The photosynthesis of ryegrass leaves grown in a simulated sward. - Ann. appl. Biol. *73* : 229 - 237, 1973.

18339 - **WOLF, F.T., KIDD, G.H.** : Effect of various gas atmospheres upon the greening of etiolated seedlings. - Z. Pflanzenphysiol. *70* : 115 - 118, 1973.

18340 - **WOLF, F.T., KIDD, G.H.** : Effects of various gas atmospheres upon the greening response of etiolated wheat seedlings. - Plant Physiol. *51* (Suppl.) : 28, 1973.

18341 - **WOLFF, C.** : Electrochromic change of the prompt fluorescence in photosynthesis. - In : Ninth International Congress of Biochemistry. Stockholm, 1 - 7 July, 1973. Abstract Book. P. 219. IUB, Stockholm 1973.

18342 - **WOLK, C.P.** : Physiology and cytological chemistry of blue-green algae. - Bacteriol. Rev. *37* : 32 - 101, 1973. [Ps.]

18343 - **WOLKEN, J.J.** : Photodynamics : The chloroplast in photosynthesis. - In : MILLER, L.P. (ed.) : Phytochemistry. Vol. I. The Process and Products of Photosynthesis. Pp. 15 - 37. Van Nostrand Reinhold Company, New York - Cincinnati - Toronto - London - Melbourne 1973.

18344 - **WOŁKOWA, E., ANTOSZEWSKI, R., POSKUTA, J.** : Gaseous exchange rates of strawberry plants during fruiting. - In : Proc. XIX Int. Hort. Congress. Vol. 1A. P. 73. Warszawa 1973.

18345 - **WONG-STAAL, F., WILDMAN, S.G.** : Identification of a mutation in chloroplast DNA correlated with formation of defective chloroplasts in a variegated mutant of *Nicotiana tabacum*. - Planta *113* : 313 - 326, 1973. [Chl.]

18346 - **WOOD, B.J.B., TETT, P.B., EDWARDS, A.** : An introduction to the phytoplankton, primary production and relevant hydrography of Loch Etive. - J. Ecol. *61* : 569 - 585, 1973. [Chl.]

*18347 - **WOOD, E.J.F., ODUM, W.E., ZIEMAN, J.C.** : Influence of sea grasses on the productivity of coastal lagoons. - In : Lagunas Costeras, un Simposio. Pp. 495 - - 502. UNAM-UNESCO, México, D.F. 1969.

18348 - **WOOD, G.B., BRITTAIN, E.G.** : Photosynthesis, respiration and transpiration of radiata pine. - New Zeal. J. Forest. Sci. *3* : 181 - 190, 1973.

18349 - **WORT, D.J., SEVERSON, J.G., PEIRSON, D.R.** : Growth stimulation by naphthenates and its metabolic bases. - Plant Physiol. *51* (Suppl.) : 46, 1973. [Ps.]

18350 - **WOŹNY, A., GWÓŹDŹ, E., SZWEYKOWSKA, A.** : The effect of 3-indolylacetic acid on the differentiation of plastids in callus culture of *Cichorium intybus* L. - Protoplasma *76* : 109 - 114, 1973.

18351 - WRIGHT, M., SIMON, E.W. : Chilling injury in cucumber leaves. - J. exp. Bot. 24 : 400 - 411, 1973. [Ps, Chl.]

18352 - WRISCHER, M. : Protein crystalloids in the stroma of bean plastids. - Protoplasma 77 : 141 - 150, 1973.

18353 - WRISCHER, M. : Ultrastructural changes in isolated plastids. I. Etioplasts. - Protoplasma 78 : 291 - 303, 1973.

18354 - WRISCHER, M. : Ultrastructural changes in isolated plastids. II. Etio-chloroplasts. - Protoplasma 78 : 417 - 425, 1973.

18355 - WRISCHER, M. : The effect of ethionine on the fine structure of bean chloroplasts. - Cytobiologie 7 : 211 - 214, 1973.

18356 - WRÓBLEWSKA, H. : Influence of water deficit and age of plant on the intensity of photosynthesis and air passage capacity in leaves of Nicotiana rustica L. - Hodowla Rośl., Aklimat. Nasienn. 17 : 387 - 411, 1973.

18357 - WRÓBLEWSKI, R. : A fine structural investigation of the chloroplasts from the root of Lemna minor L. - J. submicroscop. Cytol. 5 : 97 - 105, 1973.

18358 - WU, A., THROWER, L.B. : Translocation into mature leaves. - Plant Cell Physiol. 14 : 1225 - 1228, 1973. [Photosynthates.]

18359 - WU, J.H., SKOKUT, T., HARTMAN, M. : Ultraviolet-radiation-accelerated leaf chlorosis : prevention of chlorosis by removal of epidermis or by floating leaf discs on water. - Photochem. Photobiol. 18 : 71 - 77, 1973.

18360 - WYNN, T., BROWN, H., CAMPBELL, W., BLACK, C.C. Jr. : Dark release of $^{14}CO_2$ in higher plants. - Plant Physiol. 51 (Suppl.) : 6, 1973.

18361 - WYNN, T., BROWN, H., CAMPBELL, W.H., BLACK, C.C. Jr. : Dark release of $^{14}CO_2$ from higher plant leaves. - Plant Physiol. 52 : 288 - 291, 1973.

18362 - WYNN, T., STILLER, M. : Synchronous growth and development of Chlorella pyrenoidosa. - Plant Physiol. 51 (Suppl.) : 58, 1973. [Ps, Chl.]

18363 - YABUKI, K., KO, B. : [The dependence of photosynthesis in several vegetables on light quality.] - J. agr. Meteorol. 29 : 17 - 23, 1973. [In Jap., ab : E.]

18364 - YABUKI, K., NISHIOKA, M. : [Studies on the effect of wind speed on photosynthesis. (3) The structure of air flow in the boundary layer near leaf surface.] - J. agr. Meteorol. 29 : 173 - 177, 1973. [In Jap., ab : E.]

*18365 - YAKUBOVA, M.M., RUBIN, A.B., KHRAMOVA, G.A. : Fotokhimicheskie protsessy fotosinteza u mutantov vysshikh rasteniǐ. [Photochemical processes of photosynthesis in mutants of higher plants.] - In : Geneticheskie Aspekty Fotosinteza. Tezisy Dokladov. Pp. 93 - 94. Donish, Dushanbe 1972. [In R.]

18366 - YAKUSHKINA, N.I., DURANDIN, A.I. : Vliyanie spektral'nogo sostava sveta i gibberellina na rost prorostkov yachmenya i intensivnost' fotofosforilirovaniya v nikh. [Effect of the spectral composition of light and gibberellin on the growth and phosphorylation rate in barley seedlings.] - Sel'skokhoz. Biol. 8 : 29 - 32, 1973. [In R, ab : E.]

*18367 - YAŞNIKOV, A.A., BERSHTEĬN, B.I., VASILENOK, L.I., VOLKOVA, N.V., VOLOVIK, O.I. ZAĬTSEVA, N.A., MUSHKETIK, L.S., OKANENKO, A.S., OSTROVSKAYA, L.K., PETRENKO, S.G., POLISHCHUK, A.I., REĬNGARD, T.A., SEMENYUK, I.I. : Regulyatsiya vklyucheniya neorganicheskogo fosfata v fotofosforilirovanie (piruvatkinazoǐ i fosfatazoǐ). [Regulation of inorganic phosphate incorporation into photophosphorylation (by pyruvate kinase and phosphatase).] - In : Geneticheskie Aspekty Fotosinteza. Tezisy Dokladov. Pp. 92 - 93. Donish, Dushanbe 1972. [In R.]

18368 - YASNIKOV, A.A., BERSHTEIN, B.I., VOLKOVA, N.V., VASILENOK, L.I., DUBROVSKAYA, A.A., KANIVETS, N.P., MUSHKETIK, L.S., OKANENKO, A.S., OSTROVSKAYA, L.K., REĬNGARD, T.A. : Deǐstvie kisloǐ fosfatazy na fotofosforilirovanie v khloroplastakh gorokha. [Effect of acid phosphatase on photophosphorylation in pea chloroplasts.] - Fiziol. Biokhim. kul't. Rast. 5 : 478 - 483, 1973. [In R, ab : E.]

18369 - YASUNOBU, K.T., TANAKA, M. : The evolution of iron-sulfur protein containing organisms. - System. Zool. 22 : 570 - 589, 1973. [Ferredoxin.]

18370 - **YATSENKO-KHMELEVSKIĬ, A.A., SHIRYAEVA, G.A., GUTMAN, T.S.** : Kul'tura izoliro-
         vannykh zelenykh kletok eli i nekotorye voprosy ikh produktivnosti. [Culture
         of isolated green spruce cells and some problems of their productivity.] - I
         Upravlenie Skorost'yu i Napravlennost'yu Biosinteza u Rasteniĭ. Pp. 124 - 12
         Krasnoyarsk 1973. [In R.]

*18371 - **YENTSCH, C.S.** : The harvest - primary production. - In : HERRING, P.J., CLAR
         KE, M.R. (ed.) : Deep Oceans. Pp. 150 - 163. Praeger Publ., London 1971. [Ps

18372 - **YIM, Y.-J.** : On the growth of the surface area of isolated young trees, *Alnu*
         tinctoria SARGENT. - Korean J. Bot. *16* : 1 - 5, 1973. [Growth analysis.]

*18373 - **YIM, Y.J., OGAWA, H., KIRA, T.** : Light interception by stems in plant commun
         ties. - Jap. J. Ecol. *19* : 233 - 238, 1969.

*18374 - **YIM, Y.J., YOUNG, D.R.** :[On the growth and total nitrogen changes of *Glycine*
         max. Artificial plant communities, grown in sandy loam soil with a controlle
         moisture content.]- Korean J. Bot. *14* : 21 - 28, 1971. [Growth analysis; in
         Korean, ab : E.]

18375 - **YOCUM, C.F., SIEDOW, J.N., SAN PIETRO, A.** : Iron-sulfur proteins in photosyn
         thesis. - In : LOVENBERG, W. (ed.) : Iron-Sulfur Proteins. Vol. 1. Pp. 111 -
         - 127. Academic Press, New York - London 1973.

18376 - **YOKOHAMA, Y.** : A comparative study on photosynthesis-temperature relationshi
         and their seasonal changes in marine benthic algae. - Int. Rev. ges. Hydro-
         biol. *58* : 463 - 472, 1973.

18377 - **YOKOHAMA, Y.** : [Photosynthetic properties of marine benthic green algae from
         different depths in the coastal area.] - Bull. jap. Soc. Phycol. *21* : 70 -
         - 75, 1973. [In Jap., ab : E.]

18378 - **YOKOHAMA, Y.** : [Photosynthetic properties of marine benthic red algae from
         different depths in coastal area.] - Bull. Jap. Soc. Phycol. *21* : 119 - 124,
         1973. [In Jap., ab : E.]

18379 - **YOSHIDA, F., KOBAYASHI, T., YOSHIDA, T.** : The mineral nutrition of cultured
         chlorophyllous cells of tobacco. I. Effects of II_{salts}, $II_{sucrose}$, Ca, Cl and
         in the medium on the yield, friability, chlorophyll contents and mineral ab-
         sorption of cells. - Plant Cell Physiol. *14* : 329 - 339, 1973.

18380 - **YOSHIDA, F., SAITO, N., OGAWA, S.** : Requirement of organic constituents to
         culture habituated chlorophyllous cells of tobacco. - Bull. Fac. Agr., Tama-
         gawa Univ. *13* : 10 - 16, 1973.

18381 - **YOUN, K.B., OTA, Y.** : [Changes in the chlorophyll content and chlorophyll re
         tention of leaf segments according to the growth of various leaf blades in r
         ce plant.] - Proc. Crop Sci. Soc. Jap. *42* : 6 - 12, 1973. [In Jap., ab : E.]

18382 - **YOUN, K.B., OTA, Y.** : [Relations between the leaf senescence index and root
         activity of rice plants.] - Proc. Crop Sci. Soc. Jap. *42* : 13 - 17, 1973.
         [In Jap., ab : E.]

18383 - **YOUNG, T.C., KING, D.L.** : A rapid method of quantifying algal carbon uptake
         kinetics. - Limnol. Oceanogr. *18* : 978 - 981, 1973. [Ps.]

18384 - **YUNUSOVA, L.S., MUKHINA, O.V., GUSEV, M.V.** : O vliyanii uglerodsoderzhashchi
         soedineniĭ na khromaticheskuyu kharakteristiku sine-zelenoĭ vodorosli *Anacys*
         tis nidulans. [Influence of carbon containing compounds on the chromatic cha
         racteristics of the blue-green alga *Anacystis nidulans*.] - Nauch. Dokl. vyss
         Shkoly, biol. Nauki *16* (10) : 107 - 110, 1973. [Chl, biliproteins; in R.]

*18385 - **YURINA, N.P.** : Sravnitel'noe issledovanie ribosom khloroplastov i sine-zele-
         nykh vodorosleĭ. [Comparative study of chloroplast ribosomes and blue-green
         algae.] - In : Geneticheskie Aspekty Fotosinteza. Tezisy Dokladov. Pp. 18 -
         -- 19. Donish, Dushanbe 1972. [In R.]

18386 - **ZABIROVA, I.G., BRZHEVSKAYA, O.N., NEDELINA, O.S., KAYUSHIN, L.P., SHEKSHEEV
         E.M., MUKHIN, E.N., GINS, V.K.** : Issledovanie fotofosforilirovaniya metodom
         élektronnogo paramagnitnogo resonansa. [Investigation of photophosphorylatic
         by electron paramagnetic resonance.] - Dokl. Akad. Nauk SSSR *212* : 754 - 756
         1973. [In R.]

18387 - **ZAGDAŃSKA, B.** : Próby mierzenia fotosyntezy u zbóż w warunkach polowych.
[Field measurement of photosynthesis in cereals.] - Hodowla Rośl., Aklimat.,
Nasienn. *17* : 425 - 432, 1973. [Porometer; in Pol., ab : E, R.]

*18388 - **ZAGROMSKI, G.** : Mutanty kak ob"ekt dlya issledovaniya funktsii khlorofillov.
[Mutants as objects for investigations of the function of chlorophylls.] - In:
Geneticheskie Aspekty Fotosinteza. Tezisy Dokladov. Pp. 88 - 89. Donish, Du-
shanbe 1972. [in R.]

18389 - **ZAJĄCZKOWSKA, J.** : Gas exchange and organic matter production of Scots pine
(*Pinus silvestris* L.) seedlings grown in water culture with ammonium or nitrate form of nitrogen. - Acta Soc. Bot. Pol. *42* : 607 - 615, 1973.

18390 - **ZAKHVATAEVA, N.V.** : Vliyanie ketokislot na azotfiksatsiyu fotosinteziruyushchikh bakteriǐ. [Effect of keto acids on nitrogen fixation in photosynthetic
bacteria.] - Nauch. Dokl. vyssh. Shkoly, biol. Nauki *16* (4) : 116 - 118, 1973.
[in R.]

18391 - **ZAKRZHEVSKIǏ, D.A., KALASHNIKOV, Yu.E.** : Sravnitel'noe izuchenie predel'nogo
anaerobnogo potentsiala i ėlektrodnoǐ aktivnosti kletok rastitel'nykh organizmov, stoyashchikh na raznykh ėtapakh ėvolyutsii. [Comparative study of the maximum anaerobic potential and electrode activity of cells of plants at different stages of evolution.] - Dokl. Akad. Nauk SSSR *209* : 992 - 995, 1973. [Ps;
in R.]

18392 - **ZAKRZHEVSKIǏ, D.A., ROZONOVA, L.N., KUTYURIN, V.M.** : O spetsificheskom deǐ-
stvii atmosfery vodoroda na vydelenie kisloroda khloroplastami tradeskantsii.
[Specific effect of a hydrogen atmosphere on the evolution of oxygen by *Tradescantia* chloroplasts.] - Dokl. Akad. Nauk SSSR *213* : 980 - 982, 1973. [in
R.]

18393 - **ZANKEL, K.L.** : Rapid fluorescence changes observed in chloroplasts : their relationship to the O_2 evolving system. - Biochim. biophys. Acta *325* : 138 - 148
1973.

18394 - **ZANKER, V., RUDOLPH, E., HINDELANG, F.** : Zur Prototropie von Phaeophytin *a*. -
Z. Naturforsch. *28 B* : 650 - 655, 1973.

18395 - **ZAUSSINGER, A., RUCKENBAUER, P., FILA, R.** : Entwicklung und Bau eines photo-
elektrischen Blattflächenmeßgerätes. - Bodenkultur *24* : 352 - 361, 1973.

*18396 - **ZAVITKOVSKI, J.** : Dry weight and leaf area of aspen trees in northern Wisconsin. - In : Forest Biomass Studies. Pp. 193 - 205. Univ. Maine 1971. [Biomass
production.]

*18397 - **ZAVITKOVSKI, J., STEVENS, R.D.** : Primary productivity of red alder ecosystems.
- Ecology *53* : 235 - 242, 1972.

18398 - **ZAVODNIK, N.** : Seasonal variations in rate of photosynthetic activity and chemical composition of the littoral seaweeds common to North Adriatic. Part I.
Fucus virsoides (DON)J.AG. - Bot. mar. *16* : 155 - 165, 1973.

16399 - **ZAVODNIK, N.** : Seasonal variations in rate of photosynthetic activity and chemical composition of the littoral seaweeds common to North Adriatic. Part II.
Wrangelia penicillata C.AG. - Bot. mar. *16* : 166 - 170, 1973.

18400 - **ŻELAWSKI, W., SZANIAWSKI, R., DYBCZYŃSKI, W., PIECHUROWSKI, A.** : Photosynthetic capacity of conifers in diffuse light of high illuminance. - Photosynthetica *7* : 351 - 357, 1973.

18401 - **ZELDIN, M.H., SKEA, W., MATTESON, D.** : Organelle formation in the presence of
a protease inhibitor. - Biochem. biophys. Res. Commun. *52* : 544 - 549, 1973.
[Chl.]

18402 - **ZELENSKIǏ, M.I., CHERNYAEVA, I.I.** : Ustanovka dlya provedeniya opredelenii fo-
tokhimicheskoǐ aktivnosti khloroplastov. [Apparatus for determination of photochemical activity of chloroplasts.] - In : Metody Kompleksnogo Izucheniya Fotosinteza. Vol. 2. Pp. 218 - 223. VASKHNIL, Leningrad 1973. [in R.]

18403 - **ZELENSKIǏ, M.I., SAKHAROVA, O.V.** : Metodika issledovaniya fotofosforilirovaniya
na osnove izmerenii pH. [Method of investigation of photophosphorylation on the
basis of pH-measurements.] - In : Metody Kompleksnogo Izucheniya Fotosinteza.
Vol. 2. Pp. 182 - 217. VASKHNIL, Leningrad 1973. [in R.]

18404 - ZELITCH, I. : Alternate pathways of glycolate synthesis in tobacco and maize leaves in relation to rates of photorespiration. - Plant Physiol. *51* : 299 - - 305, 1973.

18405 - ZELITCH, I. : Plant productivity and the control of photorespiration. - Proc. nat. Acad. Sci. USA *70* : 579 - 584, 1973.

18406 - ZELITCH, I. : The biochemistry of photorespiration. - Curr. Advance. Plant Sci. *1973* (6) : 44 - 54, 1973.

18407 - ZELITCH, I., DAY, P.R. : Pedigree selection for high and low rates of photo-respiration in tobacco. - Plant Physiol. *51* (Suppl.) : 42, 1973.

18408 - ZELITCH, I., DAY, P.R. : The effect on net photosynthesis of pedigree selec-tion for low and high rates of photorespiration in tobacco. - Plant Physiol. *52* : 33 - 37, 1973.

18409 - ZENCHENKO, V.A., CHURINA, M.B., SHAKHOV, A.A. : Svetochuvstvitel'nost' organe ètiolirovannykh prorostkov pshenitsy. [Photosensitivity of organelles of etio lated wheat sprouts.] - Dokl. Akad. Nauk SSSR *212* : 1243 - 1246, 1973. [Pho-torespiration.]

18410 ZHUCHENKO, A.A., ANDRYUSHCHENKO, V.K., MEDVEDEV, V.V. : Opredelenie ploshcha-di list'ev rasteniǐ. [Determination of leaf area in plants.] - Fiziol. Biokhi kul't. Rast. *5* : 211 - 212, 1973. [In R, ab : E.]

18411 - ZHURAVLEV, Yu.N., SAVEL'EVA, T.D. : Nekotorye osobennosti rosta i razvitiya rasteniǐ tabaka, porazhennykh VTM. [Certain features of the growth and deve-lopment of TMV-infected tobacco plants.] - Tr. biol.-pochv. Inst., dal'nevost nauch. Tsentr Akad. Nauk SSSR *14* (Virusnye Bolezni Rasteniǐ Dal'nego Vostoka) 64 - 72, 1973. [Growth analysis; in R, ab : E.]

18412 - ZICKLER, H.-O., WILD, A. : Untersuchungen über den Gehalt von Ferredoxin und Plastidenchinonen bei einer Plastommutante und ihrer Normalform von *Antirrhi-num*. - Ber. deut. bot. Ges. *86* : 331 - 340, 1973.

18413 - ZIEGLER, I. : Effect of sulphite on phosphoenolpyruvate carboxylase and malat formation in extracts of *Zea mays*. - Phytochemistry *12* : 1027 - 1030, 1973.

18414 - ZINCA, N., IONESCU, P. : Influenţa atacului produs de *Agrobacterium tumefaci-ens* asupra unor procese fiziologice şi biochimice din viţa de vie. [Influence of *Agrobacterium tumefaciens* attack on physiological and biochemical proces ses in vine.] - Stud. Cercet. Biol., Ser. Bot. *25* : 87 - 93, 1973. [Ps; in Roum., ab : F.]

18415 - ZINSOU, C., COSTES, C. : Photodécoloration du β-carotène en solution. - Phy-siol. vég. *11* : 191 - 206, 1973.

*18416 - ZOSIMOVICH, V.P., BORISENKO, T.T. : Osobennosti nasledovaniya intensivnosti fotosinteza triploidnymi gibridami sakharnoǐ svekly. [Peculiarities of inhe-ritance of photosynthetic rate by triploid hybrids of sugar-beet.] - In : Ge-neticheskie Aspekty Fotosinteza. Tezisy Dokladov. Pp. 101 - 102. Donish, Du-shanbe 1972. [In R.]

18417 - ZRŮST, J. : Fotosyntetická aktivita terčíků listové čepele různě hnojených brambor. [Photosynthetic activity of discs of leaf blade of differently fert ilized potatoes.] - Věd. Práce výzk. Ústavu bramborář. Havlíčkově Brodě *5* : 137 - 149, 1973. [In Czech, ab : E, G, R.]

18418 - ZRŮST, J. : Vliv umístění listů na proměnlivost hodnot rychlosti fotosyntézy u brambor. [Effect of leaf insertion on the variability of photosynthetic ra-te in potato.] - Rostl. Výroba *19* : 243 - 252, 1973. [In Czech, ab : E, R.]

18419 - ZSOLNAY, A. : Hydrocarbon and chlorophyll : a correlation in the upwelling r(gion off West Africa. - Deep-Sea Res. *20* : 923 - 925, 1973.

18420 - ZUTE, S.O. : Pervichnaya produktsiya i destruktsiya organicheskogo veshchest· v yuzhnoǐ chasti Rizhskogo zaliva. [Primary production and destruction of or· ganic matter in the southern part of the Gulf of Riga.] - Latv. PSR Zinat. A· kad. Vēstis *1973* (1) : 32 - 35, 1973. [Ps; in R, ab : E.]

Authors' names are presented in the form in which they appear in the respective publication. The names from papers published in Cyrillic characters are transcribed as
shown on p. III of this volume. Alternative spellings and forms of the name of the
same author are usually cross-indexed. The numbers in italics refer to publications
in which the respective author acts as an editor.

A

AASE, J.K. 15553
ABBASOV, G.S. 17666
ABDULLAEV, H.A. 15368, *15369
    see ABDULLAEV, Kh.A.
ABDULLAEV, Kh.A. *18154
    see ABDULLAEV, H.A.
ABDULLAEVA, S.K. 15370, *16250
ABDURAiKHMANOVA, Z.N. *15371
ABERCROMBIE, M. *17718*
ABILOV, Z.K. 15372-3
ABOU-ZIED, E.N. 15374
ACKERMANN, W.C. *17970*
ACKMAN, R.G. 17146
ACZÉL, A. *15375
ADABRA, Y. 16960
ADAMOVA, N.P. 17829
ADAMS, M.S. 15376, 16032, 16977
ADEDIPE, N.O. 15377-9
ADEY, W.H. 15380
AFANAS'EV, E.A. 17059-60
AGALIDIS, I. 15381
AGATA, W. 16905-6
AGEEVA, O.A. 16118
AGHION, J. 15943, 15947
AGZAMOV, A. 16573
AHMADI, N. 15382
AHMADJIAN, V. *17617*
AHMED, A.M. 15383, 17553
AIKAZYAN, V.Ts. 15384
AINIS, L. 17090
AKAZAWA, T. 17339-40, 18024
AKHMEDOV, B. *15385
AKITA, S. 15386-9˙
AKIYAMA, T. 18030
AKOYUNOGLOU, G. 15390, 15439-40,
    17115
AKULOVA, A.E. 15391
AKULOVICH, N.K. 15392-4
ALBERTE, R.S. 15395
ALBERTSSON, P.-A. 16963
ALDEN, R.A. 17685
ALDERFER, R.G. 15396
ALDRICH, H.C. 17451
ALETNIKOV, I.M. 16974
ALEKSEEV, E.V. 16138
ALEXANDER, V. 15842
ALFANO, R.R. 17759
ALI, A. 16147
ALIEV, Z.Sh. *15694
ALIKOV, Kh.K. 15397
ALINA, B.A. *16137
ALLAWAY, W.G. 15398-9, 16528, 17403
ALLEN, C.F. 18109
ALLEN, H.L. 15400, 16247

ALLEN, J.F. 15401
ALLEN, M.M. 16072
ALLEN, T.D. 15402
ALLEN, T.F.H. 15403
ALLEN, W.A. 16223-4
ALLERHAND, A. 16288
ALLESSIO, M.L. 15404
ALLISON, J.C.S. 18322
AL'PEROVICH, L.I. 15470
ALPHEN-van WAVEREN, O. van 18288
AL-SANI, N. *16084
ALYPOV, V.F. 18053
AMBROSAU, A.L. 15405
AMESZ, J. 15406-11, 18190
AMICO, V. 16257
AMIRDZHANOV, A.G. 15412
AMIRSLANOV, I.A. 17666
ANDERSON, B. 16238-9
ANDERSON, I.C. 15476
ANDERSON, J.E. 15413
ANDERSON, J.L. 15414-5
ANDERSON, J.M. 15416-8
ANDERSON, J.W. 15730
ANDERSON, L.E. 15419, 17412, 17428
ANDERSON, M.M. 15420
ANDERSON, R. 17113
ANDERSON, S.M. 15421
ANDERSON, W.P. *17066*
ANDERSSON, G. 15422
ANDREENKO, T.I. 15423
ANDREEVA, T.F. 15424, 15465, 17320
ANDREI, M. -*15425
ANDREO, C.S. 15426
ANDREOLI, C. 17089
ANDREWS, T.J. 15427, 17040
ANDRIANOV, V.G. 15566
ANDRIANOV, V.K. 15723-4
ANDRYUSHCHENKO, V.K. 18410
ANISIMOV, A. 15428
ANISIMOV, A.A. 15429, *16674
ANISIMOV, B.V. 17768
ANISIMOVA, I.N. 16928
ANPILOGOVA, N.N. 15430
ANSELL, G.B. *16195*
ANSTIS, P.J.P. 15431
ANTHONY, D.S. 18139
ANTONGIOVANNI, M. 16950
ANTONYAN, A.A. 17170
ANTOSZEWSKI, R. 15432-3, 16936, 17187,
    18344
ANUFRIEVA, V.G. 17866
d'AOUST, A.L. 15434
D'AOUST, B.G. *15435
APEL, P. 15436-7

ROZIER, C. 17080
ROZONOVA, L.N. 18392
ROZOV, N.F. *17785
RUBERTÉ, R.M. 17129
RUBIN, A.B. 16844-5, 17546-8, 17647-8,
    *18365
RUBIN, B.A. 17649-50
RUBIN, L.B. 15423, 16558, 16604-5,
    17651-2, 17829
RUCKENBAUER, P. 17653, 18395
RUDIN, V.D. 16855
RUDOI, A.B. 17654-5, 17814
RUDOLPH, E. 18394
RUDOVA (ZHUKOVA), T.S. 16803-4
RUETZ, W.F. 17656
RÜGGEBERG, B. *17975
RUMSBY, M.G. 16983
RUNDEL, P.W. 17167
RUNGE, M. 17657-9
RUPPEL, H.G. 16939
RUSKA, H. *16025*
RUSKOVA, M.Kh. 16663
RUTKEVICH, N.M. 16850
RUUGE, É.K. 16612, 16910
RYAN, F.J. 17660
RYBIN, I.A. 17190, 17364, 17661-2
RYCZKOWSKI, M. 17663
RYLE, G.J.A. 17664
RYZHOVA, E.F. 17665
RZAEV, A.S. 16912
RZAEV, I.T. 17666

S

SAAKOV, V.S. 17064, 17094, 17122,
    17667-9
SABAD, I. 17670
SACKETT, W.M. 18298
SADLER, D.M. 17671
SADOVNIKOVA, L.G. *16941, 16942
SAGAR, G.R. 15798, 15847-8, 17154
SAGROMSKY, H. 17672
    see ZAGROMSKI, G.
SAHU, G. 17280
SAIFULLAH, S.M. *17673
SAIJO, Y. 17674
SAITO, K. 17675
SAITO, N. 18380
SAKALO, V.D. 17676
SAKANO, K. 16916, 17677
SAKHAROVA, O.V. 17014, *17678, 18403
SAKHIBNAZAROV, Sh. *18154
SAKSHAUG, E. 16632, 17679
SALAI, L. 17670
SALAMON, Z. 17680-1
SALE, P.J.M. 17682-3
SALEH, H.M. 16086
SALEMA, R. 17684
SALEMMÉ, F.R. 17685
SALIN, M.I. 15955, 17686-8
SALISBURY, F.B. 16774, 17689
SALVADOR, G.F. 17690
SALZER, J. 16180

SALZMANN, C. 16196
SAMEJIMA, M. 18157
SAMISH, Y.B. 17417, 17691
SAMOILOVA, L.A. 15747-8
SAMOKHVALOV, G.I. 16122
SAMPAT, T.V. 15438
SAMSEL, G.L. Jr. 17431, 17590, 17692
SAMSONOVA, I.A. 17693-4
SAMUILOV, V.D. 15517-8, 16266, 16844,
    17405, 17695
SANADI, D.R. *16145, 16715, 17833*
SÁNCHEZ, R.A. 17696
SANDY, J.D. 15916, 17310-2
SANEI, F. 15636
SANGER, J.E. 17697
SANKHLA, N. 16537, 17698
SAN PIETRO, A. 15630, 15672, 15892,
    17011, 17210, 17833, 18375
SANTARIUS, K.A. 16448, 16958, 17699-700
SANTHANAM, R. *17701
SANZ, M. 16124
SAPOZHNIKOV, D.I. 16849, 17498, 17702-3
SARAIVA, M.C. 17704
SARDA, C. 16352
SARIĆ, M. 17611
SARIĆ, M.R. 17705
SASA, T. 16559
SASTRY, P.S. 15483
SATO, K. 18025
SATOH, K. 17706
SAUER, K. 15472-3, 17138, 17382, 17466-7
SAUNDERS, V.A. 15862
SAUVEZON, R. 16018
SAVCHENKO, G.E. 15783, 17707, *17812,
    17813
SAVEL'EVA, R.P. 17708
SAVEL'EVA, T.D. 18411
SAWAMURA, M. 17709-10
SAWHNEY, R. 17711
SAXENA, J. 17712-3, 17751
SAZYKINA, N.A. 17189
SCAIFE, M.A. 17714
SCARSBROOK, C.E. 17715
SCAWEN, M.D. 17570
SCHAEDLE, M. 16151, 16465
SCHALLDACH, I. 15437
SCHAREN, A.L. 16900
SCHEGG, E. *17716
SCHEIBE, J. 15939
SCHENK, H.E.A. *17717
SCHIFF, J.A. 16024, 16276, 17718,
    *17801
    see SHIFF, D.A.
SCHILLING, N. 17719
SCHILLINGER, J.A. 16900
SCHINDLER, D.W. 16051, 17720-2
SCHLUNEGGER, U. 15527
SCHMID, G.H. 15853, 17558
SCHMID, R. 16802
SCHMIDT, A. 17723
SCHMIDT, G.W. 17724-6
SCHMIDT, K. 17727-8
SCHMIDT, L. 16957, 16959
SCHMIDT, R.E. 18273

This index contains a selection of primary items chosen according to their interest for photosynthesis researchers and to their relative importance and occurrence.
The word "Photosynthesis" is only rarely regarded as a main theme, but the individual
factors affecting photosynthesis are listed. Numbers of selected references containing
a more comprehensive information (review articles *etc.*) are printed in italics. An asterisk (*) denotes references to papers published prior to 1973.

A

Accumulation of dry matter in leaf discs see Gravimetric methods (dry-matter-accumula-
    tion methods)

Action spectra of photosynthesis 15840, 16022, *16036, 16277, 16279, 16281, 16372,
    16380, 16528, 16584, 16609, 16754-5, 16978, 16988, 17152, 18310

Aerodynamic methods, theory (*cf.* also Bioclimatological methods) 15553, 15903, 16413,
    18020

Age of tissue and plant, effect on carotenoids 16522, 16589, 16662, 16717, 16829,
    16950, 17961

Age of tissue and plant, effect on chlorophyll 15393-4, 15447, 15477, 15480, 15487,
    15507, 15783, 15990, *15998, 16109, 16123, 16224, 16292, 16339, 16411, 16418,
    16465, 16470, *16480, 16493, 16513, 16522, 16589, 16643, 16679, 16717, 16736,
    16741, 16829, 16983, 17050, 17188, 17212, 17320, 17515, 17581, 17599, 17799,
    17807, 17826, 17876, 17925, 17961, 18018, 18037, 18247-8, 18271, 18332-3,
    18356, 18381-2

Age of tissue and plant, effect on photosynthesis 15462, 15465, 15477, 15487, 15495,
    15596, 15668, 15695, 15707, 15731, 15787, 15796, 15798, 15816, 15821, 15906-7,
    15970, 15997, 16088, 16109, 16119, 16185-6, 16339, *16346, 16366, 16411, 16414,
    16418, 16570, 16578, 16589, 16598, 16626, 16643, 16653, 16679, 16736-7, 16741,
    16763, 16789, *16862, 16902, 16913, 16987, 17161, 17163, 17225, 17272-4, 17320,
    17485, *17493, 17525, 17599, 17663-4, 17686, 17688, *17785, *17797, 17799,
    17826, 17914, 17948, 17954, 18004, 18037, 18127, 18163, 18229, 18247, 18332-3,
    18338, 18356, 18387, 18408, 18411, 18418

Air flow rate, effect on photosynthesis see Wind (air-flow rate), effect on photosyn-
    thesis; Chamber, assimilation

Algae carotenoids 15552, *15923, 16172, 16330, 17346-7, 17891, 17927, 18179, 18274,
    18276

Algae chlorophylls *15923, 15993, 16072, 16167

Algae cultures, productivity see Productivity of algae cultures

Algae life cycle, photosynthesis and photosynthates 15573, 15677, 15900, 15909,
    15973, 16382, 16940, 17176, 17741, 17885, 17985, 18058, 18260, 18286, 18362

Algae life cycle, pigments in (*cf.* also Carotenoids, seasonal changes; Chlorophylls,
    seasonal changes) 15383, 15573, 15677, 15900, 15909, 15973, 16401, 16940,
    17843, 17862, 17864, 17885, 18057-8, 18362

Algae symbiosis in lichens see Lichens photosynthesis

Altitude and pressure, effect on photosynthesis 15413, 15461, 15567, 15715, 16191,
    16641, 17476, 17772

δ-Aminolaevulinic acid see Chlorophyll biosynthesis

Anatomy of leaf see Leaf properties (anatomical *etc.*) and photosynthesis

Animals, chlorophylls in see Heterotrophy ...

Antibiotics, effect on photosynthesis see Inhibitors and uncouplers of photosynthesis

Antibiotics, effect on pigments 15572, 15601, 15654, 15989, 16327, 16589, 16631,
    17366, 17447, 17581, 17789, 17808, 17811, *17812, 18054, 18145

Antigens see Serological methods

Antitranspirants (*cf.* also Transpiration and photosynthesis) 15918, 16596, 16936,
    16980, 17047-8, 17118, 17292, *17415, 18128

Carotenoids, effect of pesticides on see Pesticides, effect on pigments

Carotenoids, effect of temperature on see Temperature, effect of carotenoids

Carotenoids, effect of water supply on see Water supply and pigments

Carotenoids genetics see Genetics of pigment content

Carotenoids in algae see Algae carotenoids

Carotenoids in lipoprotein complexes see Carotenoids *in vivo*

Carotenoids in model systems see Carotenoids in photosynthesis ...

Carotenoids in photosynthesis and model photosynthetic reactions 15421, 15470, 15502,
     15517, 15585, 15751, 15773, 15881, 15909, 15968, B16034, 16082, 16138, 16183,
     16278, 16361, 16380, 16589, 16614, *16689, 16741, 16813, 16849, 17137, 17140,
     17229, 17285, 17361, 17377, 17473, 17498, 17507, 17558, *17560, 17668, 17670,
     17680-1, 17702-3, 17752, 17769, 17870, 17999, 18112, 18263, 18398-9

Carotenoids in photosynthetic bacteria see Photosynthetic bacteria, carotenoids in

Carotenoids in seeds see Seed, pigments in

Carotenoids in tissue culture see Tissue culture, pigments in

Carotenoids *in vivo* 15500, 15505, 15773, 15810, *16036, 16843, 17068, 17072, 17134,
     17668, 17800, 18114

Carotenoids, methods of analysis *15375, 15500, 15614, *15833, 17014, 17121, 17129,
     17259, 17332, 17410, 17464, 17646, 17665, 17852, 18276, 18278

Carotenoids, pathological effects on see Pathological effects on pigments

Carotenoids, seasonal changes 15383, 15564, B15671, 15680, 15720, 15984, 16039,
     16050, 16083, 16086-7, 16717, 16757-8, 17063-4, 17197, 17212, 17255, *17678,
     17866, 17869-70, 17874, 17909, 18113-4

Carotenoids, transformations of see Carotenoids in photosynthesis ...

Chamber, assimilation 15519, 15535, 15796, 15956, 16016, 16018, 16392, 16423, 16648,
     16830, 16852, 16854, 16865, 16902, 17042, 17069, 17227, 17272-3

Chemiosmotic hypothesis, proton transport 15467, 15503, *15563, 16071, 16443, 16454,
     16456, 16479, 17469, 18145-6, 18240-1

Chlorobium chlorophylls see Bacteriochlorophyll ...

Chlorophyll and plankton production see Plankton standing crop ...

Chlorophyll biosynthesis (*cf.* also Irradiance, effect on chlorophyll) 15374, 15377,
     15390, 15392-5, 15439-40, 15492, 15500, 15525, 15536-7, 15628-9, 15661, 15696,
     15759, 15775, 15780, 15783, *15916*, 16002, 16023-5, 16042,*16092, 16095-8,
     16103-4, 16122, 16147, 16152, 16184, 16190, 16196-7, 16205, 16207, 16215-7,
     16226, 16228, 16295, *16305, 16322, 16324, 16347-8, 16425, 16458-60, 16474,
     *16499, 16507, 16522, 16568, *16669, 16670, 16678, 16718, 16757-8, 16787,
     16816, 16971, 16974, 17011, 17017, 17022-3, 17065-7, 17115, 17126, 17138,
     17141, 17144, 17281, 17283, *17290, 17310-2, 17328-9, 17366, 17371, 17381,
     17389, 17447, 17485, 17505-6, 17509, 17568, 17580-2, *17584*, 17585, 17601-2,
     17616, 17654, 17707, 17730, 17732, 17734, 17789, 17799, 17806-11, *17812,
     17813-5, 17829, 17843, 17846, 17856-7, 17862, 17865, 17887, 17931, 17935,
     17955-6, 17977, 18000-1, 18054, 18102-3, 18111, 18121, 18204, 18218, 18238,
     18339, 18401

Chlorophyll(s) chemical structure 15486, 15699, *15880, 16028, 16042-3, 16080, 16288,
     16457, 16554, 16556, 16655, 16738, 17146

Chlorophyll, *Chlorobium* see Bacteriochlorophyll

Chlorophyll degradation see Pigment degradation

Chlorophyll delayed light emission see Chlorophyll fluorescence ...

Chlorophyll derivatives see Chlorophyll(s) chemical structure; Chlorophyll(s), methods
     of analysis; Pigment degradation; Chlorophyllase

Chlorophyll dimers see Chlorophyll-lipoprotein complexes ...

Chlorophyll, diurnal variation 15383, 15614, 15697, B15969, 16965

Chlorophyll, effect of air humidity on see Humidity ...

Chlorophyll, effect of antibiotics on see Antibiotics, effect on pigments

Chlorophyll, effect of CO_2 on see CO_2 and bicarbohates, effect on pigments

Chlorophyll, effect of growth regulators on see Growth regulators, effect on chloro-
      phyll

Chlorophyll, effect of ionizing radiation on see Ionizing radiation (*gamma*, X),
      effect on pigments

Chlorophyll, effect of irradiance on see Irradiance, effect on chlorophyll

Chlorophyll, effect of leaf and plant age on see Age of tissue and plant, effect on
      chlorophyll

Chlorophyll, effect of mineral elements on see Mineral elements, effect on chlorophyll

Chlorophyll, effect of oxygen on see Oxygen, effect on pigments

Chlorophyll, effect of pesticides on see Pesticides, effect on pigments

Chlorophyll, effect of photoperiod on see Photoperiod and chlorophyll

Chlorophyll, effect of temperature on see Temperature, effect on chlorophyll

Chlorophyll, effect of water supply on see Water supply and pigments

Chlorophyll energetics in model systems (*cf.* also Chlorophyll fluorescence ...;
      Chlorophyll monolayers ...) 15641, 15701, 15863, 16080, 16165, 16202, 16209,
      16351, *16457, 16539, 16567, 16714. 16822, 16839, 16866, 16999, 17577,
      17597, 17670, 17780, 17809, 17825, *17830, 17967, 17979, 18142, 18151, 18173

Chlorophyll excitation see Chlorophyll energetics in model systems; Chlorophyll
      fluorescence ...

Chlorophyll fluorescence and delayed light emission 15372-3, 15407-8, 15410-1,
      15475, 15490, 15500-1, 15506, 15532-3, 15543, 15562, 15572, 15614, 15629,
      15644, 15646, 15648, 15658, *15670, B15671, 15675, 15686, 15693, 15708, 15727,
      15737-9, 15860, 15938, 15991, 16023, 16028, B16034, 16067, 16071, 16076,
      *16092, 16120, 16145, 16160, 16165, 16167-8, 16181-2, 16188-9, 16215, *16250,
      16253, 16264, 16273-4, *16313*, 16333, 16341, 16364, 16395, 16405, 16444, 16488,
      16503, 16513, 16553, 16563-4, 16569, 16599, 16621, 16630-1, 16661, 16684,
      16686, 16688, *16689, 16690, 16743-4, 16769-71, 16815, 16822, 16843, 16874,
      16876-7, 16879-80, 16887, 16909, 16969-70, 17015, 17022, 17024, 17041, 17055,
      17061, 17110, 17144, 17147, 17191-2, 17220-3, 17269, 17271, 17276-7, 17373,
      17380, 17406, 17423, 17585, 17589, 17608, 17623, 17670, 17681, 17743-5, 17758-
      -9, 17802, 17809, 17827, 17851, 17861, 17864-5, 17884-6, 17917, 17952, 17972,
      17976, 17979, 18014-5, 18118, 18162, 18173, 18190, 18200-2, 18205, 18341,
      *18365, 18393

Chlorophyll genetics see Genetics of pigment content

Chlorophyll(s) in algae see Algae chlorophylls

Chlorophyll in photosynthesis see Photosynthesis and chlorophyll

Chlorophyll in tissue culture see Tissue culture, pigments in

Chlorophyll *in vivo* 15372-3, 15390, 15392-5, 15406-11, 15439-40, 15444, 15447, *15457,
      15472-3, 15500, 15517, 15529, 15532-3, 15543, 15645, 15680, 15693, *15694,
      15697, 15710, 15737, 15767, 15773, 15777, 15788, 15808, 15810, 15843, 15859,
      15968, 15970, 15991, 16023, *16036, 16063-4, 16093, *16112, 16153, 16160,
      16166, 16168, 16176, 16182, 16199, 16202, 16204, 16206, 16214-7, 16223-4,
      *16250, 16251, 16266, 16273, 16313, *16345, 16353, 16355, 16370-1, 16379-80,
      16448, 16452, 16459-60, 16475-6, 16488, 16513, 16531, 16569, 16573, 16582,
      16610, 16612, 16630-1, 16660, 16687, 16690, 16715, *16750, 16758-9, 16823,
      16843-5, 16874, 16897, 16926, 16932, 16986, 16988, 16995, 17011, 17018, 17022-
      -4, 17050, 17059, 17068, 17071-2, 17094, 17122, 17124, 17134, 17138, 17143,
      17168-9, 17185, 17220, 17309, 17354, 17361, 17377, 17381-2, 17406, 17436,
      17456, 17466-7, 17556, 17561, 17580, 17585, 17589, 17654, 17667, 17669, 17696,
      17757, 17759, 17764, 17800, 17803, 17809-11, 17814, 17827, 17851, 17865, 17867,

17869, 17874, 17884, 17904, 17913, 17956, 17968, 17991, 17994, 18000-1, 18014,
18021, 18028, 18060, 18062-3, 18068, 18113-4, 18118, 18181, 18193, 18218,
*18231-2, 18237, 18263-4, 18267, 18287-8, 18319, 18329, 18343, 18363, 18370,
18378

Chlorophyll-lipoprotein complexes, dimers, and solvates *in vitro* 15370, 15942-3,
15947, B16034, 17861

Chlorophyll-lipoprotein complexes *in vivo* see Chlorophyll *in vivo*

Chlorophyll(s), methods of analysis 15463, 15500-1, 15682, 15720, *15801, *15833,
15860, 15897-9, 15901-2, 16044-5, 16169, 16355, 16401, 16406, 16426, 16552,
16555-7, 16738, 16788, 17014, 17032, 17122, 17129, *17160, 17168-9, 17178,
B17180, 17269, 17277, 17474, 17504-6, 17733, 17766, 17800, 17814, 17824, 17851,
17875, 17905, 17913, 17992, 18060, 18118, 18187, 18257, 18276

Chlorophyll monolayers and multilayers *in vitro* (*cf*. also Chlorophyll energetics in
model systems) 16539, 17022, 17597, 17967, 17979, *18070, 18071

Chlorophyll, pathological effects on see Pathological effects on pigments

Chlorophyll phosphorescence see Chlorophyll fluorescence ...

Chlorophyll precursors see Chlorophyll biosynthesis; Chlorophyll(s) chemical structure; Chlorophyll(s), methods of analysis

Chlorophyll, reaction centre see P700 ...

Chlorophyll, seasonal changes (*cf*. also Age of tissue and plant, effect on chlorophyll) 15383, *15474, 15564, 15567, 15612, 15634, *15665, 15667, *15670,
B15671, 15680, 15688, 15702, 15720, 15757, 15781, 15783, 15810, *15875,
B15969, 16005, 16039, 16050, 16086-7, 16091, 16466, 16493, 16501, 16717,
16720, 16752, 16757-8, 16979, 17015, 17063, 17082, 17102, 17197, 17205, 17212,
17255, 17398, 17481, 17502, 17549, *17678, 17679, 17721, 17846, 17857, 17866,
17869-70, 17872, 17874, 17905, 17909, 17924, 17990, 18113-4, 18117, 18123,
18140, 18175-6, 18247

Chlorophyll sensibilized reactions (*cf*. also Chlorophyll energetics in model systems)
16739

Chlorophyll solvates *in vitro* see Chlorophyll-lipoprotein complexes ...

Chlorophyll unit see Photosynthetic unit

Chlorophyllase 16559, 16700, 17655, 17871-2, 17991-4

Chloroplast (and chromatophore) chemical composition 15372-3, 15382, 15401, 15420,
15481, 15483, 15730, 15740, 15785, 15850, 15877, 15971-2, 16132, *16137,
16195-6, 16210, 16238, *16248-9, 16276, 16342, 16396, 16451-2, *16469, *16482,
16483, 16505, 16507, 16530-1, 16575, 16634, 16659, 16683, 16717, 16760, 16774,
16804, 16828, 16835, 16889, 16982, 17005-6, 17034, 17121, 17127, 17140, 17143,
17155-6, 17171-3, 17209, 17216, *17238-9, 17265, 17343, 17375, 17422, 17429,
17441, 17488-90, 17515, 17532, 17569, 17602, 17633, 17639, 17753, *17889,
17893, 18036, 18084, 18087, 18090, 18109, 18143, 18170, 18184, 18281, *18385

Chloroplast development see Chloroplast ontogenesis ...

Chloroplast dimensions see Chloroplast number ...

Chloroplast (and chromatophore) fragments 15416-7, 15439, 15488, 15532, 15582,
15599, 15645-7, 15710, 15972, 16033, 16048, 16166, 16168, 16182-3, 16199,
16214, 16295, 16332, 16334, 16439, 16475, 16553, 16569, 16583, 16599, 16615,
16630, 16663, 16687-8, 16722-4, 16728, 16792, 16808, 16811, 16841, 16995,
17108, 17124, 17275, 17373, 17394, 17406, 17465, 17484, 17931, 18068, 18143,
18267, 18288

Chloroplast genetics see Genetics of photosynthetic apparatus

Chloroplast (and chromatophore) isolation 15593, 15649, 15972, 16132, 16210, *16299,
16332, 16334, 16441, *B16784, 16788, 17219, 17245, 17380, 17394, 17489-90,
17517, 17536, 17572, 17790, 18036, 18281

Chloroplast membrane transport *16298-9, 16300-1, 16443, 16735, 17350, 17459, 18262

Chloroplast membranes, localisation of electron transport chain components in
see Electron transport chain components, localisation in membranes

Ferredoxin-NADP (NAD)-reductase (*cf.* also Electron transport chain) 15582, 15601-2,
    15630, 15972, 16889, 18120, 18375

Ferredoxin, role in photosynthesis (*cf.* also Electron transport chain) 15450, 15558-
    -9, 15582, 15600-2, *15717*, 15761, 15892, 16075, 16118, 16155, 16241, 16319,
    16375-7, 16415, 16425, 16658, 16707, 16722, 16806, 16873, 16889, 16897, 16918,
    17181, 17210, 17262, 17445, 17723, 17833, 17893, 17982, 18047, 18120, 18300,
    18343, 18369, *18375*, 18412

Field studies of photosynthesis see Photosynthesis and dry matter production

Flooding see Rain ...

Fluorescence, methods of measuring see Spectrophotometric ...

Foliage formation and heterogeneity (*cf.* also Canopy structure and density) 17240,
    17384, 17906, 17938, 18012, 18042

Fraction I protein see Ribulose-1,5-bisphosphate carboxylase

Fruit photosynthesis 16046, 17526

Fucoxanthin see Algae carotenoids; Carotenoids ...

Fungi see Pathological effects ...

G

Gasometric methods, systems and circuits (*cf.* also Chamber, assimilation; CO_2 measu-
    rement ...; Infra-red gas analyser ...; Water-plant photosynthesis, methods)
    15493, 15786, 15866, B15894, 16051, 16088, 16191-2, 16598, 16788, 17025,
    B17179, 17249, 17691, B17821, 18007, 18108, 18363

Genetics of photosynthesis (*cf.* also Ecotypes, differences in photosynthesis)
    *15371, *15385, 15430, 15436, 15466, *15468, *15471, 15490, 15495, 15498,
    15568, 15585, 15602, 15607, *15608, 15623-4, *15640, 15695, 15708, *15744,
    *15807, 15851, 15873, 15904, 15907, 15930-1, 15966, B15969, *16012, 16056,
    *16092, 16131, *16159, 16237, 16246, *16305, 16306, 16403-4, 16524, 16547,
    16582, 16642, *16694, 16740-1, *16750, *16762, 16763, 16838, 16894, 16913,
    *16935, 16987, 17011, *17020, 17051, 17069, *17099, *17119, 17120, *17261,
    17268, 17280, 17285, 17291, 17294, 17321, *17349, 17398, 17417, 17437, 17529,
    *17560, 17565, 17579, 17621, 17653, 17756, 17762, 17788, 17867, 17873, 17890,
    17893, 17957, 17962, 17982, 17997, 18005, *18039, 18167, 18226, 18247, 18270-
    -2, 18335, *18365, *18388, 18407, *18416, 18418

Genetics of photosynthetic apparatus 15368-9, *15471, *15474, *15592, 15650, 15785,
    *15815, 15853, *16092, 16156, 16251, *16305, 16363, *16459*, *16469, *16708,
    16712, 16740-1, *16762, 16938, 16944, 16997, *17182-3, 17240, 17289, *17290,
    17291, *17349, 17541, 17587-8, 17592, 17694, *17778, 17826, *18022, *18075,
    18155, *18156, 18184, *18185, 18186, *18221, 18222, 18308, 18345

Genetics of pigment content 15454, *15457, *15471, 15476, 15490, 15507, *15524,
    15564, 15584-5, *15589, 15612, *15618-9, 15654, 15661, 15696, 15708, 15720,
    15767, *15801, *15833, *15839, 15873, *15875, 15908, 15920-1, *15924, 15925,
    *15926, 15984-5, 15995, *16011, 16037, 16086, 16091, *16092, *16094, *16100,
    16103, 16126, *16136, 16208, 16276, *16305, 16379, 16403, 16452, 16459, 16463,
    *16464, *16480, 16509, 16582, 16631, 16634, *16689, 16695, 16740-1, *16750,
    *16796, 16871, 16874, 16938, *16941, 16942, 16944, 16950, 16979, *17020, 17071,
    *17073, 17085, 17104, 17121, 17124, 17126, 17131, 17135, 17158, *17160, 17169,
    17191, *17261, 17268, 17278, 17285, 17289, *17290, 17295, 17327-8, *17349,
    17391-3, 17398, 17424, 17437, 17486, *17560, 17592, 17621, *17637, 17669,
    17672, *17678, 17693, 17760, 17809, *17832, 17862, 17867, 17872, 17890, *17946,
    17957, 17961, *17966, 17984, *18022, *18048, 18068, *18122, 18135, *18136,
    *18153-4, 18184, 18200, 18204, *18232, 18247-9, 18308, 18345, *18365, *18388,
    18412

Glycolate metabolism (*cf.* also Photorespiration) 15581, 15854-5, 16031, 16271,
    16391, *16764, 16778, 16789, 17081, *17177*, 17235, 17279, 17359, 17686, 17688,
    18084, 18166, 18168, 18235, 18404-6

Glyoxysome see Peroxisome

Grana and stroma lamellae differences (*cf.* also Chloroplast structure) 16199, 16214,
    16997, 17406, *17974, 18109

Gravimetric methods (dry-matter-accumulation methods) 16788

Growth analysis 15464, 15496, 15529, 15579, 15638, 15711, 15713, 15749, 15753,
    *15791, 15849, 15852, 15872, 15952, 15966, 15983, 16013, 16055-7, 16081,
    16127-8, 16142, 16178, 16270, 16287, 16367, 16400, 16412, 16495, 16517, 16520,
    16525, 16540, 16547, 16550, 16562, 16576, 16592, 16597, 16642-3, *16644, 16751,
    16783, 16810, 16821, 16824, *16893, 16902, 16906, 16934, 16952, 17052, 17078,
    17083, 17120, 17150, 17365, 17376, 17398, 17538, 17559, 17575, 17598, 17600,
    17629-30, 17645, 17653, 17682, 17714-5, 17755-6, 17762, 17788, 17855, 17923,
    17928, 17930, 17996-7, 18005, 18013, 18017, 18030, 18042, 18206, 18226-7,
    18322, 18372, *18373-4, *18396-7, 18411

Growth regulators, effect on carotenoids 15941, 16228, 16497, 16514, 16526, 17805,
    17977-8

Growth regulators, effect on chlorophyll 15447, 15454, *15471, *15484, 15754, 15911,
    15951, 16147, 16228, 16497, *16499, 16514, 16765, 16816, 16911, 17030, 17104,
    17154, 17188, 17204-7, 17213, 17215, 17317, 17398, 17437, 17443, 17495, 17808,
    17857, 17867-8, 17937, 17977-8, 17990, 18025, 18141, 18366

Growth regulators, effect on photosynthesis *15468, *15471, 15651, 15685, 16026,
    16378, 16709-10, 17035, 17149, 17317, 17398, 17437, 17564-6, 17713, 17804,
    17867, 17937, 17942, 17945, 17990, 18025, 18141, 18349, 18366

H

H_2 evolution, photosynthetic 15558-9, 16743, 17402, 17737, 18088

Hatch-Slack pathway see C_3 and C_4 plants ...; Carbon fixation pathways; Photosyntha-
    tes formation patterns

Heterotrophy, autotrophy, mixotrophy 15654, 15770, 15804, 16006, 16074, 16140, 16337,
    16579-80, 17157, 17304, 17541, 17617, 17940, 18104-6

Hill reaction   15372, 15417, 15582, 15599, 15601, 15642, 15693, *15694, 15709-10,
    15727, *15756, 15804, 15881, 15927, 15950, B15969, 15970, 15990, 16019, B16034,
    16183, 16199, 16214, 16238-9, 16264, 16279, 16309, 16332, 16404, 16439, 16444,
    16451, 16458, 16475, *16482, 16503, 16522, 16527, 16553, 16600, 16663, 16665-
    -6, 16686, 16688, *16762, 16763, 16775, 16792, 16874, 16887, 16982, 17004,
    17017-8, 17024, 17061, 17079, 17175, 17189, 17192, 17219, 17264, 17282, 17306,
    17317, 17360, 17380, 17399, 17456, 17468, 17485, 17530, 17558, *17560, 17561,
    17572, 17601-2, 17687, 17700, 17735, 17740, 17769, *17797, 17798, 17817, 17835,
    17867, 17870, 17873, 17895, 17952-3, 18036, 18099-101, 18105, 18207, 18229,
    18253, 18260, 18270-1, 18293, 18329, *18365, *18388

Hill reaction, methods 16788, 17572, 18402

Humidity of air, effect on photosynthesis and pigments (*cf.* also Water supply ...)
    15655, 16622, 16672, 16731, 16826, 17656, 17964, 18324, 18356

Hydrogen see H_2 evolution, photosynthetic

Induction phenomena in fluorescence see Chlorophyll fluorescence ...

Induction phenomena in photosynthesis see Transient phenomena in gas exchange

Infra-red gas analyser, use in photosynthesis measurement 15742, 17025, 17164,
    *17415, 17442, 18271

Inhibitors and uncouplers of photosynthesis 15382, 15411, 15414, 15418, 15426,
    15432, 15446, 15450-1, 15455-6, 15473, 15503, 15509, 15513, 15515, 15528,
    15545, 15558, 15562, 15568, 15588, 15598, 15600, 15625, 15631, 15658, 15700,
    15727, 15768, 15799, 15824-5, 15888, 15915, 15927, 15938, 15965, 15980, 16035,

16052, 16061, 16105, 16109, 16120, 16132, 16135, 16145, 16161-2, 16183, 16239,
16253, 16257, 16264, 16309-10, 16329, 16350, 16383-5, 16391, 16405, 16415,
*16419, 16420-1, *16437, 16444, 16456, 16458, 16476, 16481, 16492, 16500,
16503, 16553, 16563, 16574, 16594, 16599, 16612, 16618, 16646, 16666, *16669,
16687, 16728, 16775, 16799, 16812, 16818-20, *16870, 16874, 16909, 16970,
16982, 17005, 17049, 17061, 17067, 17095, 17103, 17108, 17136, 17191-2, 17221-
-3, 17226, *17261, 17282, 17315, 17336, 17342, 17358, 17399-400, 17408, 17440,
17444, 17450, 17468-9, 17474, 17557, 17595, 17605-7, 17623, 17662, 17667-8,
17688, 17695, 17711-2, 17718, 17735, 17740, 17745, 17747, 17751, 17754, 17761,
17791, *17794, *17801, 17802, 17811, 17813, 17816, 17822-3, 17867, 17870,
17907, 17947, 17957, 17987, 18058, 18099-101, 18128, 18137, 18146, 18166-7,
18169, 18180-1, 18205, 18228, 18254, 18275, 18289, 18375, 18401, 18405

Insertion see Age ...

Intracellular resistance see Mesophyll (intracellular) resistance

Ionizing radiation (*gamma*, X), effect on photosynthesis 15585, 16124, 16519, 16873,
     16918, 18147, 18149

Ionizing radiation (*gamma*, X), effect on pigments 15584-5, *15589, 15660, 15920,
     16304, 16656, 17131, 17134, 17760, 17781, 17843, 17905, 18054, *18136

Irradiance, effect on carotenoids 15441, 15722, 15751-2, 15804, 15932, 15985, 15987,
     16082-3, 16153, 16183, 16228, 16348, 16362, 16422, 16461, 16561, 16577, 16853,
     16932, 16960, 17000-1, 17006, 17063, 17072, 17371, 17373, 17498, 17652, 17665,
     17672, 17703, 18049, 18114, 18117

Irradiance, effect on chlorophyll 15383, 15393-4, 15411, 15417, 15441, 15476, 15487,
     15490, 15492, 15560, 15614, 15628, 15677, 15688, 15697, 15722, 15751-2, 15783,
     15804, 15810-1, 15838, *15839, 15873, 15877, 15898-9, 15932, 15951, 15987,
     15993, 16023-4, 16072, 16101, 16103-4, 16152-3, 16166, 16175, 16197, 16206,
     16222, 16228, 16257, 16266, 16295, 16319, 16324, 16327, 16339, 16348, 16352-3,
     16362, 16381, 16390, 16412, 16417, 16421, 16425, 16458, 16461, 16488, 16498,
     16507, 16513, 16561, 16564, 16568, 16577, 16590, 16662, 16670, 16686,
     16691, 16704-5, 16713, 16718, 16770-1, 16831, 16853, 16887, 16932, 16960,
     17000-1, 17006, 17011, 17063, 17072, 17102, 17118, 17141, 17185, 17201, 17204,
     17207, 17248, 17319, 17371,17373, 17471, 17481, 17510, 17527, 17568, 17580-2,
     17602, 17616, 17632, 17672, 17705, 17734, 17803, 17829, 17846, 17905, 17917,
     17925, 17931, 17976, 18001, 18049, 18053, 18063, 18083, 18102-3, 18114, 18117,
     18135, 18177, 18218, *18232, 18237-8, 18248-9, 18320, 18329, 18359, 18363,
     18384

Irradiance, effect (after-effect) on photosynthesis 15380, 15383, 15386-7, 15408,
     15417, 15424, 15452, 15459, 15477, 15487, 15490, 15495, 15498-9, 15509, 15511,
     15514, 15519, 15528, 15530, 15545, 15549, 15573, 15582, 15587, 15596, 15604-5,
     15607, 15610, 15633, 15655, 15668, 15676-7, 15695, 15700, 15703-4, 15706,
     15716, 15723, 15728, 15751-2, 15771, 15782, 15786, 15793-4, 15804, 15809,
     15826, 15837, 15840, 15873, B15894, 15907, 15927, 15949, 15957, 15959-60,
     15976, 15982-3, 15996-7, 16007-8, 16013, 16016, 16018, 16022, 16032-3, *16073*,
     16081, 16088, 16104, 16107-8, 16110-1, 16119-20, 16133, 16135, 16171, 16185,
     16194, 16218, 16230, 16252, 16257, 16260, 16270, 16277-8, 16281-2, 16317,
     16319, 16324, 16329, 16335, 16339, *16346, 16372, 16380, 16410, 16416-7,
     16423, 16447, 16450, 16458, 16473, 16478, 16489, 16511, 16519-20, 16540-1,
     16548, 16553, 16561, 16576, 16580, 16583, 16588, 16597-8, 16604, 16609, 16626,
     16628, 16647, 16654, 16659, 16661, 16672, 16678, 16698, 16713, 16729, 16754-5,
     16781, 16800, 16802, 16826-7, 16831, 16846, 16853, 16865, 16886-7, 16896,
     16902, 16932, 16973, 16977, 16991, 17016, 17035, 17037, 17054, 17074, 17078-9,
     17086, 17097, 17163-4, 17186, 17190, 17195, 17225, 17237, 17243, 17247, 17249,
     17250-1, 17268, 17280, 17286, 17320, 17322, 17338, 17351, 17355, 17358, 17373-
     -4, 17378, 17407, 17417, 17420, 17438, 17440, 17444, 17448, 17471, 17476,
     17520-2, 17538, 17559, 17574, 17576, 17591, 17600-1, 17617, 17622, 17624,
     17626, 17628, 17630, 17642, 17652, 17656-7, 17663-4, 17682-3, 17687, 17705,
     17711, 17715, 17720, 17727, 17736, 17747, 17771-2, 17777, *17784-5, 17787-8,
     17803, 17828, 17836, 17848, 17855, 17893, 17901, 17917, 17923, 17930, 17942-5,
     17957, 17962, 17964, 17972, 17976, 18026-7, 18030, 18043, 18049-50, 18053,
     18055, 18072, *18091, 18119, 18171, 18197, 18209-10, *18213, 18219, 18227,
     *18236, 18242, 18253-4, 18260, 18264, 18271, 18279, 18286-7, 18302, 18306-7,

Photosynthesis and dry matter production (*cf.* also Photosynthates, translocation, exudation and distribution) 15388-9, 15424, 15436-7, *15458, 15495, 15504, 15531, 15587, 15610, 15706, 15715-6, 15831-2, 15851, 15904, 15930, 15964, 15975, 15983, 15994, 16008, 16017, 16027, 16062, 16073, *B16157, 16211, 16225, 16244, 16369, 16496, 16524, 16550, 16585, 16598, 16643, 16827, 16833, 16900, 16905, 16912, 16972, 17018, 17028, 17042, *17119, 17132, 17167, 17219, 17240--1, 17250, 17286, 17321-2, 17334, 17420, 17538, 17550, 17563, 17565, 17591, 17626, 17657, *17785, *17863, 17906, 17923, 17999, 18152, 18182, 18199, 18207, 18216, 18223, *18225, 18287, 18302, 18321, 18333, 18335, *18388, 18389, *18396, 18400, 18405, 18407, 18417

Photosynthesis, diurnal course 15376, 15383, *15385, 15495, 15498, 15519, 15655, 15695, 15779, 15795, B15894, B15969, 16010, 16114, 16151, 16194, 16230, 16410, 16416, 16576, 16643, 16654, 16802, 16826, 16831, *16862, 16868, 17148, 17164, 17167, 17187, 17333, 17416-7, 17438, 17448, 17482, 17540, 17656, 17736, *17785, 17786, 17919, 17949, 17962, 18016, 18020, 18026-7, 18043, 18139, 18144, 18170--1, 18210-1, 18260, 18305, 18310, 18325, 18348, 18387

Photosynthesis, general considerations B16242, *B17390, *17493, 17552, B17922

Photosynthesis, seasonal course 15376, 15383, 15405, 15437, 15461, 15495, 15504, 15519, 15531, 15582, 15623-4, 15667, 15676, 15702, 15706, 15725, 15728, 15758, 15809, 15818, *15830, B15894, 15900, 15930, 15935, 15962, B15969, 16005-6, 16021, 16027, 16046, 16055, 16107, 16110, 16151, 16171, 16173, 16230, 16255, 16318, 16335, *16346, 16369, 16398, 16414, 16416, 16466, 16473, 16496, 16545, 16578, 16597, 16603, 16608, 16638, 16643, *16694, 16720, 16729, 16730, 16732, 16753, 16757, 16810, 16826, 16860, *16862, 16865, 16867, 16894, 16945, 16957, 17021, 17028, 17087, 17100, 17148, 17208, 17219, 17233, 17241-3, 17251, 17270, 17294, 17334, 17368, 17398, 17448, 17472, 17475-6, 17482, 17518, 17538, 17540, 17549, 17624, 17630, 17656, 17721, 17755, 17787, 17837, 17866, 17870, 17873, 17894, 17933, 17962, 17990, 18005, 18027, 18030, *18039, 18139, 18150, 18158, 18182, 18227, 18229, 18234, 18247, 18307, 18376, 18387, 18398-9, 18405, 18417-8

Photosynthetic bacteria, carbon fixation pathways *15807, 16140, 17157, 17247, 17713, 17878-9, 17936, 18024, 18134, 18390

Photosynthetic bacteria, carotenoids in 15502, 15773, 17405, 17728, 17752, 17858-9, 17884, 18263

Photosynthetic bacteria, chlorophylls in see Bacteriochlorophyll ...

Photosynthetic bacteria, electron transport chain 15381, 15408, 15423, 15475, 15502, 15637, 15761, 15772-3, *15843-4*, 15856, 15862-3, 15876, 15981, 16000-1, 16003, 16375-7, 16379, 16492, 16512, 16515, 16613-4, 16637, 16668, 16690, 16727, 16807, 16818-20, 16842, 16875, 16924-5, 17029, 17125, 17181, 17209, 17336, 17343-5, 17405, 17519, 17546-8, 17647, 17652, 17685, 17695, 17723, 17904, 18088, 18263

Photosynthetic quotient see Respiration and photosynthesis

Photosynthetic unit 15604, *15844*, 15948, B16034, 16163, 16741, 17373, 17436, 17757, 17884, 18062, 18264, 18321

Photosynthetically active radiation see Irradiance ...; Radiation regime in canopy

Photosystem I 15416, 15452, 15515, 15568, 15646-7, 15672, 15892, 16199, 16214, 16275, 16370, 16439-40, 16446, 16458, 16583, 16604, *16721*, 16722, 16724, 16775, 16808, 16823, 16898, 16998, 17124, *17239, 17373, 17484, 17558, 17706, *17833*, 17835, 18021, 18120, 18143, 18267

Photosystem II 15401, 15444, 15472, 15538, 15559, 15656, 15675, 15727, *15736*, 15739, 15881, 16014, 16058, 16061, *16092, 16183, 16303, 16310, 16476, 16503, 16509, 16553, 16564, 16569, 16611, 16646, 16997, 17049, 17061, 17105, 17108, 17192, 17377, 17601, *17794, 17816, *17888, 17952, 17976, 18099-101, 18194-5, 18288, 18330

Photosystems (*cf.* also Electron transport chain) *15408*, 15417-8, 15439, 15452, 15490, 15562, 15571, 15588, 15598-9, 15602, 15607, 15657, 15693, *15708*, 15710, 16025, B16034, *16036, 16214, 16239, 16259-60, 16334, 16341, 16380, 16441, 16444, 16485, 16582, 16630, 16661, 16688, *16689, *16750, 16792, 16806, 16886-7, 16897, 17016, 17335, 17399, 17406, 17471, 17485, 17554, 17604-8, 17651, 17758,

Productivity of algae cultures   17940

Productivity of aquatic communities   15376, 15380, 15400, 15403, 15422, *15425, 15443,
    15448, 15511, 15528, 15548, *15591, 15639, 15659, *15663, 15664, 15721, 15725,
    15842, 15870-1, 15893, 15895-6, 15919, 15928, 15940, 15946, 15963, 16069,
    16099, 16106, 16111, 16143, 16225, 16236, 16291, 16297, 16314, 16386-7, 16394,
    16409, 16416, 16442, 16468, 16500, 16516, 16518, *16533, 16534, 16545, 16551,
    16628, 16671, 16817, 16885, 16922, 16930-1, 16948-9, 16985, 17039, *17088,
    17089, 17117, *17202, 17308, 17325-6, 17348, 17356, 17407, 17432, *B17528,
    17635, *17716, 17720, 17724-5, 17831, 17841, 17903, 17910, 17918, 17920-1,
    17929, *17932, *17934, 17970, 18066-7, 18076-7, 18095, 18098, 18107, 18139,
    18165, 18250, 18252, 18255-6, 18294-6, *18347, 18420

Productivity of aquatic communities, methods   15403, 15578, 15842, 15893, 16041,
    16051, *16059, 16069, 16110-1, 16243, 16357, 16360, *16402, 16417, *16532,
    16627, 16629, 16696, 16949, *17088, 17159, 17174, 17256, 17433, 17478, 17480,
    17539, 17643, 17720-1, 18033, 18116, 18187, 18251, 18256, 18265, 18269, 18383

Proplastid see Chloroplast ontogenesis ...; Chloroplast structure

Protochlorophyll(ide) see Chlorophyll biosynthesis; Chlorophyll(s), methods of analysis

Proton transport see Chemiosmotic hypothesis

Protoplast photosynthesis and pigments   16019, 16681-2, 17511, 18279

Pteridines in photosynthesis   17533

Q

Quantasome see Photosynthetic unit

Quantum yield and requirement   15526, 15549, 15604, 15647, 15949, 16219, 16444,
    16447, 16892, 17411, 17606, 17915, 18264

Quinones in photosynthesis   *15409*, 15565, 15602, 15604, *15884*, 15989, 16025, 16170,
    16265, 16309, 16319, 16370-1, 16439, 16668, 16878, 16897, 17234, 17343, 17345,
    17358, 17377, 17558, 17634, 17977-8, 18021, 18192, 18412

R

Radiation see Irradiance ...; Radiation regime in canopy

Radiation, ionizing (*gamma*, X)   see Ionizing radiation

Radiation regime in canopy (*cf.* also Irradiance ...)   15396, 15496-7, 15529, 15604,
    15638, 15676, 15690, *15791, 15792, 15794, 15813, 15838, 15933, 15988, 16063,
    16307, 16315, 16317, 16412, 16511, 16576, 16595, 16779, 16825, 16838, 16892,
    16919, 16952, 16989-90, 17019, 17070, 17383-7, 17629, *17784, 17836, 17849,
    17892, 17938-9, 17965, 18018, 18094, 18129, 18321, 18338, *18373

Rain, flooding, effect on photosynthesis and pigments   16055

Reaction centre chlorophylls see P700

Resistance, carboxylation see Carboxylation resistance

Resistance to CO_2 and water vapour transfer (*cf.* also Bioclimatological methods;
    Resistance to CO_2 transfer, stomatal)   15460, 15516, 15534, 15540, *15569*,
    15604, 15713, 15779, 15786, 15976, 15979, 15983, 15997, *16073*, 16173, 16245,
    16316, 16356, 16414, 16478, 16576, 16649, 16652, 16993, 17021, 17163-4, 17195,
    17243, 17293, 16417, 17438, 17452, 17521, 17540, 17640, 17749, 17881-2, 17960,
    17980, 18020, 18044, 18050-1, 18144, 18163, 18207, 18321, 18356, 18364

Resistance to CO_2 transfer, mesophyll (intracellular)   *17797, 17880

Resistance to CO_2 transfer, stomatal (*cf.* also Porometer ...; Resistance to CO_2 and
    vapour transfer; Stomata and photosynthesis; Transpiration and photosynthesis)
    15386, 15596-7, 15607, 15979, 16060, 16450, 16652, 17045, 17048, 17231, 17292,
    17372, 17427, 17483, 17783, 17786, 17793, *17797, 17882, 17911, 18044, 18128-
    30

Temperature, effect on photosynthesis 15380, 15413, 15430, 15459, 15498, 15516,
    15519, 15528, 15530, 15540, 15583, *15587*, 15604-6, 15610, 15655, 15667,
    *15698, 15704, 15706, *15744, 15755, 15793, 15805, 15809, 15852, B15894,
    15935, *15936, 15958, 15961, 15976, 15979, 15981, 15983, 16007, 16010, 16032,
    *16059, 16081, 16108, 16171, 16185, 16194, 16218, 16229-30, 16270, 16282,
    16335, *16346, 16356, 16382, 16410, 16416, 16462, 16473, 16489, 16492, 16562,
    16576, 16578, 16596-8, 16624, 16646-7, 16672, *16694, 16753, 16760, 16800,
    16802, 16826, 16865, 16946, 16956, 16958, 17078, 17086, 17148, 17161, 17163-
    -4, 17189, 17195, 17208, 17229, 17242-3, 17249, 17262-7, 17273, 17297-8, 17361,
    17374, 17446, 17448, 17456, 17471, 17476, 17540, 17559, 17591, 17617, 17642,
    17656, 17688, 17699, 17712, 17714, 17776, 17817, 17841, *17889, 17898, 17914,
    17962, 17964, 18025, 18027, 18072-3, *18091, 18105, 18210, 18227, 18253-4,
    18273, 18287, 18293, 18315, 18321, 18327, 18338, 18348, 18351, 18376, 18405

Temperature of leaf and canopy, measurement 15535, 15812

Thermistor see Temperature of leaf and canopy, measurement

Thermocouple see Temperature of leaf and canopy, measurement

Thylakoid see Chloroplast ontogenesis ...; Chloroplast structure; Chromatophore
    structure

Tissue culture, photosynthesis in 15466, 16523, 16903, 17317, 18350

Tissue culture, pigments in 16526, *17044, 17317, 17491, 17730, 17805, *17995,
    18370, 18379-80

Transient phenomena in gas exchange 15922, 16022, 16787, 17554, 18253

Translocation of photosynthates see Photosynthates, translocation, exudation and
    distribution

Transpiration and photosynthesis (*cf.* also Stomata; Stomata and photosynthesis ;
    Resistance to CO_2 transfer, stomatal; Water supply and photosynthesis) 15386-
    -9, 15413, 15446, 15460, 15534, 15539-40, 15596, 15605, 15610, 15655, 15683,
    15728, 15796, 15918, 15935, *15945, 15967, 15979, 16060, 16191, 16218, 16316,
    16400, 16622-3, 16649, 16867, 16972, 17047, 17167, 17195, 17242-3, 17292,
    17298, 17342, 17376, *17415, 17416, 17446, 17483, 17544, 17881, B17899, 17996-
    -7, 18044, *18091, 18163, 18210, 18323, 18348, 18414

Tricarboxylic acid cycle see Carbon fixation pathways; Photosynthates formation
    patterns

U

Ubiquinone see Photosynthetic bacteria, electron transport chain; Quinones in photo-
    synthesis

Uncouplers of photosynthesis see Inhibitors and uncouplers ...

V

Vegetation changes in photosynthesis see Photosynthesis, seasonal course

Violaxanthin see Carotenoids ...

Vitamins K see Electron transport chain; Quinones in photosynthesis

Volume of photosynthetic organs see Production of dry matter measurement, methods of

Volume of plant organs, measurement see Leaf area and volume measurement

Volumetric methods see Manometric and volumetric methods

W

Warburg effect (*cf.* also Oxygen, effect on photosynthesis; Photorespiration) 17305,
    *18213, 18406

This index contains a selection of plant genera and types interesting as experimental material for physiological, ecological and agricultural studies. In general, mainly those plant names have been included which are given in the title of
the respective papers or in abstracts. The common English plant names are the main
items which present the reference numbers. An asterisk (*) denotes references to
papers published prior to 1973.

A

Abies see Fir

Acer see Maple

Acetabularia 17176, 17790, 17864, 18170-2, 18203

Alder 15846, 16410, 18372, *18397

Alfalfa 15389, 15442, 15561, *15592, 15735, 15754, 15795, 16195, 16870, 16902,
    17425, 17641, 18321

Algae (*cf.* also *Acetabularia, Chlamydomonas, Chlorella, Chrysophyta*, Diatoms, Dino-
    flagellates, *Dinophyceae, Euglena, Scenedesmus*)
    15380, 15400, 15403, 15422, 15443, 15448, 15501, 15541-3, 15545, 15551,
    15575, 15577-8, 15588, 15611, 15639, 15664, 15667, 15681, 15692, 15697, 15702-
    -3, 15705, 15710, 15714, 15721, 15725, 15757, 15761, 15781, 15842, 15845,
    15870-1, 15882, 15893, 15901, 15905, 15922, 15937-40, 15948-50, 15962-3,
    15986, 16038-9, 16047, *16059, 16065, 16099, 16101, 16106, 16193, 16195,
    16200, 16202-3, 16211, 16243, 16247, 16290, 16297, 16337, 16386-7, 16394-5,
    16407, 16429, *16435, 16436-7, 16467-8, 16472, 16488-90, *16532-3, 16543-5,
    16552, 16560-1, 16609-10, 16617, 16628, 16632, 16657, 16707, 16769-71, 16776-7,
    16834, 16843, 16860, 16872, 16882-5, 16891, 16897, 16910, 16922, 16931, 16933,
    16939, 16948-9, 16985, 16991, 16998, 17039, 17077, 17086-7, *17088, 17089-90,
    17096, 17117, 17123, 17147, 17153, 17177-8, 17247, 17252, 17256, 17269, 17281,
    17325-7, 17346-8, 17356, 17431-3, *17434, 17477-8, 17480-2, 17496, *B17528,
    17539, 17595, 17609-10, 17612, 17643, *17673, 17679, 17692, 17720-6, 17767,
    17803, *17888, 17903, 17910, 17912, 17918-21, 17929, 17933, *17934, 17970-1,
    17983, 17985, 17989, 18011, 18026-7, 18038, 18062, 18074, 18076-7, 18097-8,
    18111, 18165, 18250-1, 18264, 18294, 18296, 18393

Algae, blue-green (*cf.* also *Anabaena, Anacystis, Nostoc*)
    15422, 15485, 15545, 15550, 15558, 15560, 15568, *15591, *15663,
    15708, 15755, 15757, 15764, 15770, 15776, 15789, *15815, 15854-5, 15871, 15885,
    15898-900, 15937-8, 15958, 15961, 15968, *16036, 16053, 16075, 16099, 16102,
    16125, 16165, 16182-3, 16195, 16261-3, 16320-1, 16338, *16359, 16368, 16375,
    16377, 16380, *16532, 16579-80, 16610, 16627, 16671, 16684, 16715, 16725-6,
    16742, 16806, 16840, 16843, 16853, 16923, 16949, 16954, 16961, 16984, 16988,
    17022, 17074, 17076, 17089, 17123, 17166, 17210, 17221-3, 17269, 17308, 17330,
    17358, 17378, 17402, 17411, 17440, 17450, 17467, 17474, 17477, 17497, 17507,
    17622, 17651, *17717, 17718, 17726, 17771-2, 17779, *17794, 17838-9, 17890,
    17895, 17901, 17908, 17926, 17940, 17957-8, 18059, 18068, 18095, 18110, 18126,
    18179, 18331, 18337, 18342, 18383-4, *18385, 18391

Algae, brown 15574, 15993, 16273, 16335, 16380, 16610, 16884, 17057, 17281, 17299,
    17481, 17742, 17951, 18085, 18280, 18312-3, 18376, 18398-9

Algae, green (*cf.* also *Acetabularia, Chlamydomonas, Chlorella, Dunaliella, Scenedesmus, Ulva*)
    15376, 15422, 15572, *15591, 15602, 15648, *15663, 15709, 15717, 15757, 15764, 15799,
    15871, 15898-9, 15937, 16025, 16069, 16125, 16153, 16171, 16195, 16252-4,
    16330, 16336, 16375, 16382, 16416-7, 16437, 16468, 16500, 16510, 16519, 16610,
    16627, 16671, 16743-4, 16831, 16833, 16864, 16949, 16961, 16995, 17000-1, 17034,
    17174, 17218, 17269, 17281, 17330, 17348, 17431, 17451, 17523, 17622, 17726,
    17740, 17744, 17771, 17989, 18063, 18067-8, 18083, 18085, 18105-6, 18137,
    18269, 18276, 18337, 18376-8

Algae, red (*cf.* also *Porphyridium*)
    15422, 15547, 15595, 15648, 15751-2, 15980, 16074, 16113, 16154, 16352-3,
    16375, 16380, 16444, 16468, 16500, 16581, 16610, 16843, 16907, 17001, 17008-9,
    17223, 17269, 17281, 17330, 17363, 17431, *17458, 17744-5, 17893, 18067, 18069,
    18110, 18188-9, 18205, 18337, 18376, 18378

Allium cepa see Onion

Alnus see Alder

Alpine plants   15715-6, 16255, 16641, 16957, 16959, 16967, 17148, *17182-3, 17249,
     *17493, 17540, 17551, 18046

Amaranthus   15389, 15642, 16257, 16301, 16682, 16996, 17855, 18060, 18405

Anabaena   15854-5, 15968, 16182, 16261-2, 16715, 16725-6, 16853, 17074, 17477, 17833,
     18275

Anacystis   15938, 15958, 15961, *16036, 16072, 16391, 16583, 16624-5, 17221-2, 17411,
     17440, 17474, 17497, 17890, 18266

Ananas see Pineapple

Apple   15612, 15636, B15671, 16180, *16345-6, 16398, 16677, *16862, 16863, 16913,
     16953, 17015, 17187, 17461-2, 17518, 17850, 17874, 17924-5, 18037, 18200

Aquatic macrophytes   15415, *15425, 15527, 15548, 15579, *15665, 15879, 15982, 16008-
     -10, 16121, 16132, 16314, 16350, 16437, B16453, 16551, 17016, 17096, 17299,
     17326, 17355, 17500, 17516, 17520, 17569, 17598-600, 17893, 17923, 18018,
     18138-9, 18217, 18252, 18295-6, *18347, 18357

Arabidopsis   15368, *15369, *15371, *15468-9, *15524, 15660-1, *15924, 15925, *15926,
     16126, *16708, 16871, *17349, 17587-8, 17651, *17946, *18153-4, 18155, *18156

Arachis see Groundnut, peanut

Ash   15918, 16603, 16729, 17062-3, 17224

Aspen   17562-3, *18396

Atriplex   15607, 16431, 16434, 17403, 17881, 18093

Avena see Oat

Avocado   15883, 16223

B

Bacteria, photosynthetic (*cf.* also *Chlorobium, Chromatium, Rhodopseudomonas, Rhodospirillum*
     15381, 15410, 15423, 15475, 15502, 15517-8, 15528, 15570,
     15622, 15699, 15777, *15807, 15835-6, *15839, 15856, 15862-3, 15876, 15889,
     *15890, 15895-6, 15916-7, 15981, 16000-1, 16095-8, 16140, 16160, 16195, 16274,
     16377, 16558, 16613-4, 16676, 16727, 16807, 16817-20, 16842, 16844-5, 16875,
     16917, 17010-1, 17029, 17051, 17125-6, 17143, 17181, 17209, 17247, 17310-2,
     17336, 17343-5, 17382, 17405, 17466, 17519, 17546-8, 17597, 17652, 17695,
     17829, 18062, 18390

Bamboo   *18373

Bambussa see Bamboo

Banana   16223, 17425

Barley   15387, 15392-4, 15442, 15455-6, 15553, 15561, 15596, *15608, 15609, 15650,
     15767, 15907, 15975, 15990, 15992, 16088-9, 16104, 16119, 16124, 16168, 16195,
     16197, 16204, 16206, 16208, *16235, 16348, 16383-5, 16426, 16458-60, 16462,
     16475, 16498, 16509, 16568, 16571, 16634, *16669, 16670, 16681, *16694, 16783,
     *16796, 16911, 16945, 16974, 16987, 17007, 17018, 17065-7, 17127, 17133,
     *17160, 17219, *17238, 17250, 17255, 17292-4, 17319, 17328, 17425, 17455,
     17484-5, 17515, 17525, 17562, 17580, 17633, 17664, 17667-9, *17784, 17807-11,
     17813, 17846, 17860, 17898, 17931, 18005, 18068, 18192, 18218, 18238, 18366,
     *18388, 18403

Bean   15434, 15439, 15447, 15460, 15462, 15520, 15586, 15650, 15742, 15970, 15991,
     16020, 16025, 16033, 16073, 16075, 16081, 16191, 16195, 16215-7, 16223, 16304,
     16322-5, 16327, 16347, 16374, 16397, 16425, 16460, 16475-6, 16498, *16499,
     16521-2, 16572, 16585, 16588-9, 16596, 16681, 16787, 16859, 16901, 16926,
     17022, 17025, 17056, 17115, 17127, 17133, 17138, 17185, 17197, 17219, 17230,
     17234-6, 17250, 17277, 17279, 17282, 17381, 17439, 17449, 17453, 17513, 17570,
     17633, 17636, *17797, 17817, 17887, 17893, 17911, 17937, 17976, 18082, 18128,
     18133, 18168, *18213, 18305, 18352, 18355, 18363, 18405

Cereals see Barley, Grasses, Maize, Millet, Oat, Rye, Sorghum, Wheat *etc.*

Cherry B15671

Chicory, endive 15587

Chinese cabbage 15544

Chlamydomonas *15369, 15479, 15581-2, 15653-4, *15663, 15672, 15708, 15835-6,
        *15839, 15871, 15915, 15980-1, *16059, 16068, 16070, *16094, 16125, 16172,
        16360, 16506-7, 16516, 16594, 16630-1, 16671, 16833, 16860, 16888-90, 16935,
        16940, *16941, 16942, *16943, 16944, 16997, 17124, 17141, 17327, 17407, 17741,
        17862, 17947, *17966, *18022, *18122, 18391

Chlorella 15383, *15435, 15509, 15523, 15543, 15545, 15582, 15656, 15658, *15663,
        15677, 15686, 15804, 15826, 15853, 15949, 15957-60, 15980, 15996, *16059,
        16125, 16133, 16135, 16163, 16187, 16195-6, 16225, 16251, 16259-60, 16333-4,
        16380, 16417, 16444-5, 16500, 16558-9, 16610, 16657, 16671, 16678, 16743,
        16799, 16803-4, 16836-7, 16843, 16860, 16909, 16927, 16969-70, *17020, 17022,
        17037-8, 17054-5, 17071-2, 17103, 17157, 17170, 17223, 17226, 17229, 17304-5,
        17354, 17366, 17407, 17437, 17467, 17471, 17504-6, 17541, 17651, 17712, 17723,
        17751, 17828, 17848, 17885-6, *17975, 17979, 18041, 18053, 18057-8, 18089,
        18119, 18362, *18371, 18391

Chlorobium *15880, 15895-6, 16733, 16807, 16924, 17382

Chromatium 15761, 15772-3, 15999, 16003, 16064, 16492, 16690-1, 16723, 18024

Chrysophyta 15422, 15898, 16381, 16468, 17330, 18337

Cichorium see Chicory, endive

Citrullus see Watermelon

Citrus 15510, 16223, 16790-1, 17299, 17342, 17425, 17463-4, 17476, 17492, 17686,
        17688, 17709-10, 17949, 18125, 18158, 18257

Cladonia 16032, 16977

Clover 15389, 15754, 15852, 15978, 16195, 16639, 16905, 17322, 17641, 17877, 18321

Cocksfoot see Orchardgrass

Cocoa 15487

Coconut palm 17135, 17332

Coffea see Coffee tree

Coffee tree 15587, 17359

Coniferous plants (*cf.* also Fir, Larch, Pine, Sempervirent plants, Spruce) *15474,
        15615, 15688, 15808-11, 15840, *15875, 16729-32, 16780, 16794, 17476, 17800,
        18008

Corn see Maize

Cotton *15369, *15385, 15396, *15458, 15516, 15557, *15589, 15787, *16011-2, 16046,
        16144, 16223-4, 16396, 16412-4, 16547, 16573, 16636, 16652, 16712, *16762,
        16763, 17033, *17119, 17120, 17219, 17250, 17438, *17560, 17561, 17594, 17666,
        17783, 17793, 17880-1, 18123, 18140, 18207, *18365

Cottonwood see Poplar, cottonwood

Cowberry see Whortleberry

Cowpea 16055, 17206, 17736, 18358

Crabgrass 15828, 15955, 16535-6, 18293

Cucumber 15384, 15389, 15393-4, 15536-7, 15587, 15775, 15805, 15877, 16004, 16108,
        16123, 16147, 16218, 16428, 16484, 16570, 16578, 16598, 16800, 16816, 16821,
        17104, 17165, 17219, 17250, 17585-6, 17696, 17777, 17856, 17949, 17993, 18043,
        18182, 18351, 18363-4

Cucumis sativus see Cucumber

Cucurbita see Squash

Currant 17845

Fraxinus see Ash

Fruit trees 15636, B15671, 16180, 16746, 17646

Fungi 15446, 15763, 16900, 17112, 17234-6, 17561, 18019

G

Glycine see Soybean

Gossypium see Cotton

Gourd see Squash

Grape fruit see *Citrus*

Grape-vine, vine 15412, 15634, 15655, 15956, 16005, 16752, 16829, 16892, 16914,
        17047-8, 17052, 17250, 17804, 17867-8, 17909, 17961-2, 17990, 18202, 18212,
        18414

Grasses (*cf.* also Barley, Maize, Millet, Oat, Rice, Rye, Sorghum, Wheat, *etc.*)
        15389, 15441, 15491, 15555, 15609, 15616, 15642, *15691, 15749, *15750,
        15753, 15758, 15771, 15778, 15796, 15812, *15830, 15848, 15867-9, 15886,
        15890, 15928, 15952, 15978, 15995, 16013, 16018, 16091, 16144, 16195, 16197,
        16230, 16287, 16302, 16343, 16498, 16535, 16551, 16571, 16639-40, 16643-4,
        16753, 16768, 16810, 16848, 16861, 16959, 17031, 17101, 17111, 17254, 17258,
        17270, 17348, *17493, 17494, 17521-2, 17596, 17626, 17629-30, 17658, 17687,
        17689, 17773, 17788, 17800, 17822-3, 17837, 17853-5, 17890, 17894, 17896-7,
        17902, 18006, 18047, 18072-3, 18108, 18252, 18255, 18273, 18303-4, 18315,
        18361, 18405

Groundnut, peanut 15389, 15520, 15711, 16292, 16734, 17150, 17206, 17417, 17691

H

Halobacterium 17369

Halophilous plants 16287, 18323-4

Heath see *Ericaceae*

Hedera see Ivy

Helianthus annuus see Sunflower

Helianthus tuberosus see Jerusalem artichoke

Hemp *15684, 17868

Hibiscus cannabinus see Kenaf

Hordeum see Barley

Hornbeam 15812, *17950, 18124, 18175-6

Horsetail 15846, 16314, 16377

I

Ipomea batatas see Sweet-potato

Ivy   16278, 16446, 17793

J

Jerusalem artichoke 16989

Joint-grass see Dallis-grass

Juglans see Walnut

P

Panicum see Millet

Parasitic plants 17476

Parsley 17215-6

Paspalum see Dallis-grass

Pasture plants see Forage plants

Pea   *15369, 15384, 15391, 15418-9, *15471, 15520-1, 15587, 15599, 15613, *15618-
           -9, 15645-7, 15650, 15822, 15881, 15927, 15951, 15970, 15992, 16025, 16033,
           16035, 16049, 16100, 16118, 16122, 16131, 16134, *16137, 16195, 16199, 16226,
           *16240, 16280-1, 16306, 16333, 16336, 16424, 16434, 16452, 16462, *16480,
           16569, 16575, 16588, 16604, 16666, 16686-8, *16694, 16734, 16756, 16767,
           16828, 16859, 16887, 16973, 17024, 17050, 17053, 17098, 17114, 17154, 17175,
           17189, 17196, 17206, 17219, 17262-3, 17268, 16350, 17360, 17412, 17422,
           17428, 17465, 17530, 17532-4, 17562, 17578, 17636, 17651, 17703, 17795,
           17809, 17817, *17832, 17898, 17905, 18036, 18049, 18102-3, 18166, 18168,
           18193, *18221, 18222, 18230, 18254, 18277, 18291, 18368, *18385, 18386,
           18391

Peach   15463, 16223, 17457, 17749

Peanut see Groundnut, peanut

Pear   B15671, 17425

Peavine   16859

Pennisetum see Napier-grass

Peperomia 15722-4, 16220, 18240

Pepper, paprika 15516, 15877, 15914, 16195, 16223, 16598, 16622-3, 16938, 17299,
           17567, 17948-9

Perilla 15759, 16781, 17251

Persea gratissima see Avocado

Persica see Peach

Petroselinum see Parsley

Phaseolus see Bean

Phleum see Timothy

Photosynthetic bacteria see Bacteria, photosynthetic

Phragmites see Reed

Picea see Spruce

Pine   15628, 15690, 15766, 15983, 15988, *15998, 16026, 16150, 16195, 16210,
           16282, 16293-4, 16461, 16473, 16603, *16692, 16716, 16729, 16731, 16757-9,
           16956, 16980, 17042, 17069, 17100, 17425, 17476, 17562-3, 17573, 17866,
           18008, 18096, 18113, 18147, 18149-50, *18231, 18237-8, 18259, 18311, 18348,
           18389, 18400

Pineapple 15461, 15554, 15609, 15953, 17297, 17425

Pinus see Pine

Pirus see Pear

Pisum see Pea

Plane tree *16235

Platanus see Plane tree

Plum   15685, *16747

Poa see Meadowgrass

Poplar, cottonwood B15671, 15846, 16151, 16185-6, 16195, 16465, 16496, 16586, 16591,
           16603, 16633, 16729, 16962, 17224, 17531, 17868, 17954

Populus see Poplar, cottonwood

Populus tremula see Aspen

Porphyridium 15588, 16202, 16488, 16711, 17363

Potato   15405, 15428-9, 15431, 15442, 15480, 15504, 15652, *15744, 15763, 16025,
         16369, 16463, 16608, 16838, 16908, 16956, 17030, 17118, 17206, 17219, 17224-
         -5, 17228, 17361-2, 17476, 17564-5, 17568, 17632, 17682, 17768-9, 17796,
         *17889, 18019, 18417-8

Prune see Plum

Prunus see Cherry, Peach, Plum

Pumpkin see Squash

Q

Quercus see Oak

R

Radish   15377, 15389, 15434, 15587, 15625, 15837, 16588, 17250, 17333-4, 17527,
         17641, 17798-9, 17942, 17944-5, 17977-8, 18056, 18234

Rape   15897, 16179, 16462, 17755-6, 17876, 17898, 17981

Raphanus see Radish

Raspberry 17052

Reed   15728, 15962, 16008-10, 16127-8, 16314, 16525, 16934, 17383-7, 17599, 17836-
         -7, 18012-3, 18018, 18256

Rhodopseudomonas 15381, 15410, 15475, 15502, 15570, 15637, *15807, 15856, 15862,
         15876, 15889, 15917, 16000-1, 16095-8, 16103, *16112, 16190, 16295, 16379,
         16399, 16474, 16530-1, 16614, 16875, 17010, 17051, 17125-6, 17310-2, 17519,
         17536, 17589, 17728, 17858-9, 17884, 17936, 17969, 18031, 18121

Rhodospirillum 15517, 15699, 16166, 16266, 16340, 16379, 16512, 16515, 16590, 16637,
         16668, 16842, 16844, 16917, 17143, 17209, 17343-5, 17405, 17685, 17713,
         17737, 17752, 17878-9, 17904, 18134-5, 18263

Ribes see Currant

Rice   15388-9, 15585, 15851, 16073, 16291, 16571, 16577, 16582, 16584, 16597,
         16979, 17142, 17204-7, 17233, 17245, 17280, 17398, 17425, 17503, 17996-7,
         18025, *18136, 18381-2, 18405

Ricinus see Castor bean

Robinia see Locust

Rosa see Rose

Rose   16060, 16146, 17938

Rubus see Raspberry

Rye   15389, 15507, 15651, 15980, 16197, 16498, 16717, 16774, 16908, *17099,
         *17116, 17277, 17425, 17627, 17689, 17993, *18185, 18186, 18332-4, 18336

Ryegrass 15848, 16091, 16245, 16639, 17521, 17629-30, 18321, 18338

S

Saccharum see Sugar cane

Salix see Willow

Sambucus see Elder

Scenedesmus 15573, 15582, *15663, 15708, 15757, 15871, 15898, 15909-10, 15937,
        15973, 15996, 16360, 16377, 16417, 16621, 16671, 16864, 16927, 16937, 17007,
        17026, 17058-9, 17354, 17407, 17579, 17651, 17744, 17772, 18167, 18286-7

Secale see Rye

Sempervirent plants 15510, 15695, 15974, 16207, 16638, 16732, *16749, 16850, 17476

Selaginella 15626, 16437

Service-tree 16195

Setaria see Bristlegrass

Sinapsis see Mustard

Solanum (other than Egg plant, Potato, Tomato) 17684

Solanum melongena see Egg plant

Solanum tuberosum see Potato

Sorbus see Service-tree

Sorghum 15389, *15458, 15534, 15568, 15609, 15704, 15713, 15771, 16015, 16019,
        16073, 16075, 16197, 16223, 16498, 16535, 16571, 16682, 16753, 16894, 16950,
        16964, 17284, 17425, 17438, 17645, 17893, 17935, 18020, 18042, 18044, 18129-
        -30, 18298, 18405

Sorgum see Sorgnum

Soybean 15389, 15427, *15482, 15520, 15539-40, 15731, *15732, 15733-4, 15816-7,
        15819, 15821, 16077, 16057, 16073, 16195, 16223, 16358, 16400, 16408, 16542,
        16555-7, 16703, 16766, 16856-7, 16915, 16946, 16989, 16993, 17070, 17206,
        17233, 17288, 17351-2, 17374, 17398, 17401, *17415, 17452, 17550, 17800,
        17850, 17893, 18044-5, 18144, 18270-2, 18277, 18335, *18374

Sphagnum 15849, 16317-8

Spinach 15382, 15407-8, 15411, 15416-8, 15420, 15426-7, *15435, 15450-3, 15472-3,
        15488-9, 15505, 15532, 15538, 15546, 15559, 15562, 15582, 15587, 15631,
        15642, 15650, 15656-8, 15669, 15672, 15675, 15686-7, 15708, 15710, 15727,
        15730, 15739, *15756, 15768, 15834, 15934, 15971-2, 15980, 16025, 16031,
        16033, 16048-9, 16120, 16139, 16145, 16148, 16161-2, 16195, 16210, 16214,
        16222, 16238-9, *16240, 16241, 16256, 16271, *16299, 16301, 16308-12, 16328-
        -9, 16371, 16376-7, 16388-9, 16420, 16434, 16443-5, 16447, 16451-2, 16454-6,
        *16482, 16485, 16503, 16539, 16563-6, 16583, 16588, 16599, 16611-2, 16615,
        16645-6, *16658, 16659-60, 16666, 16683, 16775, 16792-3, 16814, 16843, 16873,
        16878-80, 16899, 16963, 16966, 16968, 16982, 17004, 17012-3, 17032, 17049-50,
        17068, 17075, 17105-9, 17136, 17139, 17149, 17155, 17157, 17162, 17192,
        17219, 17264-7, 17271, 17275-6, 17335, 17339-41, 17354, 17380, 17399, 17413,
        17429, 17444-5, 17456, 17460, 17467, 17483, 17487, 17489-90, 17508-10, 17517,
        17543, 17570-1, 17593, 17604-8, 17633, 17641, 17700, 17706, 17727, 17729,
        17735, 17761, 17800, 17833, 17893, 17979, 17986, 17988, 18021, 18028, 18036,
        *18070, 18087, 18090, 18100-1, 18109, 18120, 18180-1, 18190, 18194-5, 18254,
        18262, 18266-7, 18285, 18288-9, 18329-30, 18341, 18403, 18405

Spinach beet see Sugar beet, beet, mangold, spinach beet

Spinacia see Spinach

Spruce 15808-11, 15813, 15840, 15846, 16150, 16716, 16730, 16757, 16951-2, 17069,
        17476, 17535, 17562, 18008, 18113-4, 18148, 18370, 18400

Squash 15718, 16195, 16223, 16411, 16720, 16735, 16956, 17027, 17043, 17050, 17095,
        18127

Strawberry 15432-3, 16936, 17754, 17993, 18201, 18344

Submersed plants see Aquatic macrophytes

Succulents (*cf.* also Cacti) 15519, 15554, 15649, 15674, 15722-4, 15953, 16210,
        16218, 16681, 16736-7, 17203, 17298, 17642, 18028, 18361

Sugar beet, beet, mangold, spinach beet 15428, 15526, 15587, 15650, 15707, 15726,
        15771, 15831, 15980, 16025, 16129, 16195, 16201, 16356, 16366, *16469, 16471,

V

Vaccinium see Blueberry

Vaccinium vitis idaea see Whortleberry

Vegetables (*cf.* also Cabbage, Cauliflower, *etc.*)
      15587, 15883, 16030, 16107, 16587, 16976, 17030, 17085, 17129, 17214-6,
      17557, 17719, *17785, 18234, 18350

Vetch   15920, 17562

Vicia faba see Broadbean

Vicia sativa see Vetch

Vigna see Cowpea

Viruses   15544, 15718, 15734, 17095, 17769, *17794

Vitis see Grape-vine, vine

W

Walnut   16603

Watermelon   15587, 16218, 16223, 16664, *16709, 16971, 17389, 17857

Weeds   15913, 15978, 16037, 16470, *16644, 17277, 17495, *18385

Wheat   15384, 15386-7, 15428, 15430, 15436-7, 15442, 15449, 15481, *15492, 15495-6,
      15498-9, 15609, 15623-4, 15629, 15635, 15662, 15771, 15783, 15838, 15877,
      15904, 15907-8, 15921, *15936, 15970, 15997, 16002, 16037, 16043-5, 16073,
      *16084, 16109, 16142, 16156, 16173, 16175, 16197, 16205, 16223, 16286, 16427,
      16434, *16464, 16498, 16501-2, 16586-7, 16647-8, 16681, 16751, 16760-1,
      16787, 16826, 16868, 16900, 16975, 17019, 17024, 17078, 17128, 17164-5,
      17206, 17208, 17212, 17219, 17240, 17250, 17255, 17322, 17371, 17373, 17396,
      17421, 17425, 17441-2, 17476, 17526, 17538, 17550, 17570, 17601-2, 17633,
      17644, 17653, 17689, *17778, 17781, 17800, 17880, 17928, 18000-1, *18039,
      18040, 18049, 18064, 18115, 18242, 18277, 18284, 18299, 18336, 18339-40,
      18405, 18409

Whortleberry   15567

Willow   15413, 15567, B15671, 15846, 16230, 16794, 16956, 17868, 18303

Wolffia see *Lemna, Wolffia*

Woody plants (*cf.* also Fir, Forest ..., Larch, Pine, Spruce, *etc.*)
      15567, 15603, *16748, 17258, 17322

Z

Zea see Maize

Zebrina see *Tradescantia, Zebrina*